Design, Instrumentation, and Controls

Design, Instrumentation, and Controls

Editor:

Harinirina Randrianarisoa, Ph.D.

ArclerPress

www.arclerpress.com

Design, Instrumentation, and Controls

Editor: Harinirina Randrianarisoa, Ph.D.

© 2017 Arcler Press LLC
708 3rd Avenue, 6th Floor
New York
NY 10017
United States of America
www.arclerpress.com

ISBN: 978-1-68094-383-2

Library of Congress Control Number: 2016950944

About the Editor

Harinirina Randrianarisoa, Ph.D.

Hari Randrianarisoa obtained is Ph.D from Institute of Electronics, Microelectronics and Nanotechnology (IEMN) of the University of Science and Technology – Lille (France). His thesis focused for designing and manufacturing of thermal micro sensors. His interests span from on renewable energy especially solar energy. He has a strong knowledge in technology transfer field. He is currently working as a Business advisor for economic development organization in Québec City and as a consultant Scientist advisor.

Contents

List of Contributors

Chananchai Wutthithanyawat
Department of Instrumentation and Control Engineering, Pathumwan Institute of Technology, Bangkok 10330, Thailand

Nawadee Srisiriwat
Department of Petrochemicals and Environmental Management, Pathumwan Institute of Technology, Bangkok 10330, Thailand

Angeliki Mavrantoni
Department of Neurophysiology, Institute of Physiology and Pathophysiology, Philipps-Universität Marburg, Marburg, Germany

Veronika Thallmair
Department of Neurophysiology, Institute of Physiology and Pathophysiology, Philipps-Universität Marburg, Marburg, Germany

Michael G. Leitner
Department of Neurophysiology, Institute of Physiology and Pathophysiology, Philipps-Universität Marburg, Marburg, Germany

Daniela N. Schreiber
Department of Neurophysiology, Institute of Physiology and Pathophysiology, Philipps-Universität Marburg, Marburg, Germany

Dominik Oliver
Department of Neurophysiology, Institute of Physiology and Pathophysiology, Philipps-Universität Marburg, Marburg, Germany

Christian R. Halaszovich
Department of Neurophysiology, Institute of Physiology and Pathophysiology, Philipps-Universität Marburg, Marburg, Germany

Noe Samano
Electronics Engineering Department, Technological Institute of Celaya, 38010 Celaya, GTO, Mexico

José Alfredo Padilla-Medina
Electronics Engineering Department, Technological Institute of Celaya, 38010 Celaya, GTO, Mexico

Nimrod Vázquez
Electronics Engineering Department, Technological Institute of Celaya, 38010 Celaya, GTO, Mexico

Zhenyu Chen
Department of Control Science and Engineering, Harbin Institute of Technology, Harbin 150001, China

Yang Liu
Department of Control Science and Engineering, Harbin Institute of Technology, Harbin 150001, China
Ultra-Precision Optoelectronic Instrumentation Engineering Centre, Harbin Institute of Technology, Harbin 150001, China

Zhenxian Fu
Department of Control Science and Engineering, Harbin Institute of Technology, Harbin 150001, China

Shenmin Song
Department of Control Science and Engineering, Harbin Institute of Technology, Harbin 150001, China

Jiubin Tan
Ultra-Precision Optoelectronic Instrumentation Engineering Centre, Harbin Institute of Technology, Harbin 150001, China

Qiuzhan Zhou
State Key Laboratory of Automotive Simulation and Control, Jilin University, Changchun 130022, China
College of Communication Engineering, Jilin University, Changchun 130022, China
College of Instrumentation and Electrical Engineering, Jilin University, Changchun 130026, China

Chunhui Wang
College of Communication Engineering, Jilin University, Changchun 130022, China

Yongzhi Chen
College of Communication Engineering, Jilin University, Changchun 130022, China

Shuozhang Chen
College of Communication Engineering, Jilin University, Changchun 130022, China

Jun Lin
College of Instrumentation and Electrical Engineering, Jilin University, Changchun 130026, China

Jing Zhou
Department of Engineering Sciences, University of Agder, N-4898 Grimstad, Norway.

Jan Einar Gravdal
International Research Institute of Stavanger, N-5008 Bergen, Norway

Per Strand
Statoil ASA, Bergen, Norway

Svein Hovland
National Oilwell Varco, Stavanger, Norway

BingYou Liu
Department of Precision Machinery and Precision Instrumentation, University of Science and Technology of China, Hefei 230027, China
Key Lab of Electric and Control of Anhui Province, Anhui Polytechnic University, Wuhu 241000, China

Yi Jin
Department of Precision Machinery and Precision Instrumentation, University of Science and Technology of China, Hefei 230027, China

ChangZheng Chen
Changchun Institute of Optics, Fine Mechanics and Physics, Chinese Academy of Sciences, Changchun 130033, China

HuiCheng Yang
Key Lab of Electric and Control of Anhui Province, Anhui Polytechnic University, Wuhu 241000, China

Xiaodong Lu
School of Instrumentation Science and Opto-electronics Engineering, Beihang University, Beijing 100191, China;

Jiangwen Wan
School of Instrumentation Science and Opto-electronics Engineering, Beihang University, Beijing 100191, China;

Emerson Lazzarotto
Universidade Estadual do Oeste do Paraná - UNIOESTE, Brazil

Liliana Madalena Gramani
Universidade Federal do Paraná, Brazil

Anselmo Chaves Neto
Universidade Federal do Paraná, Brazil

Luiz Albino Teixeira Junior
Universidade Federal da Integração Latino Americana - UNILA, Brazil

Qingchang Tian
Research Center for Analytical Instrumentation, Institute of Cyber Systems and Control, State Key Laboratory of Industrial Control Technology, Zhejiang University, Hangzhou 310058, Zhejiang, PR China
College of Life Sciences, Zhejiang University, Hangzhou 310058, Zhejiang, PR China

Ying Mu
Research Center for Analytical Instrumentation, Institute of Cyber Systems and Control, State Key Laboratory of Industrial Control Technology, Zhejiang University, Hangzhou 310058, Zhejiang, PR China

Yanan Xu
Research Center for Analytical Instrumentation, Institute of Cyber Systems and Control, State Key Laboratory of Industrial Control Technology, Zhejiang University, Hangzhou 310058, Zhejiang, PR China

Qi Song
Research Center for Analytical Instrumentation, Institute of Cyber Systems and Control, State Key Laboratory of Industrial Control Technology, Zhejiang University, Hangzhou 310058, Zhejiang, PR China

Bingwen Yu
Research Center for Analytical Instrumentation, Institute of Cyber Systems and Control, State Key Laboratory of Industrial Control Technology, Zhejiang University, Hangzhou 310058, Zhejiang, PR China

Congcong Ma
Research Center for Analytical Instrumentation, Institute of Cyber Systems and Control, State Key Laboratory of Industrial Control Technology, Zhejiang University, Hangzhou 310058, Zhejiang, PR China

Wei Jin
Research Center for Analytical Instrumentation, Institute of Cyber Systems and Control, State Key Laboratory of Industrial Control Technology, Zhejiang University, Hangzhou 310058, Zhejiang, PR China

Qinhan Jin
Research Center for Analytical Instrumentation, Institute of Cyber Systems and Control, State Key Laboratory of Industrial Control Technology, Zhejiang University, Hangzhou 310058, Zhejiang, PR China

M. B. Krassovski
Carbon Dioxide Information Analysis Center, Oak Ridge National Laboratory, Oak Ridge, TN 37831-6290, USA

J. S. Riggs
Integrated Operations Support Division, Oak Ridge National Laboratory, Oak Ridge, TN 37831-6290, USA

L. A. Hook
Carbon Dioxide Information Analysis Center, Oak Ridge National Laboratory, Oak Ridge, TN 37831-6290, USA

W. R. Nettles
Climate Change Science Institute, Oak Ridge National Laboratory, Oak Ridge, TN 37831-6290, USA

P. J. Hanson
Climate Change Science Institute, Oak Ridge National Laboratory, Oak Ridge, TN 37831-6290, USA

T. A. Boden
Carbon Dioxide Information Analysis Center, Oak Ridge National Laboratory, Oak Ridge, TN 37831-6290, USA

Jian Li
Key Laboratory of Energy Thermal Conversion and Control of Ministry of Education, School of Energy and Environment, Southeast University, Nanjing 210096, China

Ming Kong
College of Metrology and Measurement Engineering, China Jiliang University, Hangzhou 310018, China

Chuanlong Xu
Key Laboratory of Energy Thermal Conversion and Control of Ministry of Education, School of Energy and Environment, Southeast University, Nanjing 210096, China

Shimin Wang
Key Laboratory of Energy Thermal Conversion and Control of Ministry of Education, School of Energy and Environment, Southeast University, Nanjing 210096, China

Ying Fan
School of Energy and Power Engineering, Lanzhou University of Technology, Lanzhou 730050, China

Alan Kardek Rêgo Segundo
Department of Control and Automation Engineering and Fundamental Techniques, School of Mines, Federal University of Ouro Preto (UFOP campus Morro do Cruzeiro, sn, 35400-000 Ouro Preto, MG, Brazil

José Helvecio Martins
Department of Agricultural Engineering, Federal University of Viçosa (UFV Avenida P.H. Rolfs, sn, 36570-900 Viçosa, MG, Brazil

Paulo Marcos de Barros Monteiro
Department of Control and Automation Engineering and Fundamental Techniques, School of Mines, Federal University of Ouro Preto (UFOP campus Morro do Cruzeiro, sn, 35400-000 Ouro Preto, MG, Brazil

Rubens Alves de Oliveira
Department of Agricultural Engineering, Federal University of Viçosa (UFV Avenida P.H. Rolfs, sn, 36570-900 Viçosa, MG, Brazil

Gustavo Medeiros Freitas
Vale Institute of Technology (ITV)—Avenida Juscelino Kubitschek, 31, Bauxita, 35400-000 Ouro Preto, MG, Brazil

Xiangyang Zhou
School of Instrumentation Science and Opto-electronics Engineering, Beihang University (BUAA), Beijing 100191, China

Beilei Zhao
School of Instrumentation Science and Opto-electronics Engineering, Beihang University (BUAA), Beijing 100191, China

Guohao Gong
Beijing Aerospace Era Laser Navigation Technology Co., Ltd., Beijing 100094, China

Robin L. Owen
Diamond Light Source, Harwell Science and Innovation Campus, Didcot, OX11 0DE, UK

Jordi Juanhuix
Alba Synchrotron, Carrer de la llum 2-26, Cerdanyola, 08192, Spain

Martin Fuchs
National Synchrotron Light Source II, Brookhaven National Lab, Upton, NY, 11973, USA

M. Alhroob
Department of Physics and Astronomy, University of Oklahoma, Norman, OK 73019, USA

M. Battistin
CERN, 1211 Geneva 23, Switzerland

S. Berry
CERN, 1211 Geneva 23, Switzerland

A. Bitadze
School of Physics and Astronomy, University of Manchester, Oxford Road, Manchester M13 9PL, UK

P. Bonneau
CERN, 1211 Geneva 23, Switzerland

G. Boyd
Department of Physics and Astronomy, University of Oklahoma, Norman, OK 73019, USA

O. Crespo-Lopez
CERN, 1211 Geneva 23, Switzerland

C. Degeorge
Physics Department, Indiana University, Bloomington, IN 47405, USA

C. Deterre
DESY, Notkestraße 85, Hamburg 22607, Germany

B. Di Girolamo
CERN, 1211 Geneva 23, Switzerland

M. Doubek
Czech Technical University, Technická 4, 166 07 Prague 6, Czech Republic

G. Favre
CERN, 1211 Geneva 23, Switzerland

G. Hallewell
Centre de Physique des Particules de Marseille, 163 Avenue de Luminy, 13288 Marseille Cedex 09, France

A. Hasib
Department of Physics and Astronomy, University of Oklahoma, Norman, OK 73019, USA

S. Katunin
B.P. Konstantinov Petersburg Nuclear Physics Institute (PNPI), 188300 St. Petersburg, Russia

D. Lombard
CERN, 1211 Geneva 23, Switzerland

A. Madsen
DESY, Notkestraße 85, Hamburg 22607, Germany

S. McMahon
Rutherford Appleton Laboratory, Science & Technology Facilities Council, Chilton, Didcot OX11 OQX, UK

K. Nagai
Department of Physics, Oxford University, Keble Road, Oxford OX1 3RH, UK

A. O'Rourke
DESY, Notkestraße 85, Hamburg 22607, Germany
School of Physics and Astronomy, University of Manchester, Oxford Road, Manchester M13 9PL, UK

B. Pearson
Department of Physics and Astronomy, University of Oklahoma, Norman, OK 73019, USA

D. Robinson
Department of Physics and Astronomy, University of Cambridge, Cambridge, UK

C. Rossi
INFN Genova, Via Dodecaneso 33, 16146 Genova, Italy

A. Rozanov
Centre de Physique des Particules de Marseille, 163 Avenue de Luminy, 13288 Marseille Cedex 09, France

E. Stanecka
Institute of Nuclear Physics PAS, ul. Radzikowskiego 152, 31-342 Kraków, Poland

M. Strauss
Department of Physics and Astronomy, University of Oklahoma, Norman, OK 73019, USA

V. Vacek
Czech Technical University, Technická 4, 166 07 Prague 6, Czech Republic

R. Vaglio
CERN, 1211 Geneva 23, Switzerland

J. Young
Department of Physics and Astronomy, University of Oklahoma, Norman, OK 73019, USA

L. Zwalinski
CERN, 1211 Geneva 23, Switzerland

Xiangyang Zhou
School of Instrumentation Science and Opto-electronics Engineering, Beihang University (BUAA), Beijing 100191, China

Yuan Jia
School of Instrumentation Science and Opto-electronics Engineering, Beihang University (BUAA), Beijing 100191, China

Qiang Zhao
Shanghai Institute of Satellite Engineering, Shanghai 200240, China

Ruixia Yu
School of Mechanical Engineering, University of Science and Technology Beijing, Beijing 100083, China

Anne Dabrowski
CERN, CH-1211 Geneva 23, Switzerland

Rock Pulak
Union Biometrica, Inc., 84 October Hill Rd., Holliston, MA 01746, USA

Preface

The instrumentation control in the engineering design process is that of the system allowing the engineers to see and understand what is going on within all the different processes being used by the process in the engineering design being tested. With regards to subject, this book discussed the design process, the elements of a successful project, the specific issues to address in a well-designed instrumentation and control system, and the engineering products that enable practical design and successful maintenance. The specifics advances discussed in this book include the experimentation, simulation and the characterization of the instrumentation behavior. Therefore, this book introduces new research idea and development to bring the systems becoming very efficient and require the very minimum of human intervention.

Editor
Harinirina Randrianarisoa, Ph.D.

CHAPTER 1

TEMPERATURE CONTROL OF HEATING ZONE FOR DRYING PROCESS: EFFECT OF AIR VELOCITY CHANGE

Chananchai Wutthithanyawat[1] and Nawadee Srisiriwat[2]

[1]Department of Instrumentation and Control Engineering, Pathumwan Institute of Technology, Bangkok 10330, Thailand

[2]Department of Petrochemicals and Environmental Management, Pathumwan Institute of Technology, Bangkok 10330, Thailand

ABSTRACT

This paper proposes a temperature control technique to adjust air temperature in a heating zone for drying process. The controller design is achieved by using an internal model control (IMC) approach. When the IMC controller parameters were designed by calculating from an actual process transfer function estimated through an open-loop step response with input step change from 50% to 60% at a reference condition at air velocity of 1.20 m/s, the performance of temperature controller was experimentally tested by varying an air velocity between 1.32 m/s and 1.57 m/s, respectively. The experimental results showed that IMC controller had a high competency for controlling the drying temperature.

INTRODUCTION

Drying process is widely used for food processing to preserve foods by removing enough moisture from foods to prevent food spoilage in which effective drying depends on: adequate heat to draw out moisture, without cooking the food; dry air to absorb the released moisture; and enough air circulation to remove the moisture [1]. Air flow rate and temperature in drying chamber affect the moisture removal rate from foods [2]. Consequently, the air flow rate and temperature control is very important to generate hot air in heating zone of food drying process.

In our previous work [3], the temperature distribution of hot air flow in heating zone of a rectangular duct was investigated by varying the power supply to a coil heater and an electrical fan in order to generate hot air entering the drying chamber that is crucial to control the desired temperature and keep the quality of air for drying process. This experiment installation associated with a data acquisition (DAQ) card and a computer was also conducted to generate user interfaces on the computer by using LabVIEW [4].

Currently, there are many control techniques used for the food drying system such as herb [5], gain [6], codfish [7-8], and fruit [9]. Nafeh et al. [5] investigated a temperature control in a PV-Wind medical herb dryer system by using on/off controller that showed the high capability of the proposed technique in controlling the drying temperature. Han et al. [6] used a model predictive control (MPC) system based on neural network which was designed and tested on two model corn dryers that the control method of the system can be used in drying process

Citation: Chananchai Wutthithanyawat and Nawadee Srisiriwat, Temperature Control of Heating Zone for Drying Process: Effect of Air Velocity Change, DOI: 10.1051/ matecconf/20166503002.

control. Boeri et al. experimentally investigated the optimized operating drying control for temperature, relative humidity and air velocity parameters by using fuzzy logic approach [7] and also proposed the multiinput-multi-output-based PID and Fuzzy logic (FLC) controller for a drying control process [8]. Geddes [9] implemented an internal model control (IMC) model to the temperature controller for fruit dryer process in which the simulation results showed that IMC controlled system could achieve the requirement of providing offset-free responses at steady state. Although there are many works concerning a controller design for the food drying process, the research relating to the experimental installation and controller design using IMC for the drying system is limited. In this paper, the drying temperature controller was designed by using IMC to adjust the desired temperature by manipulating the power supply of coil heater used for heating air in the heating zone of drying process. Because the foods are sensitive to heat, there is a suitable drying temperature for each food and the air flow rate is another important parameter varied to remove moisture to be the required value at the fixed temperature. Consequently, the temperature response with varying air velocity was investigated and the capability for temperature controlling was experimentally tested.

EXPERIMENTAL INSTALLATION

In the heating zone, the main apparatuses are composed of a 500 W coil heater, supplied with the input signal in the range of 0 to +5 VDC to generate the maximum temperature about 102.8 C, and an electrical fan (EBMPAPST, 8414N/2GH), adjusted the input signal in the range of 0 to +5 VDC to provide variable air velocity between 1.20 and 1.57 m/s corresponding to the air flowrate from 0.87 to 1.14 m^3/min. The atmospheric air forced by electrical fan flows across the coil heater to generate hot air passing through the non-insulated rectangular duct as shown in Fig. 1. The air velocity was measured by using a Panasonic SUNX Photo Sensor PMK44. The temperatures inside the rectangular duct at various positions were measured by RTD (resistance temperature detector). The bottom of apparatus is the circuits of measurement and control system and the output terminal can be used to send the signal and then monitor the temperature of three positions in rectangular duct and the air velocity via LabVIEW [4].

Figure 1. Experimental apparatus [3].

DESIGN OF TEMPERATURE CONTROL

It is known from the previous discussion that the main function of the electrical heater is to supply the heating required by the drying air. The block diagram of thermal process based on two-input single-output (TISO) system is illustrated in Fig. 2 which includes the temperature control system and the air velocity control system. When the open-loop system was provided with a step input, the process step testing was carried out to characterize three parameters, the gain K, the dead time t0 and the time constants τ, for determining the first-order plus dead-time (FOPDT) model. Accordingly, the combination of the transfer functions, GP(s), is approximated by FOPDT model for the purpose of characterizing the dynamic response of the process. In this step testing, to obtain the maximum gain value, the change of 10% of step input from 0 to 90% was tested for RTD#1 and it was found that at a step change, started at 100 sec, from 50% to 60% gave the highest value of steady-state gain in which the process variable was changed from 49.62% to 65.13% or from 62.21°C to 73.85°C for the step change in controller output of magnitude is 0.5 V. Therefore, the steady-state gain of process, K, can be calculated as 23.27°C/V corresponding to 1.47%/%. Using Fit 1 method to determine the dead time and time constant [10], the values of t0 and τ are approximately 15 sec and 150 sec, respectively. Thus, the resulting FOPDT model transfer function of actual process, GP(s), is shown in Eq. (1).

$$G_P(s) = \frac{Ke^{-t_0 s}}{\tau s+1} = \frac{1.47e^{-15s}}{150s+1} \qquad (1)$$

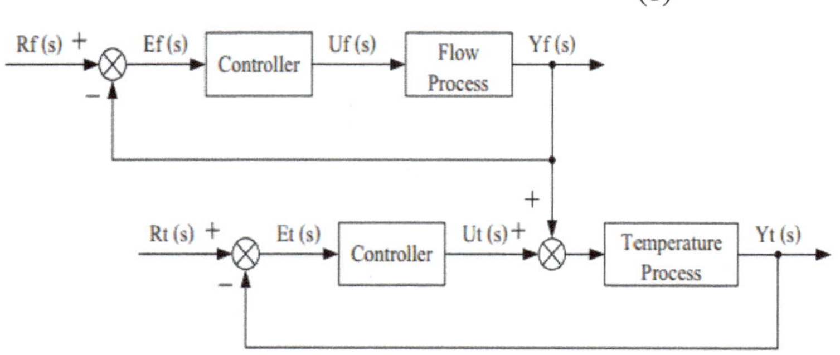

Figure 2. TISO Block diagram of the temperature control system coordinated air flow control system.

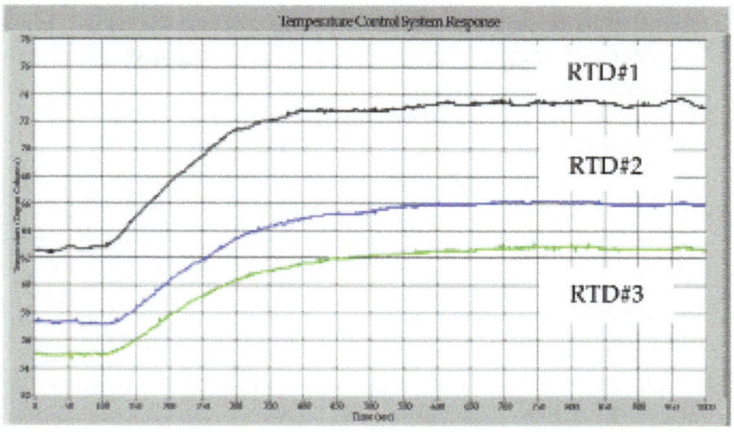

Figure 3. Open-loop step response of temperature process from 50% to 60%.

In process control applications, model based control systems are often used to track set points and reject low disturbances. Particularly, if the control structure has been developed based on the exact model of the process then perfect control is theoretically possible. For controller design in this paper, IMC based PI (proportional integral) design procedure for a first order system was applied to control air temperature at the RTD#1 position and the fixed air velocity of 1.20 m/s. The PI tuning parameters of controller were calculated by Eq. (2) [11]. Hence, the tuning parameters of K_c and T_i are 1.13 and 150 sec, respectively.

$$K_C = \frac{T}{K(T_f + t_0)} \; ; \; T_i = T \tag{2}$$

where K_c is controller gain [%/%], K process gain [%/%], Ti integral time constant [sec], T_f filter time constant [sec], t0 dead time [sec] and T is process time constant [sec]. In this paper, as a rule of thumb T_f is 75 sec chosen to be at least twice as fast as the time constant from open-loop response of process [9].

RESULTS AND DISCUSSION

The aim of the temperature control system is to adjust the temperature in heating zone of drying chamber at which the range of temperature and air velocity varying in this study was generally used in the drying foods. After the IMC controller parameters were designed, to investigate the performance of the proposed temperature control system, the response of the drying temperature was tested at different air velocities between 1.32 and 1.57 m/s corresponding to the step change of input signal from 20 to 100% of +5 VDC for supplying the electric fan.

Figs. 4(a) and 4(b) illustrate the temperature response and the control signal, respectively, in case of using IMC controller. In this test, a set point of drying temperature was nearly 62.52°C which is at 50% operating point of temperature control system. The change of air velocity entering the heating zone in drying chamber was assumed as a disturbance of drying process. As shown in Fig. 4, the step changes of input signal of the electric fan at 20, 40, 60, 80 and 100%, representing the air velocity at 1.32, 1.41, 1.49, 1.55 and 1.57 m/s, respectively, were started at 100 sec. An ambient temperature was about 25°C.

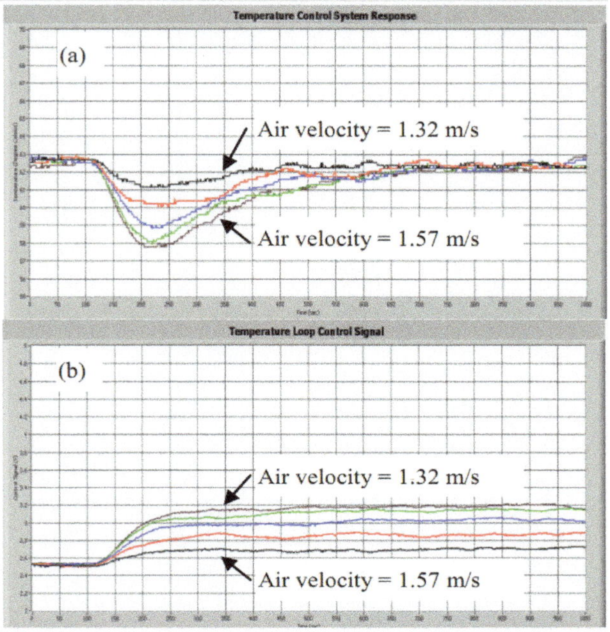

Figure 4. Effects of air velocity on (a) temperature control system response; x-axis is time [sec] and y-axis is temperature [°C] and (b) temperature loop control signal; x-axis is time [sec] and y-axis is control signal [V].

At step change of 20%, when the air velocity was changed from 1.20 m/s to 1.32 m/s, the minimum temperature was about 61.14°C (about 48.0%) which decreases to 2.0% from the set point and the control system took a settling time of 350 sec to come within the prescribed value and remain within the acceptably value as shown in Fig. 4(a). The control signal was about 54% which is more than 50% of a reference operating condition, that is, the control signal increased 4% as illustrated in Fig. 4(b). If the inlet air velocity increases, thus making a disturbance, its effect must propagate through the entire heating zone before the drying temperature changes. Once the required value of drying temperature changes, the signal from the controller also changes. An increase of air velocity decreased the minimum temperature response. At 100% step change of air velocity, the minimum temperature was at 57.85°C or about 43.8% which decreased to 6.2% from set point that the control signal was about 64%. The settling time increased with increasing the air velocity to reach the set point of drying temperature and moreover, the increase of air velocity in range of 60 to 100% used the highest settling time about 600 sec. Although the temperature control system took a long time to control the desired drying temperature because this is the worst case of a non-insulated system in order to test an ability of control system, the drying temperature was efficiently controlled nearly set point. Consequently, it can be concluded that the control of drying temperature was achieved through using IMC controller which gave a high capability.

CONCLUSION

The temperature control system of heating zone in drying chamber was performed by using internal control model (IMC). The drying temperature and air velocity in range of testing in this paper are usually applied in food drying process. The IMC based PI control was designed to adjust the drying temperature at the required value at approximately 62.52°C (50% operating condition) at 1.20 m/s of air velocity to control the operation of the electric heater. To investigate an efficiency of control system, the air velocity, assumed as a disturbance, was varied between 1.32 and 1.57 m/s. The experimental results indicated that the proposed IMC based PI controller gave a good capability in controlling the drying temperature.

REFERENCES

1. 1. P. Kendall, P. DiPersio, J. Sofos, Food and Nutrition Series/Preparation 9.308, (2012)
2. 2. M.A. Tabassum, V.K. Jindal, Pakistan Journal of Agricultural Research 13, 312-319 (1992)
3. 3. C. Wutthithanyawat, N. Srisiriwat, Applied Mechanics and Materials 627, 153-157 (2014)
4. 4. C. Wutthithanyawat, N. Srisiriwat, Applied Mechanics and Materials 619, 303-308 (2014)
5. 5. A. Shafy, A. Nafeh, E.A. Sweelem, F.H. Fahmy, Smart Grid and Renewable Energy 4, 153-160 (2013) 6. F. Han, C. Zuo, W. Wu, J. Li, Z. Liu, Mathematical Problems in Engineering Article ID 584376, (2012)
6. 7. C. Boeri, F. Neto da Silva, J. Ferreira, M. Soares dos Santos, European Drying Conference EuroDrying'2011, Spain, (2011)
7. 8. C.N. Boeri, F.N. da Silva, J. Ferreira, Acta Scientiarum. Technology 35, 279-289 (2013)
8. 9. S.A. Geddes, Internal Model Control (IMC) of a Fruit Drying System, Bachelor of Engineering dissertation, Faculty of Engineering and Surveying, University of Southern Queensland, (2006)
9. 10. C.A. Smith, A.B. Corripio, Principles and Practice of Automatic Process Control, John Wiley & Sons, Inc., Canada, (1985)
10. 11. R. Singh, R. Bala, B. Bhatia, International Journal of Advanced Research in Computer Science and Software Engineering 4, 915-922 (2014)

CHAPTER 2

A METHOD TO CONTROL PHOSPHOINOSITIDES AND TO ANALYZE PTEN FUNCTION IN LIVING CELLS USING VOLTAGE SENSITIVE PHOSPHATASES

Angeliki Mavrantoni, Veronika Thallmair, Michael G. Leitner, Daniela N. Schreiber, Dominik Oliver and Christian R. Halaszovich

Department of Neurophysiology, Institute of Physiology and Pathophysiology, Philipps-Universität Marburg, Marburg, Germany

ABSTRACT

Voltage sensitive phosphatases (VSPs), including engineered voltage sensitive PTEN, are excellent tools to rapidly and reversibly alter the phosphoinositide (PI) content of the plasma membrane *in vivo* and study the tumor suppressor PTEN. However, widespread adoption of these tools is hampered by the requirement for electrophysiological instrumentation to control the activity of VSPs. Additionally, monitoring and quantifying the PI changes in living cells requires sophisticated microscopy equipment and image analysis. Here we present methods that bypass these obstacles. First, we explore technically simple means for activation of VSPs via extracellularly applied agents or light. Secondly, we characterize methods to monitor PI(4,5)P_2 and PI(3,4,5)P_3 levels using fluorescence microscopy or photometry in conjunction with translocation or FRET based PI probes, respectively. We then demonstrate the application of these techniques by characterizing the effect of known PTEN mutations on its enzymatic activity, analyzing the effect of PTEN inhibitors, and detecting in real time rapid inhibition of protein kinase B following depletion of PI(3,4,5)P_3. Thus, we established an approach that does not only allow for rapidly manipulating and monitoring PI(4,5)P_2 and PI(3,4,5)P_3 levels in a population of cells, but also facilitates the study of PTEN mutants and pharmacological targeting in mammalian cells.

INTRODUCTION

Phosphoinositides (PIs) are membrane-bound signaling molecules essential for many cellular processes like migration, proliferation, membrane trafficking, and regulation of ion channels (Balla et al., 2009). Various experimental techniques have been developed to control PI levels in living cells, ranging from pharmacological approaches and genetic manipulations such as over-expression or knock-down of PI-metabolizing enzymes, to more dynamic approaches that allow temporal control over PI levels. Three distinct approaches for the acute manipulation of PI concentrations have been developed, all displaying certain limitations that need to be considered.

First, the FKBP-FRB heterodimerization system recruits a PI metabolizing enzyme (e.g., PI(4,5)P_2-5-phosphatase) to the membrane when rapamycin is applied (Suh et al., 2006; Varnai et al., 2006). The timing of recruitment and consequently the change in, e.g., PI(4,5)P_2 concentration can be controlled with

Citation: Mavrantoni A, Thallmair V, Leitner MG, Schreiber DN, Oliver D and Halaszovich CR (2015) A method to control phosphoinositides and to analyze PTEN function in living cells using voltage sensitive phosphatases. Front. Pharmacol. 6:68. doi: 10.3389/fphar.2015.00068

precision on the order of seconds. However, recruitment and hence the alteration of PI levels is irreversible. Furthermore, the use of the rapamycin-induced dimerization system is limited to cell types that do not express large amounts of FKBP or FRB protein domains endogenously (Coutinho-Budd et al., 2013) and unwanted effects via endogenous targets of rapamycin have to be considered (Brown et al., 1994).

The second technique is similar, being based on the recruitment of PI-phosphatases to the membrane controlled by light-induced dimerization of CRY2 and CIBN domains (Idevall-Hagren et al., 2012). This approach allows for PI(4,5)P_2 depletion that is reversible within minutes, but it has limitations concerning the analysis of PI levels using standard fluorescent biosensors, as the system is highly sensitive to blue and green light.

A third strategy to manipulate PI levels in living cells utilizes voltage sensitive phosphatases (VSPs) that have been isolated from several chordates (Murata et al., 2005; Hossain et al., 2008; Ratzan et al., 2011). VSPs consist of a voltage-sensing transmembrane domain (VSD) linked to a cytoplasmic phosphatase domain. Membrane depolarization activates the protein resulting in dephosphorylation of PIs. Although their biological function remains largely unknown, VSPs have proven to be highly useful to manipulate PI levels experimentally. For example, the *Ciona intestinalis* VSP (Ci-VSP), which primarily acts as a 5-phosphatase toward PI(4,5)P_2 and PI(3,4,5)P_3 (Iwasaki et al., 2008; Halaszovich et al., 2009), has been used to study regulation of ion channels by PI(4,5)P_2 (Lindner et al., 2011; Yudin et al., 2011).

The potency and versatility of this strategy has recently been expanded by the development of engineered voltage sensitive enzymes (Lacroix et al., 2011; Halaszovich et al., 2012). One such enzyme is the chimeric VSP $PTEN_{CiV}$ (termed "Ci-VSPTEN16" in (Lacroix et al., 2011)), where the PI(3,4)P_2/PI(3,4,5)P_3 3-phosphatase PTEN (phosphatase and tensin homolog deleted from chromosome 10) replaces the phosphatase domain of Ci-VSP. $PTEN_{CiV}$ is activated by depolarization of the membrane potential while fully retaining enzymatic properties of wild-type PTEN including substrate specificity (Lacroix et al., 2011). Accordingly, $PTEN_{CiV}$ can be used as a tool to manipulate PI(3,4)P_2 and PI(3,4,5)P_3 levels.

Of note, PTEN itself has been the focus of extensive research efforts, since it is one of the most frequently mutated tumor suppressor genes in human cancer (Chalhoub and Baker, 2009) and in hamartoma-related syndromes (Hollander et al., 2011); PTEN dysfunction is also linked to autism spectrum disorders (ASDs) (Varga et al., 2009). Despite the significance of PTEN mutations in pathological conditions, there are only limited techniques to evaluate the effects of these mutations on the phosphatase activity of the protein. These include *in vitro* phosphatase assays and the functional analysis of PTEN in a yeast survival model (Cid et al., 2008). There is a need for methods to fully characterize PTEN activity in living mammalian cells. When combined with established fluorescence-based methods for reading out PI concentrations, the voltage-controlled $PTEN_{CiV}$ chimera may provide an ideal platform for analyzing PTEN and its mutants in living cells.

One major drawback in using VSPs for analyzing PI functions and PTEN properties in living cells is the need for rather sophisticated methods and expertise for the control of the membrane potential, usually by patch clamp electrophysiology (e.g., Halaszovich et al., 2009; Sakata et al., 2011). Other disadvantages are the time consuming single cell technique and perturbation of the cell interior by whole-cell patch clamping. Here, we report the development of a novel experimental strategy to activate VSPs and $PTEN_{CiV}$ without the use of electrophysiological instrumentation, enabling rapid, reversible, scalable, and broadly applicable manipulation of PI levels in a population of intact cells. Thus, we show that chemical or light induced activation of cation channels can be employed to depolarize the membrane potential and thereby activate Ci-VSP or $PTEN_{CiV}$. Similarly, we present $PTEN_{CiV}$ as a novel approach to study PTEN mutations and inhibitors with high precision. Furthermore, we demonstrate that $PTEN_{CiV}$ can also be used to study the ubiquitous PI(3,4,5)P_3/Protein kinase B (PKB, Akt) pathway on a time scale not easily accessible to biochemical approaches.

MATERIALS AND METHODS

Molecular Biology

The chimeric construct $PTEN_{CiV}$ was described previously (Lacroix et al., 2011). The Frubby construct was created from the Pippi-PI(4,5)P_2 sensor kindly provided from Dr. M. Matsuda, Japan (Yoshizaki et al., 2007), by replacing the PLCδ_1-PH domain with the tubbyC domain (Santagata et al., 2001). To remove PLCδ_1-PH from the original plasmid we used the *BspE*1 and *Kpn*1 restriction sites. The same restriction sites were utilized to insert tubbyC into the construct. The amino acid sequence of this novel construct is provided. The AktAR construct was kindly provided from Dr. J. Zhang, Baltimore, USA (Gao and Zhang, 2008). Mutagenesis was performed using the QuickChangeII XL Site-Directed mutagenesis kit (Agilent Technologies, Santa Clara, CA).

Plasmids and Transfection

CHO and HEK 293 cells were plated on glass coverslips or—for TIRF experiments—glass bottom dishes (WillCo Wells B. V., Amsterdam, The Netherlands). For HEK 293 cells the glass was coated with poly-D-lysine. Cells were transfected 24–48 h later using JetPEI (Polyplus Transfection, Illkirch, France) as described (Halaszovich et al., 2009). For the used expression vectors. Bovine phosphatidylinositol 3-kinase p110α (constitutively active mutant K227E) was cotransfected with $PTEN_{CiV}$ to increase PI(3,4,5)P_3 levels except for experiments involving AktAR (Rodriguez-Viciana et al., 1996). In the latter experiments, PI(3,4,5)P_3 levels were elevated by incubating the cells with 0.4 ng/μl insulin-like growth factor-1 (Sigma, St.Louis, MO) for 15 min (Halaszovich et al., 2009). HEK cells were preferred for confocal imaging, since they present a stronger membrane localization of PI markers. Experiments were performed at room temperature 24–48 h after transfection.

Fluorescence Microscopy

Confocal images were acquired using a Zeiss LSM710 Examiner upright microscope and a W-Plan-Apochromat 20 × 1.0 DIC M27 objective (Carl Zeiss AG; Jena; Germany). For excitation, laser lines of 561 nm for RFP and 488 nm for GFP were used. For detection, wavelength ranges were 582–754 nm for RFP and 493–582 nm for GFP. ChR2 was activated using an HBO lamp with excitation band-pass filters 470/40 nm (blue) and 436/20 nm (violet).

TIRF (Total Internal Reflection Fluorescence) imaging was performed by using a B×51WI upright microscope with a TIRF-condenser (Olympus, Hamburg, Germany) and a 488 nm laser (Picarro, Sunnyvale, CA). The setup was equipped with a LUMPlan FI/IR 40x/0.8- numerical aperture water immersion objective (Olympus), an IMAGO-QE cooled CCD camera (TILL Photonics GmbH, Grafelfing, Germany) and a monochromator (Polychrome IV, TILL Photonics).

Widefield illumination FRET (Förster Resonance Energy Transfer) was gained using a Nikon Eclipse TE2000-U inverted microscope equipped with a 40 × 0.55 LWD objective and coupled with an Oligochrome (TILL Photonics) light source. CFP and YFP were excited using a 430/24 nm (F49-430, AHF Analysentechnik, Tubingen, Germany) filter or a 500/20 nm (F49-500, AHF) filter respectively. Emission signal passed first through a tripleband beamsplitter CFP/YFP/mCherry (F68-017, AHF) and next detection area was selected using a viewfinder and a uEye camera (TILL Photonics). For CFP and YFP/FRET emission detection, a 470/23 nm (F49, AHF) and a 535/30 nm filter respectively, were placed before the photodiodes (TILL Photonics). Photocurrent was amplified by Oligochrome Imaging Control Unit, transmitted to an EPC-10 amplifier (HEKA electronics, Lambrecht, Germany) and data acquisition and analysis were performed with PatchMaster software (HEKA).

Patch Clamping

Whole-cell patch clamping was performed in CHO cells using an EPC-10 amplifier and PatchMaster (both HEKA). Patch pipettes' resistance was typically 1.5–4.0 MΩ, recordings were sampled at 20 kHz and low-pass-filtered at 5 kHz. For current clamp recordings, cells were held at 0 pA and changes of membrane voltage were recorded upon 1 s blue light exposure. $PTEN_{CiV}$ was activated by 30 s voltage steps from a holding potential of −60 to +80 mV. Sensing currents from $PTEN_{CiV}$ were measured as a result of a voltage jump to +120 mV from a holding potential of −70 mV with subtraction of linear leak and capacitive transient currents.

Solutions and Chemicals

In most experiments, cells were kept in extracellular solution containing (mM) 144 NaCl, 5.8 KCl, 0.9 $MgCl_2$, 1.3 $CaCl_2$, 0.7 NaH_2PO_4, 5.6 D-glucose, and 10 HEPES, pH 7.4 (with NaOH), 305–310 mosm/kg. Solutions with increased K^+ concentration were made by replacing NaCl with KCl. For maximal depolarization, a 150 mM K^+ solution was required prepared as follows (mM): 150 KCl, 0.9 $MgCl_2$, 1.3 $CaCl_2$, 0.7 NaH_2PO_4, 5.6 D-glucose, and 10 HEPES, pH 7.4 (with KOH). For the TRPV1 and ChR2(H134R) experiments Ca^{2+} was omitted from the standard extracellular solution to prevent the Ca^{2+} dependent activation of phospholipases-C (PLC) (Ca^{2+}-free extracellular solution) (Rebecchi and Pentyala, 2000). Intracellular solution used for patch clamp experiments contained (mM) 135 KCl, 2.41 $CaCl_2$, 2.5 $MgCl_2$, 5 EGTA, 5 HEPES, and 3 Na_2-ATP, pH 7.3 (with KOH), 290–295 mosm/kg.

Capsaicin and insulin-like growth factor-1 were purchased from Sigma-Aldrich, the bisperoxovanadate from Santa Cruz Biotechnology (Dallas, TX), DTT from Fermentas (St. Leon-Rot, Germany).

Data Analysis and Statistics

Confocal microscopy data acquisition was performed using ZEN2009 (Zeiss) software. For time series recordings, the fluorescence was averaged for regions in the cytosol (F_{cyt}) and normalized to the baseline fluorescence prior to stimulation (F_0). TIRF image analysis was performed using TILLvision software, and the background-corrected TIRF signal was normalized to F_0. For FRET recordings, fluorescence intensity from a wide area including several cells—except for AktAR experiments, were fluorescence was collected from a single cell—was acquired and processed with PatchMaster. FRET ratios, calculated as fFRET/fCFP, were normalized to the baseline before stimulation. fCFP and fFRET were the signals from the CFP and FRET emission channels, corrected for the dark current recorded without illumination. IgorPro (Wavemetrics, Lake Oswego, OR) software was used for data analysis and presentation. To calculate the maximal response to K^+or capsaicin, values acquired during the last 30 s of the application were averaged. For ChR2-mediated Ci-VSP activation, the maximal response represents the first value acquired after the blue light exposure. To acquire the initiation kinetics using KCNQ4 and TRPV1, we calculated the time required to reach 50% of the maximal response [$t_{(1/2)}$ of initiation], for each individual cell. For recovery kinetics, the time required to reach 50% of the baseline after the end of the application was calculated [$t_{(1/2)}$ of recovery]. For ChR2 initiation kinetics analysis, we utilized the blue light exposure time curve (Figure 1H) to calculate $t_{(1/2)}$ of initiation. Only cells that demonstrated a fluorescence intensity change greater than 4% upon application of the substance were used for the analysis. Error bars show the S.E.M, *n* represents the number of cells and *e* represents the number of experiments. For statistical analysis, a two-tailed Dunnett's test was performed to compare the single control to different experimental groups, two-tailed Student's *t*-test was used to compare two groups and Scheffe's test for multiple comparisons. Statistical significance was assigned at $P < 0.05$ and is marked as * for the Dunnett's and Scheffe's test and as # for Student's *t*-test.

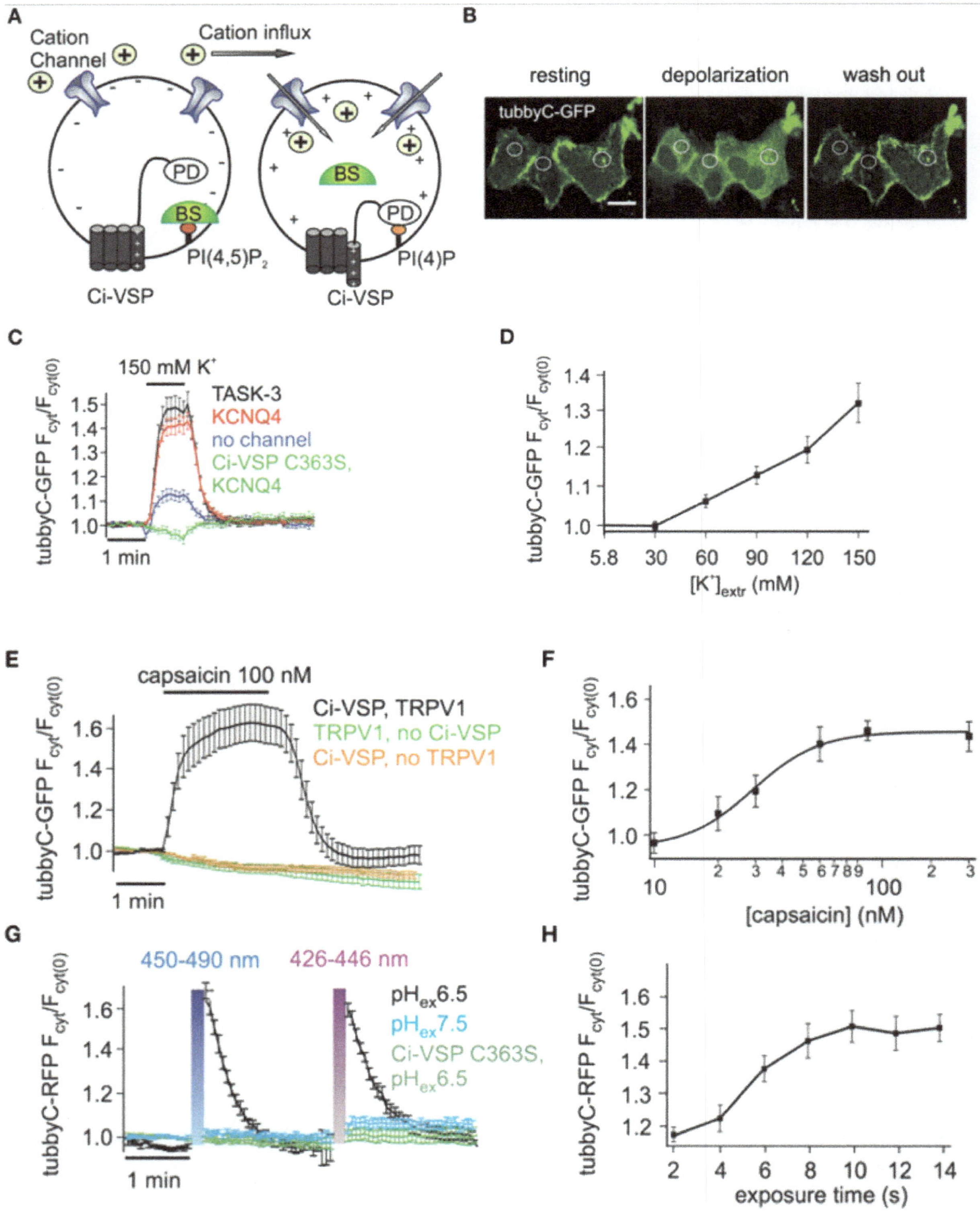

Figure 1. Activation of Ci-VSP by membrane depolarization through overexpressed cation channels. (A) Scheme representing the general strategy used to activate Ci-VSP. Cells co-express Ci-VSP, a PI biosensor and a cation channel. Cation influx into the cell results in membrane depolarization, activation of Ci-VSP, depletion of PI(4,5)$P_2$and lastly translocation of the sensor to the cytoplasm. BS, biosensor; PD, phosphatase domain. (B) Confocal images of HEK cells co-expressing Ci-VSP, KCNQ4 and tubbyC acquired before, during and after application of 150 mM K^+. Fluorescence signal was recorded from cytoplasmic regions. Scale bar 20 μm. (C) Time series of relative increase of cytoplasmic fluorescence calculated from CHO cells expressing Ci-VSP, tubbyC-GFP and the K^+ permeable channels TASK-3 (n = 48, e = 16), KCNQ4 (n = 82, e = 16) or no channel (n = 28, e = 9). The catalytically dead mutant Ci-VSP C363S (n = 23, e = 8) was used as negative control. tubbyC translocates to the cytoplasm upon 150 mM K^+ application for 1 min. (D) K^+ concentration-response curve from HEK cells co-expressing Ci-VSP, tubbyC-GFP, and KCNQ4 (n = 15, e = 6). (E) Application of 100 nM capsaicin in HEK cells co-expressing TRPV1 channel, Ci-VSP and tubbyC-GFP resulted in Ci-VSP activation and tubbyC translocation (n = 25, e = 8). No effect was observed in the absence of Ci-VSP (n = 19, e = 7) or TRPV1 (n = 16, e = 6). (F)Capsaicin concentration-response curve (EC_{50} = 29 ± 3 nM; at least 10 cells per concentration). (G) Activation of ChR2 with violet (426–446 nm) or blue (450–490 nm) light exposure for 10 s in HEK cells co-expressing tubbyC-RFP and Ci-VSP. tubbyC translocation was observed only at extracellular pH 6.5 (n = 12, e = 4), but not at pH 7.5 (n = 11, e = 4) or with the catalytically dead Ci-VSP C363S mutant (n = 23, e = 4). (H) Exposure time-response curve showing that 10 s of exposure with blue light resulted in maximal tubbyC translocation.

RESULTS

Activation of Ci-VSP in Intact Cells by Membrane Depolarization via Cation Channels

Activation of VSPs requires a depolarization of the membrane potential. To facilitate cell-based assays using VSPs we aimed at eliminating complex electrophysiological instrumentation such as patch clamp or two-electrode voltage clamp amplifiers from the experimental setting. To this end, we developed the experimental strategy depicted in Figure 1A, where cation channels are used to depolarize the cell membrane, thereby activating the VSP. The activity of the VSP can be monitored using fluorescence-tagged biosensor specific for the VSP's substrate (Varnai and Balla, 2006).

We started by exploring cation channel based strategies to activate the prototypical Ci-VSP, a $PI(4,5)P_2$ 5-phosphatase. First, the K^+ channels TASK-3 or KCNQ4 (Kv7.4) were used to depolarize the cells. These channels are expected to be largely or partially open under basal conditions, since TASK-3 is a constitutively open "leak" channel (Enyedi and Czirjak, 2010), whereas KCNQ4 is a voltage gated channel open at potentials above −50 mV, when expressed in CHO cells (Chambard and Ashmore, 2005). Given that both channels provide a substantial K^+ conductance, cell depolarization can be simply achieved by increasing the extracellular K^+ concentration.

Indeed, we observed $PI(4,5)P_2$ depletion as reported by translocation of the $PI(4,5)P_2$ biosensor tubbyC-GFP (Santagata et al., 2001) when 150 mM K^+ was applied to cells co-expressing Ci-VSP and either KCNQ4 or TASK-3 (Figures 1B,C). Strong and rapid activation of Ci-VSP was achieved with either channel. Slightly more cells co-transfected with KCNQ4 (92%) than cells co-transfected with TASK-3 (80%) responded to application of K^+. Therefore KCNQ4 was used for further experiments. When co-expression of channels was omitted, only 50% of the cells responded to application of K^+ and changes in $PI(4,5)P_2$ concentration were weak (Figure 1C), indicating that in CHO cells endogenous K^+ channels were not sufficient to strongly drive Ci-VSP activity. The translocation of tubbyC-GFP was completely absent when catalytically dead Ci-VSP C363S was used (Figure 1C), confirming that K^+-induced changes of $PI(4,5)P_2$ were indeed mediated by activation of Ci-VSP.

VSPs can be used to deplete PIs in a gradual fashion by imposing graded membrane depolarization (Halaszovich et al., 2009). Membrane potential has a roughly logarithmic dependence on extracellular $[K^+]$ when the K^+ conductance is dominant, suggesting that membrane potential and consequently Ci-VSP activity could be gradually increased by elevating extracellular $[K^+]$. Indeed, when various K^+ concentrations were washed in, Ci-VSP activity measured as the degree of translocation of the biosensor was also graded (Figure 1D). Maximum activity occurred at 150 mM K^+, which was the maximum concentration we could achieve while keeping physiological osmolarity.

As an alternative approach to achieve membrane depolarization we utilized the capsaicin-activated TRPV1 channel (Figure 1E), a non-selective cation channel that depolarizes the membrane when active (Caterina et al., 1997). Because TRPV1 is also Ca^{2+} permeable and Ca^{2+} influx could potentially activate PLC (Rebecchi and Pentyala, 2000) causing $PI(4,5)P_2$ cleavage, we used Ca^{2+}-free extracellular solutions in these experiments. Application of capsaicin to HEK cells co-expressing TRPV1, Ci-VSP and tubbyC-GFP resulted in readily detectable $PI(4,5)P_2$ depletion as indicated by tubbyC-GFP translocation, whereas no effect was observed in the absence of TRPV1 or Ci-VSP (Figure 1E). 100 nM of capsaicin was sufficient to induce the full response, as demonstrated by the capsaicin concentration-response curve (Figure 1F).

Over the last years, light-activated prokaryotic ion channels have been established as optogenetic tools for manipulation of the membrane potential (Lin, 2011). Among these channels, Channelrhodopsin-2 (ChR2), has been studied extensively and used widely to depolarize cells. ChR2 is a non-selective cation channel activated by visible light with a response spectral peak at around 450 nm (Nagel et al., 2003; Lin et al., 2009). Since activation by light could enable enhanced temporal control and may facilitate experimentation, we also attempted to use ChR2 to activate VSPs,

We first explored conditions for sufficient depolarization by ChR2. Extracellular acidification (pH 6.5) turned out to strongly increase the depolarization of cells expressing ChR2 when stimulated with blue light, most likely due to the high H^+ permeability of ChR2 (Nagel et al., 2003). This finding suggested low extracellular pH as the optimal condition to activate Ci-VSP via ChR2. Indeed, Ci-VSP was successfully activated only when extracellular solution of pH 6.5 was used but not at the standard extracellular pH 7.4 (Figure 1G). As in the case of TRPV1, we used Ca^{2+}-free solution to prevent activation of PLC by Ca^{2+} influx. 10 s of blue or violet light exposure were sufficient for maximal stimulation of Ci-VSP, measured via translocation of tubbyC-RFP to the cytoplasm (Figure 1H). An RFP-tagged sensor was used in these experiments, because GFP excitation interferes with ChR2 activation.

To compare the performance of the above methods of Ci-VSP activation, we evaluated the maximal response to each stimulus as well as the time required for $PI(4,5)P_2$ depletion (i.e., sensor translocation) and the time required for $PI(4,5)P_2$ recovery after cessation of the stimulus. Activation of all three channel types resulted in robust sensor translocation. Specifically, Ci-VSP activation induced by TRPV1 and ChR2 was slightly stronger (sensor translocation 1.63 ± 0.09- and 1.65 ± 0.06-fold, respectively), compared to the KCNQ4-mediated response (1.51 ± 0.04-fold increase; Figure 2A). However, these differences did not reach statistical significance. The fastest response (Figure 2B) was observed using ChR2 [$t_{(1/2)}$ = 4.4 ± 0.4 s], followed by KCNQ4 [$t_{(1/2)}$ = 18.2 ± 1.8 s] and TRPV1 [$t_{(1/2)}$ = 25.0 ± 3.7 s]. For the recovery after stimulation, $t_{(1/2)}$ was 12.4 ± 1.2 s for ChR2, 19.9 ± 2.7 s for KCNQ4 and 27.9 ± 3.1 s for TRPV1 (Figure 2C).

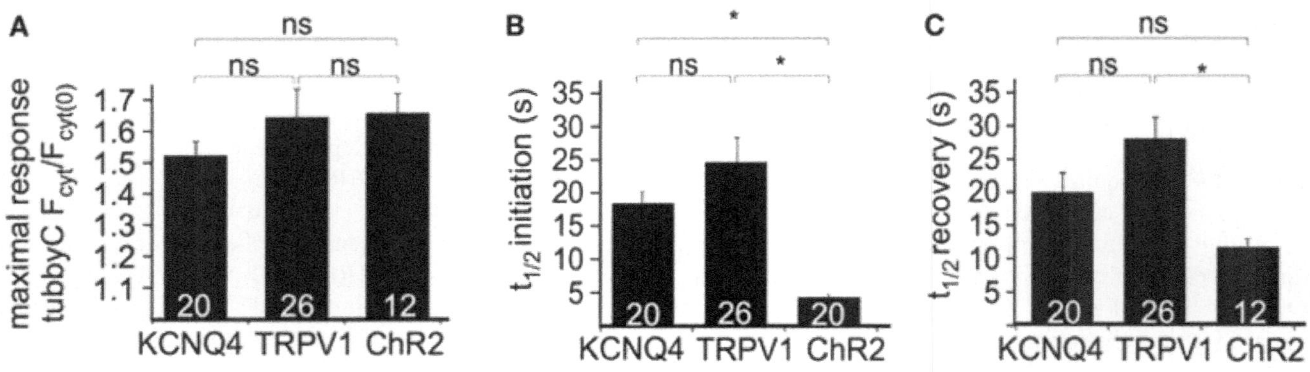

Figure 2. Comparison of the three depolarization methods in HEK cells cotransfected with Ci-VSP, tubbyC, and one of the different channels. (A) Bars representing the maximal response observed by overexpression of KCNQ4, TRPV1, and ChR2. For KCNQ4 we used the tubbyC translocation value from experiments as in graph Figure 1C, for TRPV1 from graph Figure 1E, and for ChR2 the maximal value from the blue light activation in graph Figure1G. (B) Bars representing $t_{(1/2)}$ of initiation, meaning the time required to reach 50% of the maximal response, for the three channels, calculated from data summarized in Figures 1C,E,H, respectively. (C) Bars showing $t_{(1/2)}$ of recovery, the time required for 50% recovery to baseline. Numbers in columns represent the number of cells. * Indicates statistically significant difference; ns, not significant.

Monitoring Ci-VSP Activity without Imaging

The above method for monitoring PI depletion is based on the translocation of PI sensors between cytosol and plasma membrane. Detection of translocation requires imaging-based experiments, usually involving confocal microscopy or advanced microscopy techniques such as TIRF microscopy (van der Wal et al., 2001; Varnai and Balla, 2008; Halaszovich et al., 2009). Such approaches may not be optimal in many experimental settings including high-throughput assays, because experiments and analysis are based on single-cell data and thus are time-consuming, may be difficult to automatize and involve costly and complex equipment (e.g., confocal microscope, laser, TIRF-lenses). One way to circumvent these disadvantages is to employ PI sensors based on Förster resonance energy transfer (FRET): previously, pleckstrin homology (PH) domains of phospholipase C δ_1 (PLCδ_1-PH) fused to CFP and YFP have been used to detect PI(4,5)P_2concentration dynamics; dissociation of the domains into the cytosol increases their distance, thus reducing FRET efficiency (Figure 3A) (Varnai and Balla, 1998; Falkenburger et al., 2010; Hertel et al., 2011). As shown in Figures 3B,C, activation of Ci-VSP with capsaicin in cells co-expressing TRPV1, PLCδ_1-PH-CFP, and PLCδ_1-PH-YFP induced an increase of the CFP signal and a decrease of the FRET signal, indicating the decrease in FRET efficiency. Here, we used photodiodes to measure both CFP and YFP fluorescence emission, demonstrating that the FRET-based approach eliminates the need for microscopic spatial separation of membrane-bound and cytosolic fluorescence. We further sought to improve this approach, because consistency and reliability of this method may suffer from variable expression stoichiometry of both PH domains and from changes in fluorescence intensity resulting from sensor translocation. We therefore developed an intramolecular FRET sensor where binding of a PI(4,5)P_2 binding domain to the membrane affects FRET efficiency between CFP and YFP flanking this domain (Figure 3A). Such a sensor based on PLCδ_1-PH and anchored to the plasma membrane via a CaaX-box has been developed previously and termed Pippi-PI(4,5)P_2 (Yoshizaki et al., 2007). However, dual sensitivity of the PLCδ_1-PH sensor domain to both PI(4,5)P_2 and IP_3 may be a problem under some experimental conditions (Varnai and Balla, 2006). In contrast, the tubbyC domain used in the present study combines insensitivity to IP_3 (Szentpetery et al., 2009) with higher sensitivity to changes in PI(4,5)P_2 levels than PLCδ_1-PH, as shown previously (Halaszovich et al., 2009). We thus designed a PI(4,5)P_2sensor, which we termed "Frubby," where the PLCδ_1-PH domain of the original Pippi PI(4,5)P_2sensor (Yoshizaki et al., 2007) was replaced by tubbyC as the PI(4,5)P_2-binding domain. Depletion of PI(4,5)P_2 using Ci-VSP under voltage clamp conditions showed that Frubby yields a robust, rapid, and readily reversible change in FRET ratio as well as a graded response across a useful dynamic range.

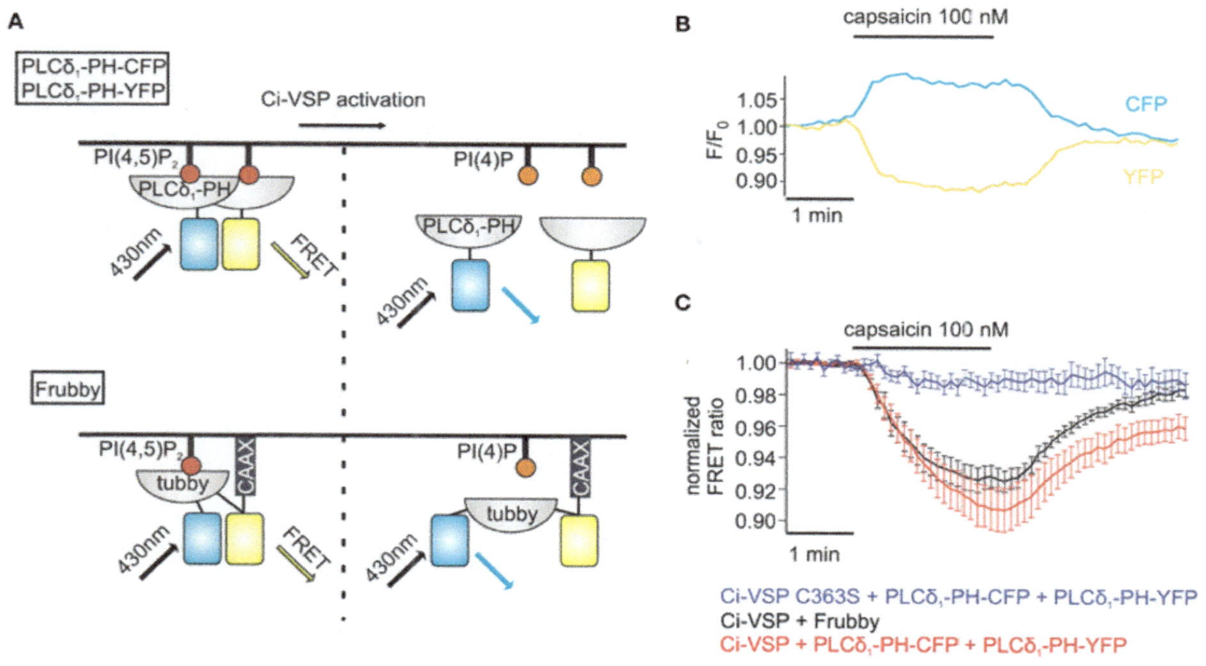

Figure 3. FRET sensors to monitor Ci-VSP activation. (A) Schematic representation of two different FRET sensors used to monitor Ci-VSP activity using photometry. PLCδ$_1$-PH-CFP and PLCδ$_1$-PH-YFP are located at the membrane before Ci-VSP activation and act as FRET donor and acceptor, respectively. FRET is disturbed when the sensors translocate to the cytoplasm after activation of Ci-VSP. Frubby is a construct that allows for intramolecular FRET and as shown in the schema it is held at the plasma membrane via a CaaX motif. (B) Representative trace of fluorescence signal from a cell expressing Ci-VSP, TRPV1 channel, PLCδ$_1$-PH-CFP, and PLCδ$_1$-PH-YFP. Treatment with capsaicin results in an increase of CFP and a decrease of FRET signal. (C) FRET ratio traces of the two FRET sensor systems. Both the PLCδ$_1$-PH–CFP and PLCδ$_1$-PH–YFP system (e = 15) and Frubby (e = 13) show a robust decrease of FRET ratio. No decrease in FRET is observed when the catalytically dead mutant Ci-VSP C363S (e = 6) is used.

We went on to combine detection of PI(4,5)P$_2$ changes using Frubby with activation of Ci-VSP by TRPV1. As shown in Figure 3C, application of capsaicin to cells co-expressing Frubby, Ci-VSP and TRPV1 induced a strong FRET ratio change. Notably, signal amplitude was similar, but signal variability was smaller and reversibility was more complete compared to the bi-molecular FRET sensor (Figure 3C).

It should be noted that experiments using Ci-VSP to suppress PI(4,5)P$_2$ dependent activity of ion channels such as GIRK and KCNQ2/KCNQ3 have demonstrated that the apparent PI(4,5)P$_2$affinity of these effectors lies within the dynamic range we demonstrated for Frubby and the degree of PI(4,5)P$_2$ depletion can be expected to be effective on those targets of PI(4,5)P$_2$ signaling (cf. e.g., Murata et al., 2005; Iwasaki et al., 2008; Villalba-Galea et al., 2009; Rjasanow et al., submitted).

In conclusion, the combination of ligand-controlled activation of Ci-VSP together with FRET detection using the Frubby sensor allowed for the temporally precise depletion of PI(4,5)P$_2$ in intact cells and concomitant monitoring of PI(4,5) P$_2$ changes without the need for single-cell imaging.

VSP-PTEN Chimera for Functional Analysis of the Tumor Suppressor PTEN

PTEN is one of the most frequently mutated tumor suppressors in human cancer and exerts its tumor suppressor activity via direct antagonism of the phosphatidylinositol 3-kinase (PI3K) signaling pathway by counteracting the PI(3,4,5) P$_3$ production via PI3K (Chalhoub and Baker, 2009; Varga et al., 2009; Hollander et al., 2011).

Development of the chimeric protein PTEN$_{CiV}$, which fully reproduced the substrate specificity of wild-type PTEN and could be activated by membrane depolarization, provided a novel approach for controlling the activity of PTEN and studying this important tumor suppressor (Lacroix et al., 2011). Similarly to the approach we used for Ci-VSP, we attempted to activate PTEN$_{CiV}$ without electrophysiological methods and compare the phosphatase activity of different PTEN mutants. We selected TRPV1 for activation of PTEN$_{CiV}$ because high extracellular K^+ concentrations interfered with some GFP-tagged biosensors, precluding the use of KCNQ4 or TASK-3, and activation of ChR2 with blue light is incompatible with GFP-tagged sensors for PI(3,4,5)P$_3$.

Since PTEN is a PI(3,4)P$_2$/PI(3,4,5)P$_3$ 3-phosphatase, we first tested various sensor domains that specifically bind to these 3-phosphorylated PIs for their suitability to quantitatively assess activity of PTEN$_{CiV}$. Akt-PH-GFP (Servant et al., 2000) and Btk-PH-GFP (Salim et al., 1996) showed larger responses than GRP-PH-GFP (Klarlund et al., 1997). Because response kinetics were faster for Akt-PH than for Btk-PH, we chose Akt-PH-GFP as the most suitable sensor for subsequent experiments.

For examination of PTEN mutations by these methods, we focused on Lysine 125, located in the catalytic pocket of the protein (Figure 4A). PTEN mutations at this position are frequently found in cancer, according to COSMIC (Catalog of Somatic Mutations in Cancer,http://cancer.sanger.ac.uk/cancergenome/projects/cosmic/). In addition to oncogenic K125R and K125N, we also analyzed K125L because all three PTEN mutants were previously examined with a yeast survival model (Rodriguez-Escudero et al., 2011), thus allowing for direct comparison of methods.

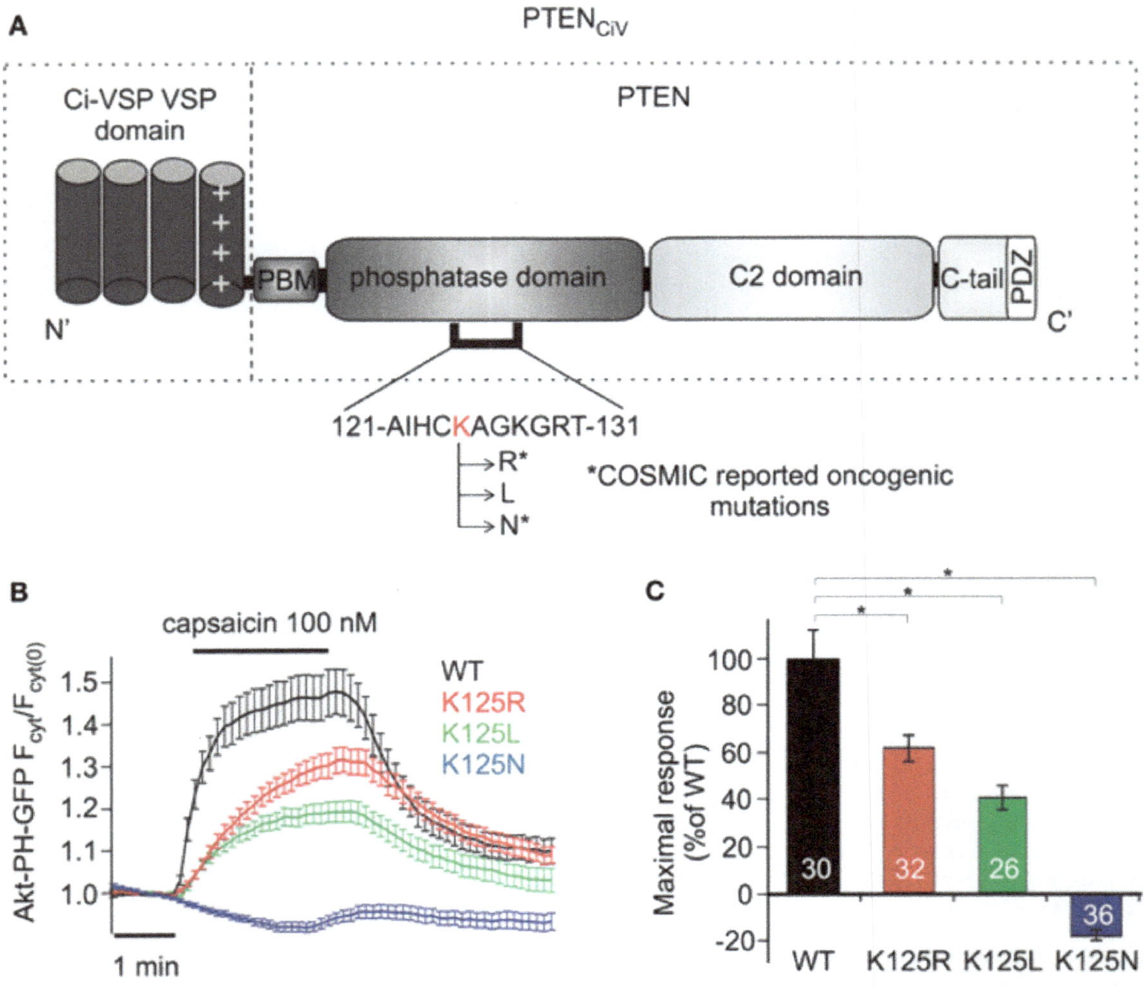

Figure 4. Analyzing PTEN mutations using PTEN$_{CiV}$ **and TRPV1 mediated membrane depolarization. (A)** Schematic representation of PTEN$_{CiV}$. The VSD of Ci-VSP, consisting of four transmembrane domains, was fused to the N terminal region of PTEN. Point mutations were introduced at K125 located in the P-loop of the catalytic domain. Asterisks mark oncogenic mutations as reported at COSMIC database (http://cancer.sanger.ac.uk/cancergenome/projects/cosmic/). PBM, PIP_2 binding module; C-tail, the carboxyl-terminus domain; PDZ, postsynaptic density protein–*Drosophila* disc large tumor suppressor–zonula occludens 1 protein domain; P-loop, phosphatase binding loop. (B) Time series traces of Akt-PH-GFP translocation upon activation with capsaicin in cells co-expressing TRPV1 channel and WT or mutated PTEN-$_{CiV}$. (C) Mean of Akt-PH-GFP translocation calculated for the last 30 s of capsaicin application, normalized to the average response measured with WT PTEN$_{CiV}$. Numbers in columns represent the number of cells. Statistically significant difference between WT and mutants is marked with *.

When challenged by capsaicin-triggered depolarization, all three mutants showed impaired phosphatase activity compared to "wildtype" PTEN$_{CiV}$ (Figure 4B). K125R and K125L presented significantly reduced enzymatic activity, yielding maximum sensor translocation of 61.17 ± 5.48% and 40.43 ± 5.11% relative to "wildtype" PTEN$_{CiV}$ activity respectively, while the K125N mutation resulted in a complete loss of enzymatic function (Figure 4C).

PTEN$_{CiV}$ Can be Exploited to Test PTEN Inhibitors

Bisperoxovanadium (bpV) compounds are specific PTEN inhibitors and have been widely used to assess the role of PTEN in cellular processes (Schmid et al., 2004). They have been characterized using *in vitro* phosphatase assays and Western blot analyses of the phosphorylation of PKB (Protein Kinase B, Akt), a downstream target of the PTEN pathway. We considered these compounds as a useful example for exploring the potency of the PTEN$_{CiV}$/live-cell approach in analyzing the effect of PTEN modulators in intact cells. We examined the effects of two compounds, bpV(HOpic) where bpV is attached to a polar functional group and bpV(phen) where bpV is attached to a neutral functional group.

TRPV1 was used to activate PTEN$_{CiV}$ as described above. Here, we measured membrane association of the PTEN activity sensor Akt-PH-GFP by TIRF microscopy (Lacroix et al., 2011). TIRF selectively captures fluorescence from the cell membrane, thus dissociation of the PI sensor used to monitor PTEN$_{CiV}$ activation results in a decrease of the fluorescent signal.

First, cells were incubated with either bpV compound overnight and throughout the experiment. As shown in Figure 5A, both compounds inhibited PTEN activity in a dose-dependent manner. Nearly complete inhibition was observed with 10 μM of bpV(phen) or bpV(HOpic). Inhibition was partially reversible when cells that were incubated overnight with 10 μM of bpV(phen) were washed and incubated with 5 mM of the reducing agent Dithiothreitol (DTT) for 3 h.

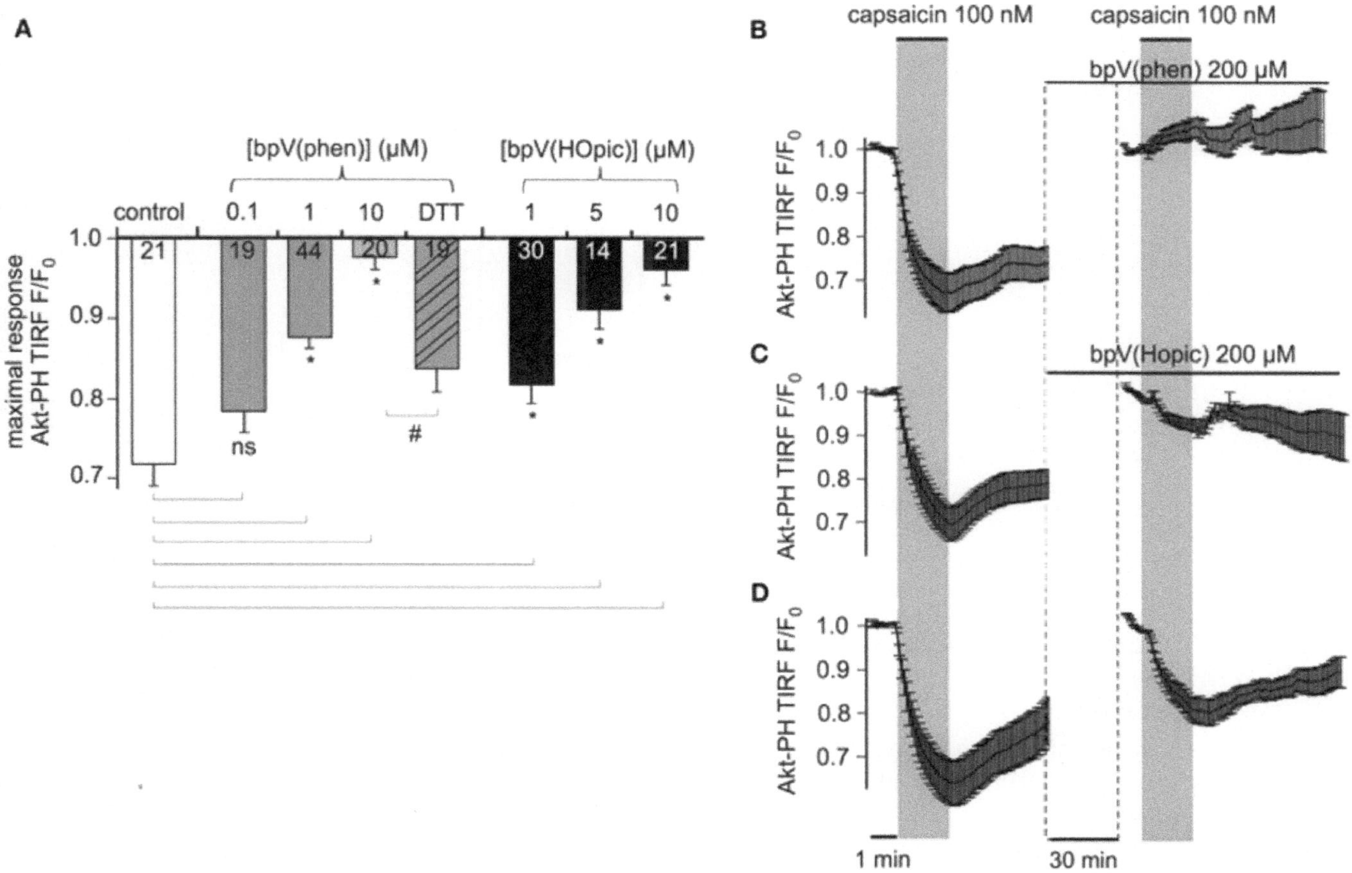

Figure 5. PTEN$_{CiV}$ **is inhibited by bisperoxovanadate compounds. (A)** Bars representing the effect of overnight incubation of bisperoxovanadate compounds on PTEN$_{CiV}$ activity. Fluorescence intensity was acquired with TIRF microscopy of HEK cells expressing PTEN$_{CiV}$, TRPV1 channel, and Akt-PH-GFP. Decrease in the fluorescence signal recorded from the membrane is observed upon capsaicin application. Cells were treated with different concentrations of the PTEN inhibitors bpV(phen) or bpV(Hopic). Cells that after an overnight incubation with 10 μM bpV(phen) were washed out and incubated for 3 h with 5 mM of the reducing agent DTT showed a partial recovery of phosphatase activity. (B–D) Graph demonstrating the effect of the inhibitors in real time. Data is normalized to the time intervals immediately preceding each capsaicin application. For a presentation of the data without re-normalization for the second agonist application.(B) HEK cells that had initially responded to a 100 nM capsaicin activation were incubated for 30 min with 200 μM of the inhibitor bpV(phen). After the 30 min incubation they presented no response to a second capsaicin application (n = 11, e = 4). (C) Application of the inhibitor bpV(Hopic) also impairs PTEN$_{CiV}$activity (n = 10, e = 4). (D) Graph showing a control experiment where no inhibitor was applied between the two capsaicin applications (n = 9, e = 3). Statistically significant difference between control and treated samples is marked with *, while the # sign indicates significant difference between the 10 μM bpV(phen) and the DTT treated samples. Numbers in columns represent the number of cells.

To investigate the time course of the inhibition, which is less easily addressed by conventional biochemical methods, we monitored the same cells before and after application of inhibitors. Cells responsive to a first stimulation by capsaicin were incubated for 30 min with 200 μM bpV(phen) (lower concentrations failed to inhibit PTEN$_{CiV}$ in acute application, data not shown). These cells did not respond to a second application of capsaicin indicating the inhibition of PTEN$_{CiV}$. Similarly, 30 min application of 200 μM of bpV(HOpic) greatly decreased the phosphatase activity. Control experiments were performed where no inhibitor was applied between subsequent capsaicin treatments. In that case, the second response—although decreased—remained pronounced. This decreased response might be explained by incomplete recovery of PI(3,4,5)P_3levels. The much lower efficacy of bpV in acute application when compared to overnight incubation may indicate that permeation into the cell membrane is slow. Indeed, when the inhibitor was applied intracellularly, the inhibitory concentrations were similar to the concentrations required with overnight incubation.

PTEN$_{CiV}$ Activation Allows for Rapid Control of PI3K Downstream Pathways

The major role of PTEN is to antagonize the phosphatidylinositol 3-kinase (PI3K) signaling pathway. This pathway is controlled by the serine-threonine protein kinase Akt (also known as protein kinase B) which is recruited and activated by plasma membrane PI(3,4,5)P_3 and in turn regulates a plethora of downstream targets through a cascade of protein kinases (Brazil et al., 2004) (Figure 6A).

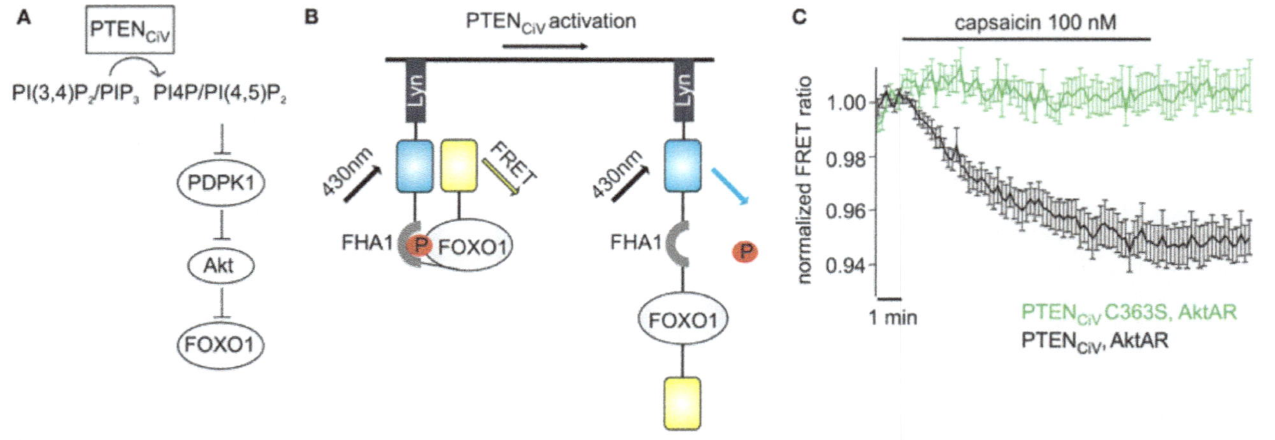

Figure 6. Activation of PTEN$_{CiV}$ **leads to rapid dephosphorylation of FOXO1. (A)** PTEN$_{CiV}$ activation decreases PI(3,4)P$_2$/PI(3,4,5)P$_3$ levels which results in Akt's displacement from the plasma membrane. This displacement prevents the phosphorylation and activation of Akt by PDPK1 (3 phosphoinositide dependent kinase 1). Akt is then unable to phosphorylate downstream targets like FOXO1. (B) Schematic representation of the Akt activation reporter, AktAR. FOXO1 at the phosphorylated state dimerizes with the Forkhead associated (FHA) domain and results in approach of the fluorophores. Activation of PTEN$_{CiV}$ should disrupt this interaction. (C) FRET ratio traces of cells expressing TRPV1, PTEN$_{CiV}$, and AktAR. Activation of PTEN$_{CiV}$ with 100 nM of capsaicin for 10 min results in a decrease of the FRET signal, indicating the rapid dephosphorylation of FOXO1 ($n = 11$, $e = 4$ for PTEN$_{CiV}$; $n = 9$, $e = 3$ for PTEN$_{CiV}$ C363S).

We wondered if our method to rapidly change PTEN activity might have the potential to interrogate the dynamics of PI3K signaling. To assess the effect of PTEN$_{CiV}$ on downstream targets of the Akt signaling pathway we utilized the Akt activity reporter, AktAR, described by Gao and Zhang (2008). AktAR is a membrane-anchored FRET sensor to monitor the kinase activity of Akt that is composed of part of the Akt-recognition sequence of FOXO1 and a forkhead-associated domain (FHA1). Akt-dependent phosphorylation of FOXO1 leads to dimerization of the two domains and an increase in the FRET signal, i.e. the FRET ratio directly correlates to the kinase activity of Akt (Figure 6B). In cells co-expressing TRPV1, PTEN$_{CiV}$, and AktAR, application of capsaicin for 10 min indeed led to a decrease of the FRET signal indicating decreased Akt activity. In contrast, no decrease was observed when catalytically-dead PTEN$_{CiV}$ C363S was co-expressed (Figure 6C). In conclusion, ligand-induced activation of PTEN$_{CiV}$ does not only allow manipulation of PI(3,4,5)P$_3$ at the plasma membrane, but also control and analysis of the downstream signaling cascade, including phosphorylation of Akt substrates in intact living cells.

DISCUSSION

We developed a fast, simple, reversible, and scalable method that can be used not only as a tool to control PI levels in living cells, but also to study the tumor suppressor PTEN. The approach is based on well characterized VSPs, namely the PI(4,5)P$_2$ 5-phosphatase Ci-VSP (Iwasaki et al., 2008;Halaszovich et al., 2009) and the engineered PI(3,4)P$_2$/PI(3,4,5)P$_3$ 3-phosphatase PTEN$_{CiV}$(Lacroix et al., 2011). To activate these enzymes in living cells, the membrane potential must be depolarized. So far, this required the use of electrophysiological techniques, usually whole-cell patch clamping for mammalian cells. Since such methods are not available to many laboratories, either because of lack of equipment or electrophysiological expertise, we sought to find a simpler and broadly applicable method for membrane depolarization. Our data demonstrate that activation of overexpressed cation channels is a simple and powerful method to activate Ci-VSP or chimeric VSPs and thereby manipulate PI(4,5)P$_2$ and PI(3,4,5)P$_3$ concentrations in a temporally well-controlled manner.

Currently, widely used methods for acutely altering PI concentrations are based on the recruitment of PI metabolizing enzymes to the membrane through a dimerization strategy, triggered either by rapamycin (Suh et al., 2006; Varnai et al., 2006) or via an optogenetic approach (Idevall-Hagren et al., 2012). Compared to these methods, the main advantage of the VSP-based approach is the reversibility of PI changes with times for 50% recovery of 12 to 28 s. In contrast, the rapamycin system is virtually irreversible and with the optogenetic method half-maximal recovery of PI(4,5)P$_2$, measured as either sensor domain translocation or KCNQ2/3 current inhibition, required several minutes (Idevall-Hagren et al., 2012). One additional disadvantage of the optogenetic approach is that its sensitivity to blue and green light places restrictions on which fluorophores can be used for imaging / monitoring purposes. On the other hand it allows for subcellular manipulation of PI concentrations, whereas VSP activation in our system always affects the entire plasma membrane of the cell. We further note that activation of VSPs by each of the tested ion channels allowed for a graded depletion of PI(4,5)P$_2$. This high level of control over PI(4,5)P$_2$ levels may be used to compare the apparent affinity of intracellular PI(4,5)P$_2$ effectors in a semi-quantitative manner, as we have demonstrated in earlier work (Halaszovich et al., 2009).

Each of the three classes of channels explored here activated Ci-VSP with similar efficacy, which was assessed as degree of translocation of the PI(4,5)P$_2$ sensor tubbyC, tagged with either GFP or RFP. Although not being a direct quantification of actual [PI(4,5)P$_2$], this method provides a sufficient means of comparison of degree of PI(4,5)P$_2$ depletion. While we cannot exclude the possibility that the choice of the tag, which was based on the technical requirements discussed below, influences the degree of fluorescence increase seen in the experiments, we expect the main determinant to be the properties of the $PI(4,5)P_2$ binding domain itself. Therefore a meaningful comparison of the degree of Ci-VSP activity can be drawn from experiments even when differently tagged sensors are used.

Advantages and limitations of each channel type mostly relate to the technical aspects of channel activation. Depolarization through constitutively open K^+ channels requires full exchange of bath solution to alter the K^+ concentration; in some cases, high extracellular K^+ concentrations interfered with GFP fluorescence. Activation of ligand gated channels does not necessarily require complete bath exchange, since the ligand can be added to the bath solution, especially when small concentrations can be used; for the TRPV1 channel used here, an agonist concentration as low as 100 nM was sufficient for maximal effect. For plate-reader systems, there is equipment available to perform such an application of a ligand to the assay. A disadvantage of using TRPV1 is the requirement for a Ca^{2+}-free bath solution, although in many cases this will not be a severe restriction. Also, use of TRPV1 was the method showing the slowest kinetics, which is probably related to ligand permeation and binding/unbinding. Manipulations of the bath solution can be avoided altogether when the optogenetic approach using a light gated channel (ChR2) is used. Here, we achieved the fastest response times. Disadvantages of using ChR2 are that a low pH (6.5) and Ca^{2+}-free bath solution had to be used and that the spectral characteristics of ChR2 preclude the use of certain wavelengths for excitation of fluorophores. Notably, we were not able to use GFP-tagged PI-sensors and used RFP instead. We would like to point out that naturally we could only study a subset of ion channels that potentially can be useful for VSP activation. Further candidates like e.g., the FMRFamide-gated sodium channel FaNaC (Lingueglia et al., 1995), which has been used to depolarize tsA201 cells as well as to activate mammalian neurons (Scheel et al., 2005; Schanuel et al., 2008), might be better suited for specific applications and should be considered accordingly. One caveat when using ion channels for VSP activation is the $PI(4,5)P_2$ dependence of some of these channels. Regarding those used in this study, KCNQ4 shows inhibition following $PI(4,5)P_2$ depletion, in contrast to the $PI(4,5)P_2$ insensitive TASK-3 (Lindner et al., 2011), also TRPV1 has been described as $PI(4,5)P_2$ sensitive (Klein et al., 2008). While the $PI(4,5)P_2$ sensitivity of these two channels was not sufficient to make them unsuitable for the purpose of VSP activation, for more sensitive channels or in cell systems showing lower baseline $PI(4,5)P_2$ concentrations this might not be the case. One such example might be the highly sensitive R555C mutant of KCNQ1, where even basal Ci-VSP activity is reported to be sufficient for complete channel deactivation (Coyan et al., 2014). Therefore when deviating from the ion channels or cell systems used in the present study their suitability should be carefully assessed.

Beyond manipulation of $PI(4,5)P_2$ and $PI(3,4)P_2$/$PI(3,4,5)P_3$ levels, the availability of $PTEN_{CiV}$ considerably extends the range of useful applications of the method presented here. The role of PTEN in pathological conditions like tumors of sporadic and germline origin and ASDs is well established (Cantley and Neel, 1999; Varga et al., 2009; Hollander et al., 2011). Biochemical *in vitro* techniques or a heterologous yeast reconstitution system have usually been used to investigate the effect of mutations on protein function or to study chemical PTEN inhibitors and activators.

The method we present should provide a useful complementation of the current spectrum of methods. We note several potential advantages of this approach. First, our method proved to be more sensitive than the published yeast-based method (Rodriguez-Escudero et al., 2011) in assessing the effect of PTEN mutations (see below). Second, in the living mammalian cell, PTEN acts on its substrates in the native membrane environment, which can differ substantially from the composition of vesicles used to present the potential substrates or even soluble substrate analogs used in *in vitro* assays. This might affect activity or substrate specificity inferred from such experiments. For example, the human VSP ortholog hVSP1 (also known as TPIP or TPTE2) was first described as a $PI(3,4,5)P_3$ 3-phosphatase based on *in vitro* data (Walker et al., 2001). However, recent experiments in mammalian cells using a voltage sensitive hVSP1 chimera revealed that it is a $PI(4,5)P_2$/$PI(3,4,5)P_3$ 5-phosphatase (Halaszovich et al., 2012). Finally, the new method is applicable to a wide range of cell types, including cell lines derived from human tumors, once heterologous expression of the molecular components (VSP and cation channel) can be achieved in that cell line.

For demonstration purposes, we studied PTEN mutations, including known oncogenic mutations. For some of these PTEN mutations, results obtained with the $PTEN_{CiV}$ approach were consistent with previously published findings from *in vitro* and yeast assays (Rodriguez-Escudero et al., 2011). Thus, $PI(3,4,5)P_3$ phosphatase activity was strongly decreased for $PTEN_{CiV}$ K125L and abrogated for $PTEN_{CiV}$ K125N (Figures 4B,C). However, the yeast survival method did not detect any defect of the PTEN K125R, although the mutant is annotated as an oncogenic mutation. Our approach indeed detected a decrease of phosphatase activity by 40%. This finding might explain the tumorigenic effect of that mutation and also highlights the sensitivity of our approach. As ASD-related mutations have been shown to decrease but not abrogate PTEN phosphatase activity (Rodriguez-Escudero et al., 2011), this high sensitivity should render our method a well-suited tool for testing ASD- and tumor-associated PTEN mutations.

Similarly, the $PTEN_{CiV}$ method enabled analysis of pharmacological manipulation of PTEN. *In vitro* and *in vivo*, bpV compounds inhibit PTEN at nanomolar concentrations with minimal cytotoxicity (Schmid et al., 2004). In that study, more indirect methods like assessing the phosphorylation state of Akt/PKB were used, whereas we directly monitored PTEN's substrate $PI(3,4,5)P_3$ to determine the effect of these inhibitors in living cells. The bpV compounds have been used to study the involvement of PTEN in signaling, proliferation, apoptosis, and neurogenesis (Band et al., 1997; Batty et al., 2007; Pi et al., 2012; Tian et al., 2012; Becker et al., 2013); however, higher concentrations (200 nM–100 μM) were required. We found comparable concentrations to effectively inhibit PTEN activity in our live cell assay. This technique holds promise for high-throughput screens for novel substances with pharmacological activity toward PTEN. Such substances could also include enhancers of PTEN activity, which might be used for anticancer therapy.

These findings demonstrate that $PTEN_{CiV}$ can be used to study mutations and compounds affecting the catalytic domain of PTEN. While this technique has clear advantages in terms of simplicity and through-put over existing techniques, it might be unable to detect the effect of mutations or interactions outside the catalytic domain which might affect PTEN activity, e.g. by interfering with membrane association. Therefore we view the VSP-based method not as a complete replacement for, but as a complement to existing methods.

Moreover, because the $PTEN_{CiV}$ method is essentially a real-time approach in living cells, it can yield insights into the temporal cellular dynamics of $PI(4,5)P_2$ and $PI(3,4,5)P_3$ signaling that are difficult to obtain with a purely biochemical approach. As an example, we showed that down-regulation of $PI(3,4,5)P_3$ (and $PI(3,4)P_2$) concentration resulted in a surprisingly rapid switching-off of signaling by dephosphorylation of intermediates within the Akt signaling cascade, which (to our knowledge) has not been appreciated before. This finding points to a rather high activity of the protein phosphatase(s) that antagonize Akt kinase activity.

In summary, we presented a series of simple to use and broadly applicable methods that provide temporal control of VSPs. This allows for manipulation of PI levels and subsequently manipulation of their downstream targets. Moreover, using the engineered voltage sensitive enzyme $PTEN_{CiV}$ we suggest a novel method for charactering this critical tumor suppressor. Further development of a method combining $PTEN_{CiV}$ with an optimized FRET sensor, could create a powerful automated tool to screen for pathological PTEN mutations and for substances affecting activity of PTEN.

ACKNOWLEDGMENTS

The authors thank T. Balla for the gift of PLCδ1-PH, Btk-PH, Akt1-PH, and GRP1-PH, L. Shapiro for tubbyC, M. Matsuda for Pippi-$PI(4,5)P_2$, J. Zhang for AktAR, T. Jentsch for KCNQ4, J. Daut for TASK-3, T. Plant for TRPV1, K. Deisseroth for ChR2 H134R, Y. Okamura for Ci-VSP and J. Downward for p110α K227E.

ABBREVIATIONS

ASDs, autism spectrum disorders; bpV, bisperoxovanadium; C-tail, the carboxyl-terminus domain; ChR2, channelrhodopsin-2; Ci-VSP, Ciona intestinalis voltage sensitive phosphatase; COSMIC, Catalogue of Somatic Mutations in Cancer; FHA1, forkhead-associated domain; FOXO1, forkhead box 1; FRET, Förster resonance energy transfer; P-loop, phosphatase binding loop; PBM, PIP2 binding module; PDZ, postsynaptic density protein-Drosophila disc large tumor suppressor-zonula occludens 1 protein domain; PH, pleckstrin homology; PI, phosphoinositide; PI3K, phosphatidylinositol 3-kinase; PKB, protein kinase B; PTEN, phosphatase and tensin homolog deleted from chromosome 10; PTENCiV, PTEN with Ciona intestinalis voltage sensor domain; TIRF, total internal reflection fluorescence; VSD, voltage sensor domain; VSP, voltage sensitive phosphatase.

REFERENCES

1. Balla, T., Szentpetery, Z., and Kim, Y. J. (2009). Phosphoinositide signaling: new tools and insights. *Physiology* 24, 231–244. doi: 10.1152/physiol.00014.2009
2. Band, C. J., Posner, B. I., Dumas, V., and Contreres, J. O. (1997). Early signaling events triggered by peroxovanadium [bpV(phen)] are insulin receptor kinase (IRK)-dependent: specificity of inhibition of IRK-associated protein tyrosine phosphatase(s) by bpV(phen). *Mol. Endocrinol.* 11, 1899–1910. doi: 10.1210/mend.11.13.0041
3. Batty, I. H., van der Kaay, J., Gray, A., Telfer, J. F., Dixon, M. J., and Downes, C. P. (2007). The control of phosphatidylinositol 3,4-bisphosphate concentrations by activation of the Src homology 2 domain containing inositol polyphosphate 5-phosphatase 2, SHIP2. *Biochem. J.* 407, 255–266. doi: 10.1042/BJ20070558
4. Becker, L., Peterson, J., Kulkarni, S., and Pasricha, P. J. (2013). *Ex vivo* neurogenesis within enteric ganglia occurs in a PTEN dependent manner. *PLoS ONE* 8:e59452. doi: 10.1371/journal.pone.0059452
5. Brazil, D. P., Yang, Z. Z., and Hemmings, B. A. (2004). Advances in protein kinase B signalling: AKTion on multiple fronts.*Trends Biochem. Sci.* 29, 233–242. doi: 10.1016/j.tibs.2004.03.006
6. Brown, E. J., Albers, M. W., Shin, T. B., Ichikawa, K., Keith, C. T., Lane, W. S., et al. (1994). A mammalian protein targeted by G1-arresting rapamycin-receptor complex. *Nature* 369, 756–758. doi: 10.1038/369756a0
7. Cantley, L. C., and Neel, B. G. (1999). New insights into tumor suppression: PTEN suppresses tumor formation by restraining the phosphoinositide 3-kinase/AKT pathway. *Proc. Natl. Acad. Sci. U.S.A.* 96, 4240–4245. doi: 10.1073/pnas.96.8.4240
8. Caterina, M. J., Schumacher, M. A., Tominaga, M., Rosen, T. A., Levine, J. D., and Julius, D. (1997). The capsaicin receptor: a heat-activated ion channel in the pain pathway. *Nature* 389, 816–824. doi: 10.1038/39807
9. Chalhoub, N., and Baker, S. J. (2009). PTEN and the PI3-kinase pathway in cancer. *Annu. Rev. Pathol.* 4, 127–150. doi: 10.1146/annurev.pathol.4.110807.092311
10. Chambard, J. M., and Ashmore, J. F. (2005). Regulation of the voltage-gated potassium channel KCNQ4 in the auditory pathway. *Pflugers Arch.* 450, 34–44. doi: 10.1007/s00424-004-1366-2
11. Cid, V. J., Rodriguez-Escudero, I., Andres-Pons, A., Roma-Mateo, C., Gil, A., den Hertog, J., et al. (2008). Assessment of PTEN tumor suppressor activity in nonmammalian models: the year of the yeast. *Oncogene* 27, 5431–5442. doi: 10.1038/onc.2008.240
12. Coutinho-Budd, J. C., Snider, S. B., Fitzpatrick, B. J., Rittiner, J. E., and Zylka, M. J. (2013). Biological constraints limit the use of rapamycin-inducible FKBP12-Inp54p for depleting PIP2 in dorsal root ganglia neurons. *J. Negat. Results Biomed.* 12:13. doi: 10.1186/1477-5751-12-13
13. Coyan, F. C., Abderemane-Ali, F., Amarouch, M. Y., Piron, J., Mordel, J., Nicolas, C. S., et al. (2014). A long QT mutation substitutes cholesterol for phosphatidylinositol-4,5-bisphosphate in KCNQ1 channel regulation. *PLoS ONE* 9:e93255. doi: 10.1371/journal.pone.0093255
14. Enyedi, P., and Czirjak, G. (2010). Molecular background of leak K+ currents: two-pore domain potassium

channels. *Physiol. Rev.* 90, 559–605. doi: 10.1152/physrev.00029.2009

15. Falkenburger, B. H., Jensen, J. B., and Hille, B. (2010). Kinetics of PIP2 metabolism and KCNQ2/3 channel regulation studied with a voltage-sensitive phosphatase in living cells. *J. Gen. Physiol.* 135, 99–114. doi: 10.1085/jgp.200910345

16. Gao, X., and Zhang, J. (2008). Spatiotemporal analysis of differential Akt regulation in plasma membrane microdomains. *Mol. Biol. Cell* 19, 4366–4373. doi: 10.1091/mbc.E08-05-0449

17. Halaszovich, C. R., Leitner, M. G., Mavrantoni, A., Le, A., Frezza, L., Feuer, A., et al. (2012). A human phospholipid phosphatase activated by a transmembrane control module. *J. Lipid. Res.* 53, 2266–2274. doi: 10.1194/jlr.M026021

18. Halaszovich, C. R., Schreiber, D. N., and Oliver, D. (2009). Ci-VSP is a depolarization-activated phosphatidylinositol-4,5-bisphosphate and phosphatidylinositol-3,4,5-trisphosphate 5'-phosphatase. *J. Biol. Chem.* 284, 2106–2113. doi: 10.1074/jbc.M803543200

19. Hertel, F., Switalski, A., Mintert-Jancke, E., Karavassilidou, K., Bender, K., Pott, L., et al. (2011). A genetically encoded tool kit for manipulating and monitoring membrane phosphatidylinositol 4,5-bisphosphate in intact cells. *PLoS ONE* 6:e20855. doi: 10.1371/journal.pone.0020855

20. Hollander, M. C., Blumenthal, G. M., and Dennis, P. A. (2011). PTEN loss in the continuum of common cancers, rare syndromes and mouse models. *Nat. Rev. Cancer* 11, 289–301. doi: 10.1038/nrc3037

21. Hossain, M. I., Iwasaki, H., Okochi, Y., Chahine, M., Higashijima, S., Nagayama, K., et al. (2008). Enzyme domain affects the movement of the voltage sensor in ascidian and zebrafish voltage-sensing phosphatases. *J. Biol. Chem.* 283, 18248–18259. doi: 10.1074/jbc.M706184200

22. Idevall-Hagren, O., Dickson, E. J., Hille, B., Toomre, D. K., and De Camilli, P. (2012). Optogenetic control of phosphoinositide metabolism. *Proc. Natl. Acad. Sci. U.S.A.* 109, E2316–E2323. doi: 10.1073/pnas.1211305109

23. Iwasaki, H., Murata, Y., Kim, Y., Hossain, M. I., Worby, C. A., Dixon, J. E., et al. (2008). A voltage-sensing phosphatase, Ci-VSP, which shares sequence identity with PTEN, dephosphorylates phosphatidylinositol 4,5-bisphosphate. *Proc. Natl. Acad. Sci. U.S.A.* 105, 7970–7975. doi: 10.1073/pnas.0803936105

24. Klarlund, J. K., Guilherme, A., Holik, J. J., Virbasius, J. V., Chawla, A., and Czech, M. P. (1997). Signaling by phosphoinositide-3,4,5-trisphosphate through proteins containing pleckstrin and Sec7 homology domains. *Science* 275, 1927–1930. doi: 10.1126/science.275.5308.1927

25. Klein, R. M., Ufret-Vincenty, C. A., Hua, L., and Gordon, S. E. (2008). Determinants of molecular specificity in phosphoinositide regulation. Phosphatidylinositol (4,5)-bisphosphate (PI(4,5)P2) is the endogenous lipid regulating TRPV1. *J. Biol. Chem.* 283, 26208–26216. doi: 10.1074/jbc.M801912200

26. Lacroix, J., Halaszovich, C. R., Schreiber, D. N., Leitner, M. G., Bezanilla, F., Oliver, D., et al. (2011). Controlling the activity of a phosphatase and tensin homolog (PTEN) by membrane potential. *J. Biol. Chem.* 286, 17945–17953. doi: 10.1074/jbc.M110.201749

27. Lin, J. Y. (2011). A user's guide to channelrhodopsin variants: features, limitations and future developments. *Exp. Physiol.* 96, 19–25. doi: 10.1113/expphysiol.2009.051961

28. Lin, J. Y., Lin, M. Z., Steinbach, P., and Tsien, R. Y. (2009). Characterization of engineered channelrhodopsin variants with improved properties and kinetics. *Biophys. J.* 96, 1803–1814. doi: 10.1016/j.bpj.2008.11.034

29. Lindner, M., Leitner, M. G., Halaszovich, C. R., Hammond, G. R., and Oliver, D. (2011). Probing the regulation of TASK potassium channels by PI4,5P(2) with switchable phosphoinositide phosphatases. *J. Physiol.* 589, 3149–3162. doi: 10.1113/jphysiol.2011.208983

30. Lingueglia, E., Champigny, G., Lazdunski, M., and Barbry, P. (1995). Cloning of the amiloride-sensitive FMRFamide peptide-gated sodium channel. *Nature* 378, 730–733. doi: 10.1038/378730a0

31. Murata, Y., Iwasaki, H., Sasaki, M., Inaba, K., and Okamura, Y. (2005). Phosphoinositide phosphatase activity coupled to an intrinsic voltage sensor. *Nature* 435, 1239–1243. doi: 10.1038/nature03650

32. Nagel, G., Szellas, T., Huhn, W., Kateriya, S., Adeishvili, N., Berthold, P., et al. (2003). Channelrhodopsin-2, a directly light-gated cation-selective membrane channel. *Proc. Natl. Acad. Sci. U.S.A.* 100, 13940–13945. doi: 10.1073/pnas.1936192100

33. Pi, W., Guo, X., Su, L., and Xu, W. (2012). BMP-2 up-regulates PTEN expression and induces apoptosis of pulmonary artery smooth muscle cells under hypoxia. *PLoS ONE* 7:e35283. doi: 10.1371/journal.pone.0035283

34. Ratzan, W. J., Evsikov, A. V., Okamura, Y., and Jaffe, L. A. (2011). Voltage sensitive phosphoinositide phosphatases of Xenopus: their tissue distribution and voltage dependence. *J. Cell Physiol.* 226, 2740–2746. doi: 10.1002/jcp.22854

35. Rebecchi, M. J., and Pentyala, S. N. (2000). Structure, function, and control of phosphoinositide-specific phospholipase C.*Physiol. Rev.* 80, 1291–1335. Available online at: http://physrev.physiology.org/content/80/4/1291

36. Rodriguez-Escudero, I., Oliver, M. D., Andres-Pons, A., Molina, M., Cid, V. J., and Pulido, R. (2011). A comprehensive functional analysis of PTEN mutations: implications in tumor- and autism-related syndromes. *Hum. Mol. Genet.* 20, 4132–4142. doi: 10.1093/hmg/ddr337

37. Rodriguez-Viciana, P., Warne, P. H., Vanhaesebroeck, B., Waterfield, M. D., and Downward, J. (1996). Activation of phosphoinositide 3-kinase by interaction with Ras and by point mutation. *EMBO J.* 15, 2442–2451.

38. Sakata, S., Hossain, M. I., and Okamura, Y. (2011). Coupling of the phosphatase activity of Ci-VSP to its voltage sensor activity over the entire range of voltage sensitivity. *J. Physiol.* 589, 2687–2705. doi: 10.1113/jphysiol.2011.208165

39. Salim, K., Bottomley, M. J., Querfurth, E., Zvelebil, M. J., Gout, I., Scaife, R., et al. (1996). Distinct specificity in the recognition of phosphoinositides by the pleckstrin homology domains of dynamin and Bruton's tyrosine kinase. *EMBO J.* 15, 6241–6250.

40. Santagata, S., Boggon, T. J., Baird, C. L., Gomez, C. A., Zhao, J., Shan, W. S., et al. (2001). G-protein signaling through tubby proteins. *Science* 292, 2041–2050. doi: 10.1126/science.1061233

41. Schanuel, S. M., Bell, K. A., Henderson, S. C., and McQuiston, A. R. (2008). Heterologous expression of the invertebrate FMRFamide-gated sodium channel as a mechanism to selectively activate mammalian neurons. *Neuroscience* 155, 374–386. doi: 10.1016/j.neuroscience.2008.05.052

42. Scheel, O., Zdebik, A. A., Lourdel, S., and Jentsch, T. J. (2005). Voltage-dependent electrogenic chloride/proton exchange by endosomal CLC proteins. *Nature* 436, 424–427. doi: 10.1038/nature03860

43. Schmid, A. C., Byrne, R. D., Vilar, R., and Woscholski, R. (2004). Bisperoxovanadium compounds are potent PTEN inhibitors.*FEBS Lett.* 566, 35–38. doi: 10.1016/j.febslet.2004.03.102

44. Servant, G., Weiner, O. D., Herzmark, P., Balla, T., Sedat, J. W., and Bourne, H. R. (2000). Polarization of chemoattractant receptor signaling during neutrophil chemotaxis. *Science* 287, 1037–1040. doi: 10.1126/science.287.5455.1037

45. Suh, B. C., Inoue, T., Meyer, T., and Hille, B. (2006). Rapid chemically induced changes of PtdIns(4,5)P2 gate KCNQ ion channels. *Science* 314, 1454–1457. doi: 10.1126/science.1131163

46. Szentpetery, Z., Balla, A., Kim, Y. J., Lemmon, M. A., and Balla, T. (2009). Live cell imaging with protein domains capable of recognizing phosphatidylinositol 4,5-bisphosphate; a comparative study. *BMC Cell Biol.* 10:67. doi: 10.1186/1471-2121-10-67

47. Tian, Y., Daoud, A., and Shang, J. (2012). Effects of bpV(pic) and bpV(phen) on H9c2 cardiomyoblasts during both hypoxia/reoxygenation and H2O2-induced injuries. *Mol. Med. Rep.* 5, 852–858. doi: 10.3892/mmr.2011.737

48. van der Wal, J., Habets, R., Varnai, P., Balla, T., and Jalink, K. (2001). Monitoring agonist-induced phospholipase C activation in live cells by fluorescence resonance energy transfer. *J. Biol. Chem.* 276, 15337–15344. doi: 10.1074/jbc.M007194200

49. Varga, E. A., Pastore, M., Prior, T., Herman, G. E., and McBride, K. L. (2009). The prevalence of PTEN mutations in a clinical pediatric cohort with autism spectrum disorders, developmental delay, and macrocephaly. *Genet. Med.* 11, 111–117. doi: 10.1097/GIM.0b013e31818fd762

50. Varnai, P., and Balla, T. (1998). Visualization of phosphoinositides that bind pleckstrin homology domains: calcium- and agonist-induced dynamic changes and relationship to myo-[3H]inositol-labeled phosphoinositide pools. *J. Cell Biol.* 143, 501–510. doi: 10.1083/jcb.143.2.501

51. Varnai, P., and Balla, T. (2006). Live cell imaging of phosphoinositide dynamics with fluorescent protein domains. *Biochim. Biophys. Acta* 1761, 957–967. doi: 10.1016/j.bbalip.2006.03.019

52. Varnai, P., and Balla, T. (2008). Live cell imaging of phosphoinositides with expressed inositide binding protein domains.*Methods* 46, 167–176. doi: 10.1016/j.ymeth.2008.09.019

53. Varnai, P., Thyagarajan, B., Rohacs, T., and Balla, T. (2006). Rapidly inducible changes in phosphatidylinositol 4,5-bisphosphate levels influence multiple regulatory functions of the lipid in intact living cells. *J. Cell Biol.* 175, 377–382. doi: 10.1083/jcb.200607116

54. Villalba-Galea, C. A., Miceli, F., Taglialatela, M., and Bezanilla, F. (2009). Coupling between the voltage-sensing and phosphatase domains of Ci-VSP. *J. Gen. Physiol.* 134, 5–14. doi: 10.1085/jgp.200910215

55. Walker, S. M., Downes, C. P., and Leslie, N. R. (2001). TPIP: a novel phosphoinositide 3-phosphatase. *Biochem. J.* 360, 277–283. doi: 10.1042/0264-6021:3600277

56. Yoshizaki, H., Mochizuki, N., Gotoh, Y., and Matsuda, M. (2007). Akt – PDK1 complex mediates epidermal growth factor-induced membrane protrusion through ral activation. *Mol. Biol. Cell* 18, 119–128. doi: 10.1091/mbc.E06-05-0467

57. Yudin, Y., Lukacs, V., Cao, C., and Rohacs, T. (2011). Decrease in phosphatidylinositol 4,5-bisphosphate levels mediates desensitization of the cold sensor TRPM8 channels. *J. Physiol.* 589, 6007–6027. doi: 10.1113/jphysiol.2011.220228

CHAPTER 3

CHARACTERIZATION OF PHOTOVOLTAIC PANELS BY MEANS OF THERMOGRAPH ANALYSIS

Noe Samano, José Alfredo Padilla-Medina, and Nimrod Vázquez

Electronics Engineering Department, Technological Institute of Celaya, 38010 Celaya, GTO, Mexico

ABSTRACT

Solar panels have become attractive in order to generate and supply electricity in commercial and residential applications. Their increased module efficiencies have caused not only a massive production but also a sensible drop on sale prices. Methods of characterization, instrumentation for in situ measurements, defect monitoring, process control, and performance are required. A temperature characterization method by means of thermograph analysis is exposed in this paper. The method was applied to multicrystalline modules, and the characterization was made with respect to two different variables, first a thermal transient and seconds a characterization with respect to the current. The method is useful in order to detect hot spots caused by mismatch conditions in electrical parameters. The description, results, and limitations of the proposed method are discussed.

INTRODUCTION

The energy availability in photovoltaic (PV) panel strongly depends on temperature and solar irradiation [1]. Actually, the temperature is one of the main factors which affect seriously the efficiency of solar cells. It is well known that the generated voltage with a solar cell varies with temperature [2]. Dependence on the temperature in a typical PV module characteristic is illustrated in Figure 1.

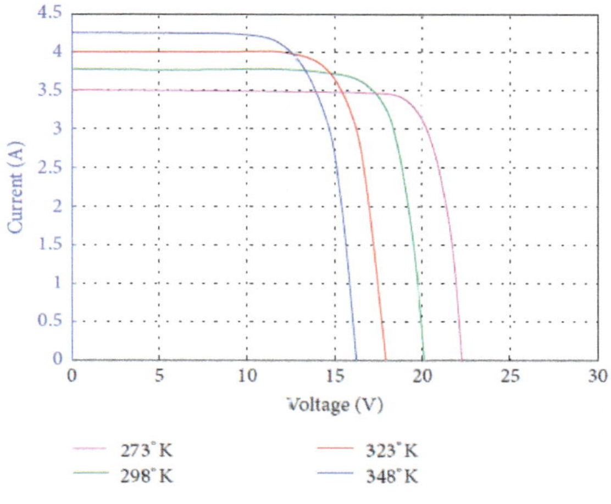

Figure 1: Typical PV solar curve and the effect of temperature.

Citation: Noe Samano, José Alfredo Padilla-Medina, and Nimrod Vázquez, "Characterization of Photovoltaic Panels by means of Thermograph Analysis," International Journal of Photoenergy, vol. 2016, Article ID 8560728, 7 pages, 2016. doi:10.1155/2016/8560728

When temperature increases, the generated voltage decreases. Apart from this temperature effect on the cell voltage, there exists another phenomenon which is related to the temperature known as a "hot spot." A hot spot is a specific well-located region with the higher temperature than the other parts of the cell. While temperature value causes voltage affections, the hot spot phenomenon causes a wide variety of cell problems, as lower efficiency and physical damage.

A thermal transient analysis is reported in the literature [1]: such study was obtained by registering the temperature for a single amorphous silicon module and a couple of monocrystalline modules every minute after a change in environmental conditions. Although the main purpose of this paper was not transient thermal characterization, authors used this for explaining the effect of the module heat in a capacitance during described conditions.

This reported method [1] began by the whole covering of the module in order to reach the equilibrium temperature with the environment. Once steady state temperature is reached, the module was uncovered and a measure of temperature was taken every minute with the aid of thermocouples. Measures were recorded during a period of 24 minutes. Results are shown in Figure 2.

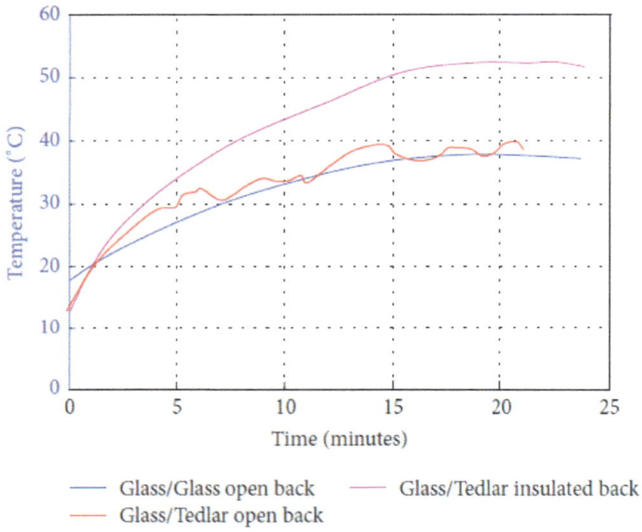

Figure 2: Transient of three PV modules as shown in [1].

A hot spot may be produced due to the next three reasons: a shadow projected in the module [3–5], a mismatch condition between cells [6–8], or a manufacturing defect [9–11].

Traditionally, the cells in a module are connected in series, and the current at the maximum power point I_{mpp} becomes critical [12]. When a cell in a serial string produces less power than the rest (due to a hot spot), the maximum power of the string cannot match the maximum power of that "poor" cell. This means that the cells cannot give more energy to the output load, although they are able to produce it.

Therefore, the generated current on excess by any "good" cell is forced to circulate through the internal diode by itself. When module works near short-circuit condition, then voltage may exhibit this as a "poor" cell. In these circumstances, the cell dissipates power instead of generating it.

Hot spots detection is important because it is easier for problem identification or finding photovoltaic module defects. Once a hot spot is detected in a module or array, it is evident that an amount of the total power produced by the array is being dissipated in that cell instead of the output load.

Since temperature and hot spots problems are the main factors which affect photovoltaic modules, research on this issue has become important, in order to have a better understanding of these phenomena and, therefore, improve PV system designs.

The PV panel's characterization by means of thermograph analysis for in situ measurements, defect monitoring, process control, and performance is proposed in this paper. Technique on infrared temperature measurement is used for detecting hot spots and establishing the thermal behavior of solar cells.

This paper focused on characterizing thermal behavior. Thermal transients in PV panels were studied at first; and, later, temperature characterization with respect to the module current was made. These two studies allow us to detect and identify the existence of hot spots in the solar module.

The paper is organized as described next: some facts of PV panels are mentioned in Section 2; the work bench for the characterization is detailed in Section 3; in Section 4, the thermal transient is discussed; the current characterization is presented in Section 5; and some final conclusions are given at the end.

Model and Facts in PV Panels

When a PV module is under the sunlight, it produces energy. When the generated energy is not required, it is consumed as heat by the panel itself, since the PV panel is not operating at the maximum power point, its life time is deteriorated no matter if energy is not delivered. It has to be kept in mind that the possible coldest temperature is reached once the PV panel is operating at the maximum power point.

The hot spots existence in a PV panel decreases the life time for cells [13, 14]; also, as a consequence, the amount of generated energy becomes reduced. The effect of a hot spot is similar to the PV panel behavior under partially shaded areas [15, 16], so that multiple maximum power points start to appear, which seriously affect energy production.

A simplified model for the PV panel is found in [16]:

$$i_{\mathrm{PV}} = \lambda I_{\mathrm{sc}} - \lambda I_s \left(e^{(q v_{\mathrm{PV}}/AKT)} - 1\right), \tag{1}$$

where A is the ideality factor of the diode, K is the Boltzmann constant, q is the electron charge, λ is the percentage of

irradiance ($1 = 100\%$), I_{sc} is the short-circuit current of the PV panel, I_s is the saturation current of the diode, T is the temperature of the ambient in ∘K, V_{PV} is the voltage of PV panel, and i_{PV} is the current of PV panel.

It should be noticed that the only variables to be considered are the irradiance, the temperature, the cell voltage, and current, once the PV panel has been built. The other terms are constants. The PV panel power is then

$$\frac{d}{dt} p_{\mathrm{PV}} = \left(\lambda I_{\mathrm{sc}} + \lambda I_s - \left(\lambda I_s + \frac{q v_{\mathrm{PV}}}{AKT}\right) e^{(q v_{\mathrm{PV}}/AKT)}\right) \frac{d}{dt} v_{\mathrm{PV}}. \tag{2}$$

The maximum power point may be obtained by making the derivative equal to zero, and neglecting the irradiance and temperature variations, the next equation is obtained:

$$\frac{d}{dt}p_{PV} = \left(\lambda I_{sc} + \lambda I_s - \left(\lambda I_s + \frac{qv_{PV}}{AKT}\right)e^{(qv_{PV}/AKT)}\right)\frac{d}{dt}v_{PV}. \quad (3)$$

From this last expression, it may be deduced that the maximum power point occurs at specific panel voltage (v_{PV}), and the value depends on the construction of the PV cell; then, if a mismatching of cells during construction could happen, the maximum power point would be different and, therefore, a cell temperature variation would also be registered between cells. This is demonstrated by the characterization proposed in this paper.

WORK BENCH AND SPECIFICATIONS

The analysis was made for two mc-Si modules with the following characteristics:(i)Module type 17(110) P1470X680, of Yingli Solar.(ii)Peak power 110 W.(iii)Peak voltage 17.5 V.(iv)Peak current 6.3 A.(v)36 cells connected in series.(vi)Glass top surface.(vii)Polymer rear surface.

Both modules were mounted in a structure oriented 20° to the south, in correspondence with the latitude of the installation place.

An infrared camera was used in order to measure the temperature. It was able to detect wavelengths from 7.5 μm to 13 μm in the region of the far infrared.

Since the front surface of the modules is reflective, the using of the rear surface of the module is a better choice. The temperature on the front side was compared with temperature on the back side of the module by using an infrared thermometer; it was found that practically both data sets have the same value. When thermographs are taken at the back side, it is avoided to have the shadows which normally occur on the front surface of the module due to weather conditions.

A temperature gradient between the back surface and the cell was considered since the temperature measured on the back surface of the modules is not exactly the actual temperature of the cells. Particularly for mc-Si cells in thermal equilibrium, irradiance is near to 1000 W/m², wind speed is less than 3 m/s, and the temperature gradient between both surfaces is about 2 or 3°C, which may just be added as an offset [3].

The thermographs were taken at 1 m distance from the back surface because this is the maximum distance, at which the camera was able to focus clearly. This limitation, added to the lens of 17 mm connected to the camera, made reading the temperature of the whole module in a single image impossible; for solving this issue, 36 cells from the module were measured in order to complete the module thermograph. According to this configuration, only a couple of cells fitted perfectly in the view field from the camera.

The two modules were operated at three different operating points in order to detect the anomalies in the temperature distribution. Then, the thermographs were taken, when the modules generate 6 W, 30 W, and 60 W. The resulting concatenated cells are shown in Figure 3.

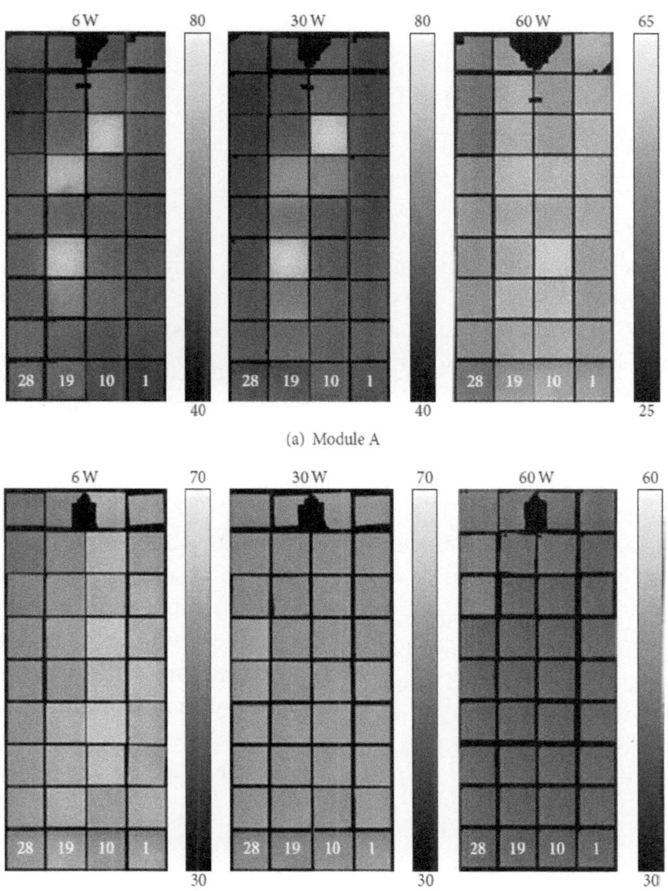

(a) Module A

(b) Module B

Figure 3: Temperature of two mc-Si modules at three operating points.

It is clearly shown in Figure 3 that some cells in module A are hotter than some other cells while the module generates 6 W and 30 W. On the other side, module B shows a homogenous distribution at the three points of operation, while module A has much contrast. It is also easily seen that module B was colder when it worked near the maximum power point; this occurs because only a portion of the total amount of irradiated energy in the module is converted to heat; the rest is supplied to the external load. On the opposite, when the module works in an operation point far from the maximum power point, then the irradiated energy in the module mainly becomes on heat.

It may also be easily seen in Figure 3 that module A contains some cells at a temperature which does not match the cells located around it. It is an alert for a hot spot problem in those cells. Due to this, cells 16 and 25 were chosen as the sample for subsequent tests, where cell 16 represents an abnormal situation and cell 25 exhibits a normal condition. Since these cells are neighbors, both were selected to be captured in a single image.

A DC/DC boost converter was used as the load for the PV modules (Figure 4); this was employed in every developed test in this paper. Since the duty cycle for the converter may be changed as wished, this means that the operation point for the module could be changed as fast as needed.

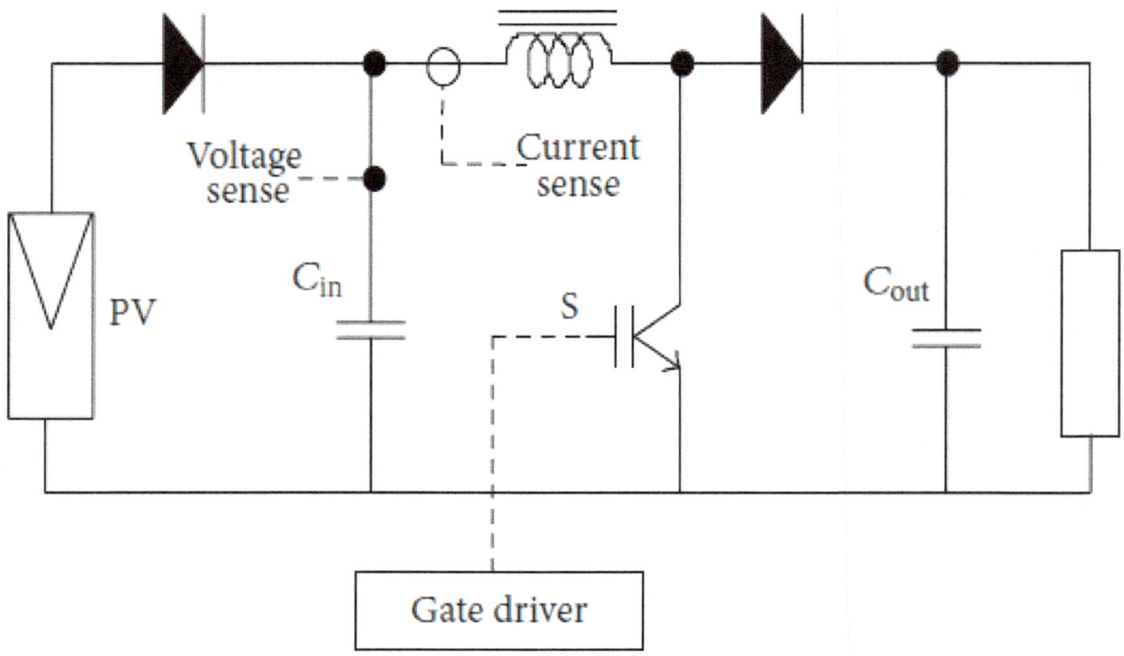

Figure 4: PV panel and power stage.

All measures were made with calibrated equipment, model E-25, of FLIR, with a resolution of 160 × 120 pixels, the spectral range of 7.5 μm and 13 μm, and a precision of ±2%. The tests were made by considering more than a single experiment; however, for a matter of simplicity, the reported tests are for just one case.

THERMAL TRANSIENT

As mentioned before, cells 16 and 25 were chosen in module A for temperature analysis and characterization data.

First of all, characterization with respect to time was made, because the obtained information for this case should be of transcendental relevance with respect to the characterization for the operating point of the module. It was found in the literature [1] that the temperatures of a c-Si module reaches steady state condition until 24 minutes have passed. For analysis purposes, a mc-Si solar cell was employed in this paper, so that more time was considered for the thermal transient characterization.

The first test was developed with a current value of 0.6 A as operating point for the module, once the thermal equilibrium was reached; the operating point was suddenly changed to a current value of 6 A. The temperature was recorded every minute during a 50-minute period.

A temperature graph for cell 16 is shown in Figure 5(a); it should be observed that although the technique for measuring transient temperature was a little bit different, the behavior during the first 24 minutes is almost the same as that reported in the literature [1]. Although temperature seems to reach its maximum value in approximately 25 minutes, however, the steady state takes more time (see Figure 5(a)). It is worth mentioning that this test was done at midday.

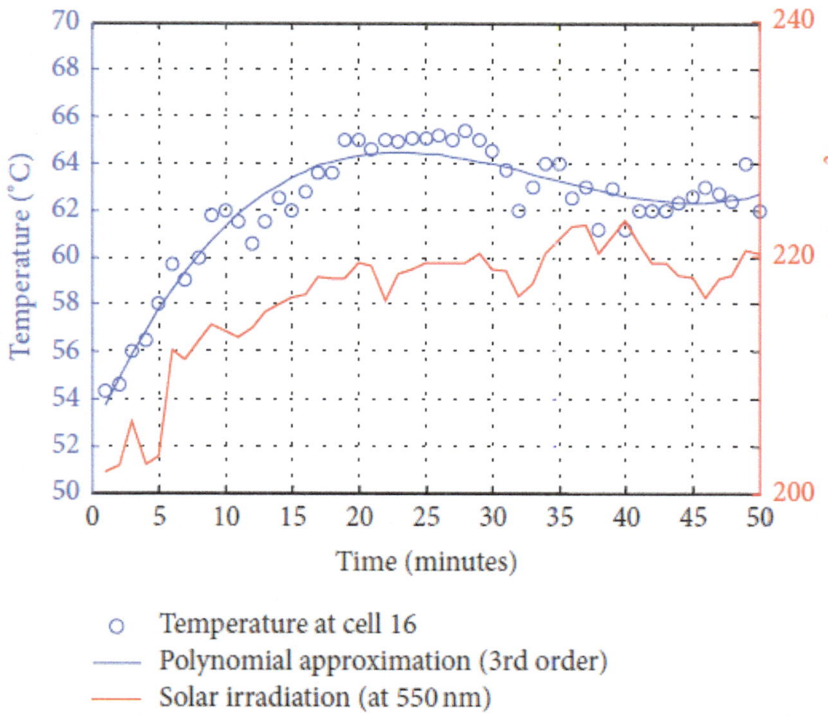

(a)

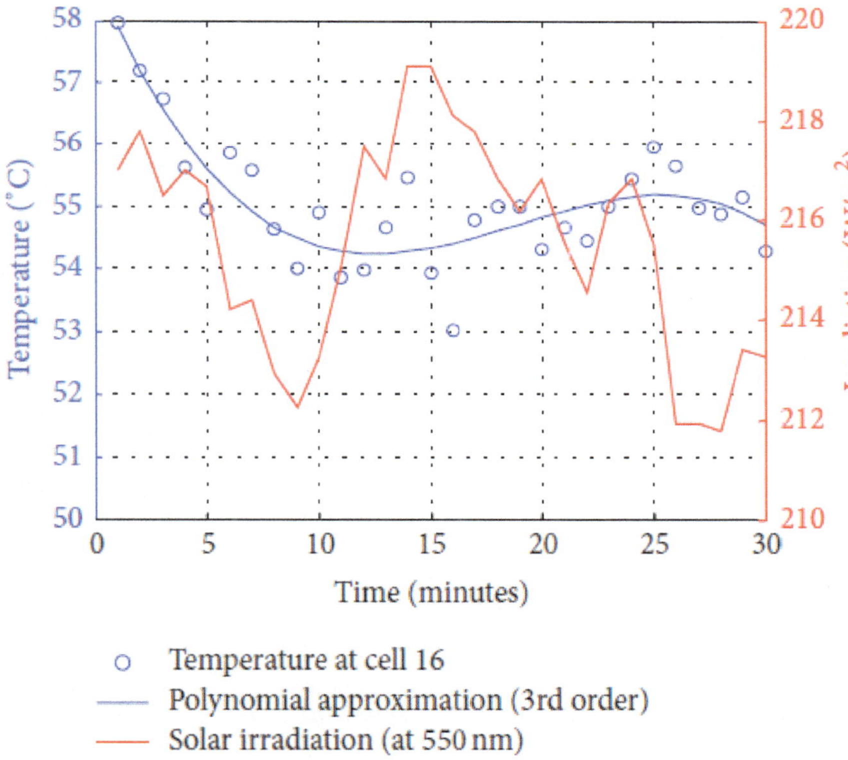

(b)

Figure 5: Thermal transient of cell 16 of module A. (a) For a current step up. (b) For a current step down.

Another test was developed, where the transient was carried out starting from a current value of 6 A and finishing with a current value of 0.6 A. The temperature graph is illustrated in Figure 5(b); particularly for this case, the recorded time was only for a period of 30 minutes; as it may be observ,ed the curve is very similar to that in Figures 2 and 5(a) but inverse.

Finally, the temperature evolution for cells 16 and 25 was recorded simultaneously, with an cperating point for the module by passing from a current of 5 A to a current of 6 A. Conditions for this test were established in order to analyze if the behavior, which is shown in Figure 5(a), is reproducible in spite of having a lower step change. The resulting curve is illustrated in Figure 6(a).

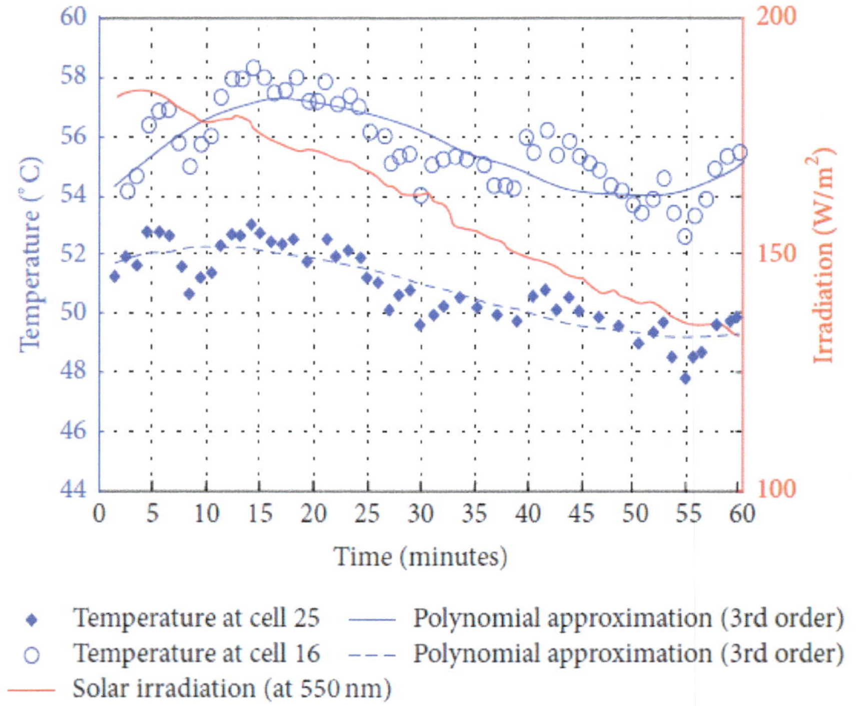

(a)

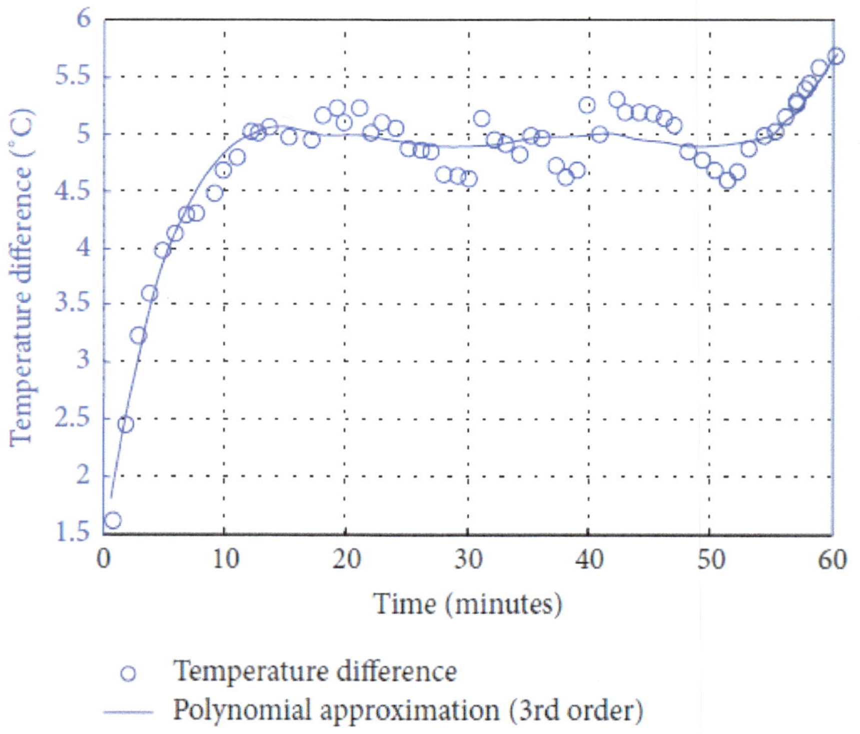

(b)

Figure 6: Thermal transient of screens 16 and 25 of module A. (a) For a small current step up. (b) Temperature difference.

Some interesting facts are implicitly revealed in this figure. In the first place, the shape of the temperature curve is similar to those obtained before, which means that the behavior of transient temperature does not depend on the step change for the operating point or weather conditions. In the second place, it should be noticed that cell 16 becomes hotter while cell 25 gets colder, under the same current step conditions. This is explained because the presence of hot spots produces mismatch conditions for the cells.

The temperature difference between cells 16 and 25 is shown in Figure 6(b). It may easily be observed how in 10 minutes the cells reach their bigger temperature difference, and later it is kept constant.

According to tests described before, the temperature transient takes around 30 minutes, so that any experiment for characterization should at least wait for 30 minutes between every capture.

The statistical analysis has an error of ±1.3, for cells 16 and 25. The difference between both cells was statistically significant ($t = -55\text{–}13$, df $= 59$, and $p = 1.85e - 52$).

CURRENT CHARACTERIZATION

An analysis for the current was developed as a test in order to obtain the best possible characterization. A period of time of 30 minutes was established in order to let the temperature reach the steady state conditions while the applied current to the module was increased at that period of time.

Unfortunately, changes on weather conditions made it very hard to carry out the tests under exactly the same conditions; however, this factor may be diminished. All the tests were developed from 11:00 a.m. to 4 p.m. in order to minimize the effects due to weather conditions; this procedure certainly meant such a great amount of time in order to be able not only to conclude and validate all cells but also to establish all the possible module defects.

The characterization for the current was done for cells 16 and 25, due to a mismatch found from these cells on previous analysis. The temperature was registered while the module current was increased every 30 minutes; the starting value for current was at 0.5 A, with step increments of 0.5 A, until 6 A value was reached. In other words, the duty cycle for the DC/DC boost converter was changed according to the specified current demand. The test began at 11:00 a.m. and finished at 4:00 p.m. It is evident that the reason for this long term of time is due to the restriction of having to wait 30 minutes for every change of load.

Obtained results from the test are illustrated in Figure 7(a); at first sight, it may seem that no relevant information was found. However, very different conclusions may be drawn from the graph.

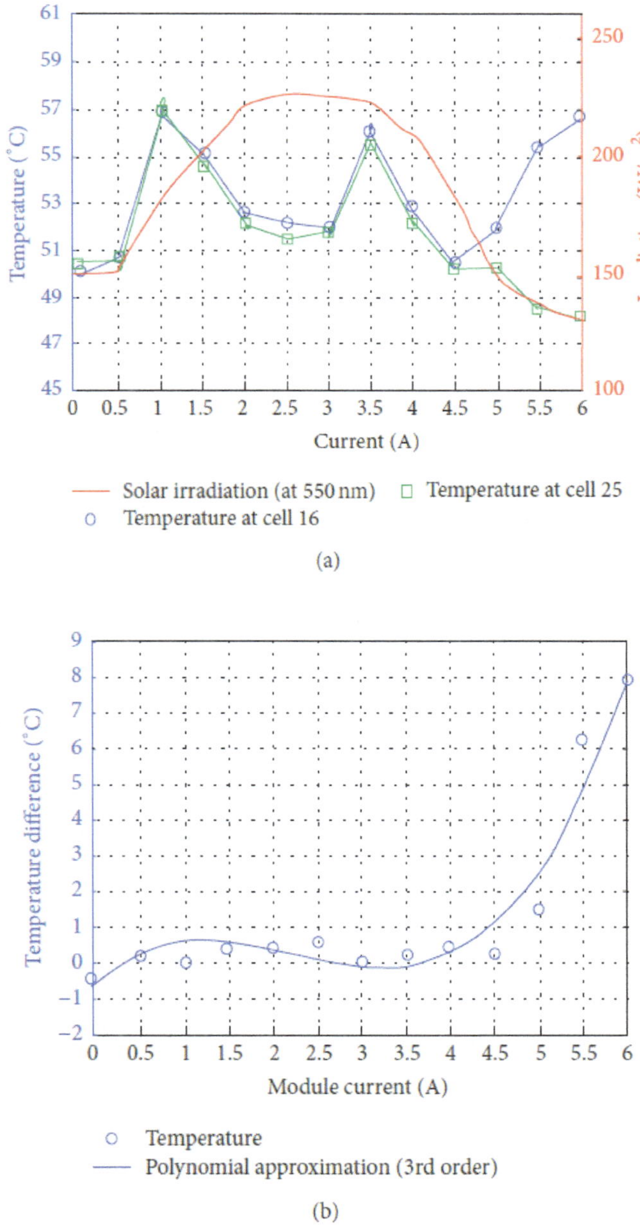

Figure 7: Current characterization under 0.5 A increments: (a) temperature graph and (b) temperature difference.

It is easily seen in Figure 7(a) that there exists more than a minimum value for temperature, so that it may be assumed that there exist multiple maximum power points. This, then, may imply that there may exist partially shaded areas or mismatch conditions at the PV panel. Since the tests were carried out under no shaded areas, it means that a mismatch occurs.

A better look to the graphs allows us to observe that the temperature for cells 16 and 25 suffers an important difference. Very often, these differences are due to weather conditions; however, thier influence should almost be the same on both cells due to the fact of being neighbors, so that a different cause should be considered forward in this particular case.

Data information would be clearer if temperature differences were graphed and compared for both cells (16 and 25). The environmental influence factor is significantly reduced when information is illustrated as it is shown in Figure 7(b). It should also be noticed that, at lower temperature conditions, both cells have almost the same temperature; however, at higher currents, the curves start to deviate from each other since cell 16 becomes hotter and cell 25 becomes colder.

The obtained result, which is illustrated in Figure 6(a), has to be confirmed and validated. For this purpose, the previous test had to be repeated; however, it was run with a current interval of 1 A at this time. This let us carry out the test in just three hours instead of the six required hours with a step of 0.5 A. Certainly, this represents a lost in curve resolution but shows minor changes due to weather conditions.

A graph for the second test is illustrated in Figure 8(a). It is easily seen that still there are significant variations; however, temperature differences on both cells exhibit similar evolution, as shown in Figure 8(b). Clearly, cell 16 becomes hotter than cell 25, when current increases. As a consequence, cell 16 has a lower capability for generating power compared to other cells. Finally, it is possible to identify mismatch conditions by measuring the temperature at the back of a PV panel; this would also let us identify manufacture defects.

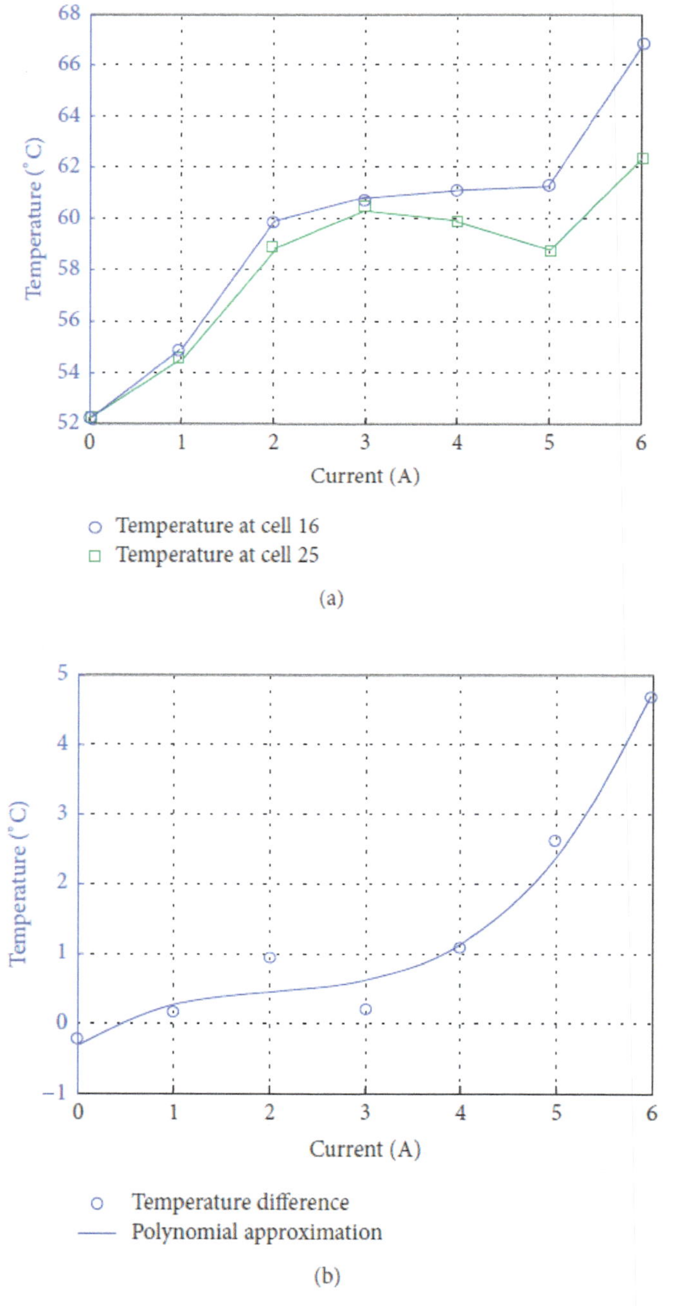

Figure 8: Current characterization under 1 A increments: (a) temperature graph and (b) temperature difference.

CONCLUSION

The characterization methods and instrumentation for in situ measurements, defect monitoring, and process control and performance are required in order to assure the quality.

A temperature characterization for a mc-Si module with time and current was developed by thermographic analysis. On the one hand, temperature characterization with respect to time made it possible to identify the temperature variations when the operation point changes for the module. On the other hand, temperature characterization with respect to PV panel current made it possible to establish cell mismatching conditions.

This method may effectively be applied for detecting defects in PV panels and hot spots regions produced by mismatch conditions in electric module features.

The main importance of the defects detection is because the PV panel will have low efficiency, but also multiple maximum power points, and then a concern to the converter and algorithm of the power extraction.

REFERENCES

1. D. L. King, J. A. Kratochvil, and W. E. Boyson, "Temperature coefficients for PV modules and arrays: measurement methods, difficulties, and results," in Proceedings of the IEEE 26th Photovoltaic Specialists Conference, pp. 1183–1186, Anaheim, Calif, USA, October 1997.
2. P. Singh and N. M. Ravindra, "Temperature dependence of solar cell performance—an analysis," Solar Energy Materials and Solar Cells, vol. 101, pp. 36–45, 2012.
3. C. Deline, "Partially shaded operation of a grid-tied PV system," in Proceedings of the 34th IEEE Photovoltaic Specialists Conference (PVSC '09), Philadelphia, Pa, USA, June 2009.
4. V. Quaschning and R. Hanitsch, "Numerical simulation of current-voltage characteristics of photovoltaic systems with shaded solar cells," Solar Energy, vol. 56, no. 6, pp. 513–520, 1996.
5. E. Díaz-Dorado, A. Suárez-García, C. Carrillo, and J. Cidrás, "Influence of the PV modules layout in the power losses of a PV array with shadows," in Proceedings of the 14th International Power Electronics and Motion Control Conference, Ohrid, Macedonia, 2010.
6. J. W. Bishop, "Computer simulation of the effects of electrical mismatches in photovoltaic cell interconnection circuits," Solar Cells, vol. 25, no. 1, pp. 73–89, 1988.
7. N. D. Kaushika and A. K. Rai, "An investigation of mismatch losses in solar photovoltaic cell networks,"Energy, vol. 32, no. 5, pp. 755–759, 2007.
8. K. Wilson, D. De Ceuster, and R. A. Sinton, "Measuring the effect of cell mismatch on module output," inProceedings of the 4th World Conference on Photovoltaic Energy Conference, pp. 916–919, IEEE, Waikoloa, Hawaii, USA, May 2006.
9. A. M. Gabor, M. Ralli, S. Montminy et al., "Soldering induced damage to thin SI solar cells and detection of cracked cells in modules," in Proceedings of the 21st European Photovoltaic Solar Energy Conference, Dresden, Germany, 2006.
10. J. W. Bishop, "Microplasma breakdown and hot-spots in silicon solar cells," Solar Cells, vol. 26, no. 4, pp. 335–349, 1989.
11. J. Bauer, J.-M. Wagner, A. Lotnyk et al., "Hot spots in multicrystalline silicon solar cells: avalanche breakdown due to etch pits," Physica Status Solidi—Rapid Research Letters, vol. 3, no. 2-3, pp. 40–42, 2009.
12. C. Honsberg and S. Bowden, PVeducation.org, http://www.pveducation.org/pvcdrom/solar-cell-operation/spectral-response.
13. W. Herrmann, W. Wiesner, and W. Vaaben, "Hot spot investigations on PV modules—new concepts for a test standard and consecuences for module design with respect to bypass diodes," in Proceedings of the Conference Record of the Twenty Sixth IEEE Photovoltaic Specialists Conference, pp. 1129–1132, Anaheim, Calif, USA, 1997.
14. E. Molenbroek, D. W. Waddington, and K. A. Emery, "Hot spot susceptibility and testing of PV modules," in Proceedings of the 22nd IEEE Photovoltaic Specialists Conference, pp. 547–552, Las Vegas, Nev, USA, October 1991.
15. V. Quaschning and R. Hanitsch, "Numerical simulation of photovoltaic generators with shaded cells," inProceedings of the 30th Universities Power Engineering Conference, pp. 583–586, Greenwich, UK, September 1995.
16. G. Liu, P. Wang, W. Wang, and Q. Wang, "MPPT algorithm under partial shading conditions," inElectrical, Information Engineering and Mechatronics, vol. 1, pp. 91–98, Springer, London, UK, 2011.

CHAPTER 4 CALIBRATION OF THE HALL MEASUREMENT SYSTEM FOR A 6-DOF PRECISION STAGE USING SELF-ADAPTIVE HYBRID TLBO

Zhenyu Chen [1], Yang Liu [1,2], Zhenxian Fu [1], Shenmin Song [1] and Jiubin Tan [2]

[1]Department of Control Science and Engineering, Harbin Institute of Technology, Harbin 150001, China

[2]Ultra-Precision Optoelectronic Instrumentation Engineering Centre, Harbin Institute of Technology, Harbin 150001, China

ABSTRACT

To determine the planar motion of a 6-DOF precision stage, a measurement system based on three Hall sensors is adopted to obtain the X, Y, Rz motions of the stage. The machining and assembly errors in the actual mechanical system, which are difficult to measure directly, cause the parameters in the model of the Hall measurement system to deviate from their designed values. Additionally, the vertical movement of the stage will render the measurement model nonlinear. To guarantee the accuracy of the measurement, the parameters in the measurement model should be estimated and the nonlinearity compensated. In this paper, a novel approach based on self-adaptive hybrid TLBO (teaching-learning-based-optimization) is proposed to estimate the parameters in the Hall measurement model. The influences of zero deviations and vertical movements on the measurement accuracy are analyzed and compensated. The effectiveness of the proposed method is validated by experimental results obtained on a 6-DOF precision stage. Thanks to parameter estimation and calibration, the measurement error of the Hall sensor array is reduced to 6 micrometers.

INTRODUCTION

Precision motion stage is the kernel part of an ultra-precision system which has been playing a critical role in Integrated Circuit (IC) manufacturing, optical components production, Liquid Crystal Display (LCD) fabrication, IC encapsulation, *etc.* Increased demand is putting stricter requirements on the precision stage, especially on the accuracy of its multiple degree-of-freedoms (DOFs) movement. To achieve high precision positioning of the stage, a sufficiently accurate measurement system is indispensable, for the measurement error in the feedback channel of the control loop will directly affect the system performance and cannot be suppressed by the controller in the forward channel [1]. Since the 6-DOF displacements of a stage cannot be measured by a single sensor, a measurement system composed of multiple sensors is necessary. Mura measured a 6-DOF Stewart platform stage by constructing a sensor array, composed of wire extensometers or MEMS displacement sensors [2,3]. By solving the kinematic equations, the 6-DOF moving information could be estimated. Similarly, Allred used a simplified Stewart platform stage to measure the helicopter blade [4]. Kim made use of some cooperative targets

Citation: Zhenyu Chen, Yang Liu, Zhenxian Fu, Shenmin Song and Jiubin Tan, alibration of the Hall Measurement System for a 6-DOF Precision Stage Using Self-Adaptive Hybrid TLBO, doi:10.3390/s16060872.

to design a measurement coordinate system, and the system parameters were determined by optimization. Then, the 6-DOF displacements could be calculated by coordinate transform [5]. However, these methods are not effective for a non-cooperative and floating stage. As noncontact sensors [6,7], Hall sensors are widely used in various industrial applications, including position, velocity and magnetic fields measurement [8,9,10,11,12,13]. Also, it is shown to be insensitive to harsh environments. For measurement of a multi-DOF stage using Hall sensor, Zhao obtained an elliptic fitting function of the magnetic field for compensation, in the case that the height between the hall probe and permanent magnet is fixed [14]. To estimate the maglev stage in planar DOFs, Kawato adopted two-axis Hall-effect sensors, constructed a model of magnetic field distribution according to the Halbach magnet matrix, and developed a Gaussian least-squares differential-correction (GLSDC) algorithm [15,16]. Sun established the model of the magnetic field and displacement using finite element analysis, and presented a 3-DOF measurement algorithm based on symmetry and linearization [17].

Note that the study in reference [14,15,16,17] is based on the assumption that the distance between the Hall probe and the permanent magnetic is fixed which, however, could not be guaranteed in practice. The slope of the Hall sensor output would vary due to the change of magnetic flux. According to [14], a variation of 1 mm in distance will cause a considerable change in the output slope, and a large error will be introduced to the Hall measurement system if only the designed parameters are employed and the influence of the relative distance variation is not calibrated.

In practice, the measurements of the precision stage by the array of Hall sensors are coupled, and should be decoupled before being utilized. However, the decoupling is subject to errors in its decoupling matrix caused by the uncertainty of mechanical parameters of the precision stage, since these parameters inevitably deviate from their designed values due to machining and assembling errors. These errors will degrade the positioning accuracy of the nano-or-micro-precision stage.

Since such mechanical parameters are hard to directly measure due to limitation of space and measurement principle, a "soft sensor" method is proposed. The so-called soft sensor technology makes use of the mathematical relationship between parameters that are directly measurable and those that are not, obtaining the concerned parameters by numerical calculation based on optimal approximating theory [18]. After nearly 40 years of development, the soft measurement technology has been widely employed by process control systems in the chemical industry, metallurgy, biology, *etc*. Up to now, modeling methods for soft sensors fall into two major categories: the mechanism based and the data driven. Using the mechanism based approach, Prasad [19] obtained the model of a styrene polymerization process, and then estimated the average molecular weight of the polymer performance indexes using an extended Kalman filter. By the data driven method, Zamprogna performed principal component analysis for a batch distillation process using the regression process model and the temperature measurement [20]; Gonzaga measured the viscosity of polymer using a feed-forward artificial neural network model [21]; Liu solved the melt viscosity measurement problem in a polymer extrusion process using a nonlinear observer-based finite impulse response model, with the temperature and screw speed as its input [22]. In mechatronic systems, the concept of soft sensors has been rarely mentioned. However, some parameter estimation methods based on the prior model structures, such as Luenberger observer, Kalman filter, and adaptive observer, have also been well developed and applied in practice. Elmas adopted a full-order Luenberger observer to estimate the rotator position of a switched reluctance machine, with satisfactory effect [23]. Szabat estimated the internal states of a string-linked double mass system using a fuzzy Luenberger observer, to control the driven system [24]. Sim designed a novel Kalman filter to estimate the parameters of PMSM online [25]. Rajamani applied an adaptive observer in a vehicle suspension system [26]. Lim proposed an online frequency parameter estimation method based on statistics, to measure the rigidity and damping coefficients in the electromagnet suspension system [27].

The measurement model of the Hall sensor array, which is a mechatronic system, could be established based on mechanism analysis, which contains mechanical assembly errors and some unknown coefficients. As the relationship between these immeasurable parameters and the measurable ones is of high dimension, high order and nonlinear, the aforementioned state space-based methods are no longer suitable to this problem.

Meta-heuristic methods, inspired by natural activities, provide another approach to parameter estimation. As long as the fitness function is determined, they could be easily applied in practice without additional constraints. Owing to their simplicity, various kinds of meta-heuristic algorithms have been proposed and effectively applied in many fields, such as mechanical design and trajectory planning of robotics. The most popular meta-heuristic methods include Genetic Algorithm (GA) [28], Differential Evolution (DE) [29], Particle swarm optimization (PSO) [30], and Teaching-Learning-Based Optimization (TLBO) [31]. TLBO is newly proposed and is used for optimizing constrained mechanical design problems. Compared with other heuristic methods, the biggest advantage of TLBO is that fewer parameters are to be adjusted, with only the population size and maximum iterations to be determined [32,33,34]. However, the TLBO is prone to prematurity for some practical problems [35]. To address this issue, Ghasemi combined the traditional TLBO with the Gaussian-Based TLBO (GBTLBO), balancing the convergence rate and the solution quality [36]. In this paper, the GBTLBO is further improved using an adaptive updating strategy for the parameter p, the mutation probability. Applying this novel self-adaptive hybrid TLBO to the parameter estimation for the Hall measurement system, the measurement accuracy could be improved.

The rest of this paper is organized as follows: Section 2 describes the configuration of the investigated precision stage. Section 3 establishes the measurement and the calibration model of the Hall sensor array, which contains a mechanical assembly error. Section 4 proposes a self-adaptive hybrid TLBO method for estimating parameters in the aforementioned model. Then,Section 5 conducts experiments and analyzes the experimental results in different situations.Section 6 concludes the paper.

SYSTEM CONFIGURATION

The structure of the 6-DOF precision stage investigated in this paper is shown in Figure 1, whose coordinate system follows the right-hand rule.

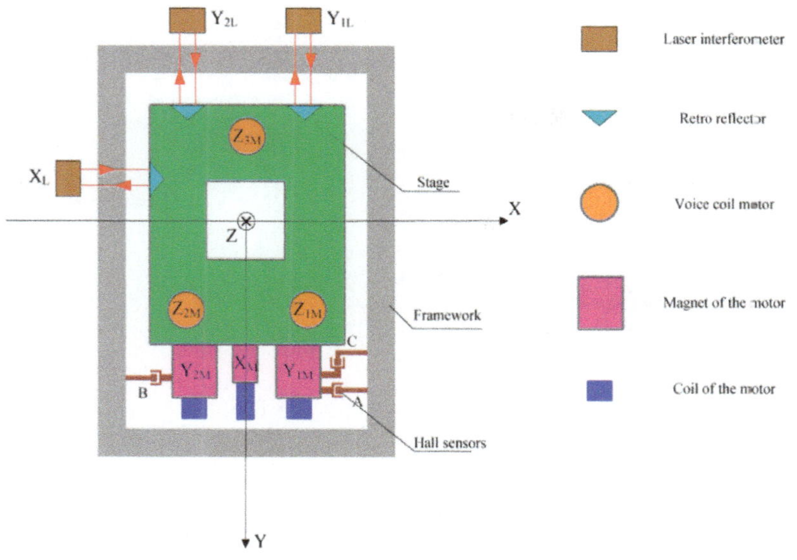

Figure 1. The configuration of the 6-DOF precision stage.

The stage is driven by six voice coil motors. Three of them, Z1M, Z2M, Z3M, are in the vertical direction, constituting an isosceles triangle. Driven by these motors, the stage could realize translation in *Z* axis and rotation around *X*, *Y* axis. The driving centers of Z1M and Z2M are located in the first and second quadrants of the coordinate system respectively, symmetric to the *Y*axis. The driving center of Z3M is on the negative *Y* axis. The forces of these motors are parallel to the *Z* axis and have the same positive direction. In the horizontal plane, two symmetrically distributed voice coil motors, Y1M and Y2M, provide driving force in Y direction. While motor XM, located in the middle between the two Y-direction motors, provides the driving force in X direction. The driving point of motor XM is located on the positive *Y* axis, taking positive X direction as driving direction. The driving points of Y1M and Y2M are located in the first and second quadrants, symmetric to the *Y* axis, taking the positive direction of *Y* axis as driving direction. Since the driving points of these horizontal linear motors do not coincide, the stage can achieve horizontal rotation and translation.

The motion of the stage is measured by the hall sensor array and the capacitance sensor. Three hall sensors, A, B, C, are responsible for measuring the displacement of the stage relative to the framework in the horizontal plane. Specifically, sensors A and B measure the relative displacements in the Y direction, and sensor C in the X direction. The output voltage of a Hall sensor is in direct proportion to the magnetic flux density where the sensor is placed, and zero output is desired in the middle of its measurement range. The structure of the Hall sensor used in the stage is shown in Figure 2, in which the structure of the magnetic steel is simplified for better versatility and easier installation. The Hall probe is fixed on the stage, while the permanent magnets are installed on the mounting plate. The dimension of the magnet is 12.7 mm × 6.4 mm × 6.4 mm, and the installation mount is designed to be 43 mm × 25.5 mm × 6.5 mm. With this configuration, an arc-shape magnetic field is generated above the upper surface of the magnet. The Hall sensor array is shown in Figure 3.

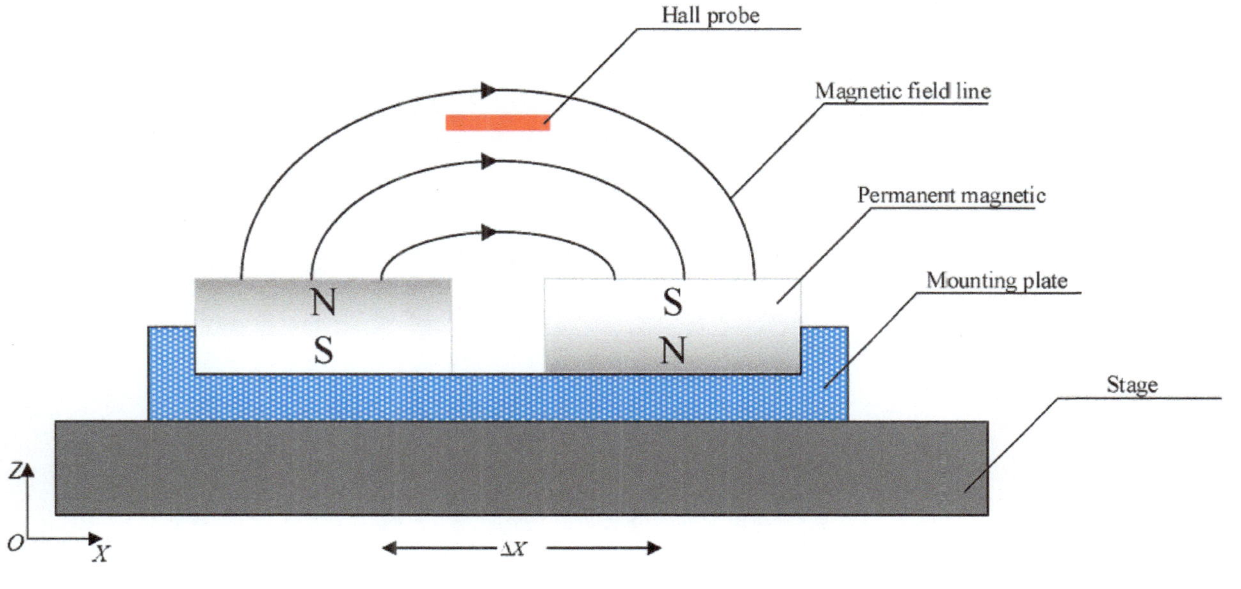

Figure 2. The structure of the designed Hall sensor.

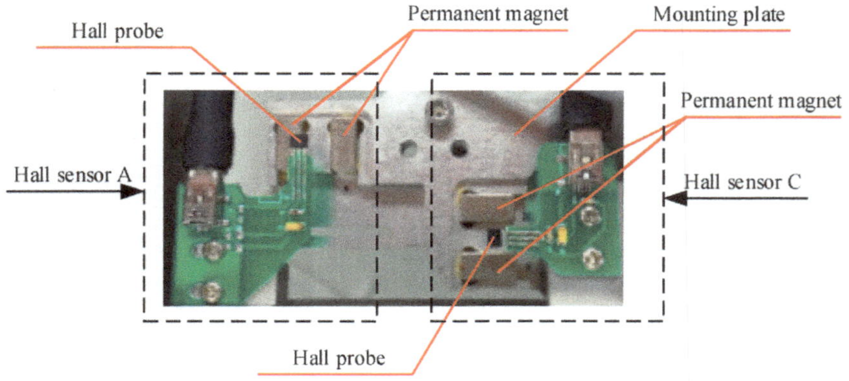

Figure 3. Part of Hall sensor array.

The vertical 3-DOF movement of the stage is measured by the capacitance sensors, whose layout is as shown in Figure 4. The probe of capacitance sensor is fixed on a disc-shaped base below the stage. Two capacitance sensors are installed on the X axis, symmetric to the origin, while the third capacitance sensor is located off the axis X. Each capacitance sensor could obtain the displacement of the probe relative to the framework, and the 3-DOF vertical movement of the stage relative to the framework can be calculated.

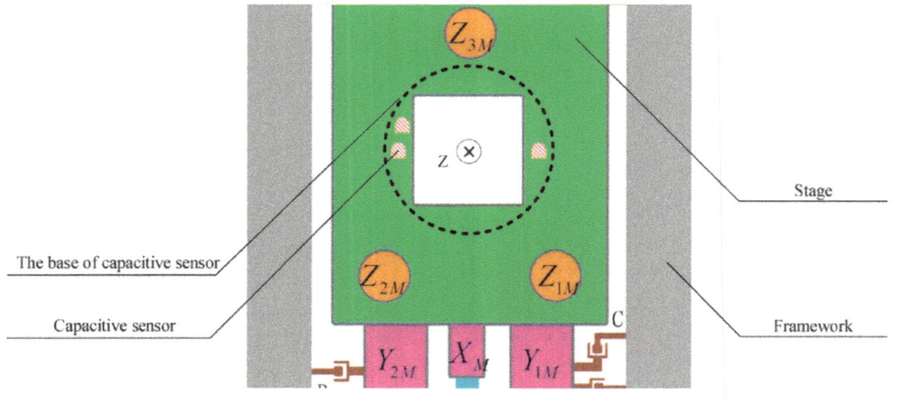

Figure 4. The layout of the capacitance sensors.

To calibrate the Hall sensor array, laser interferometers are adopted to measure the planar 3-DOF motion of the stage, acting as a reference for Hall sensor array. The investigated precision stage is shown in Figure 5.

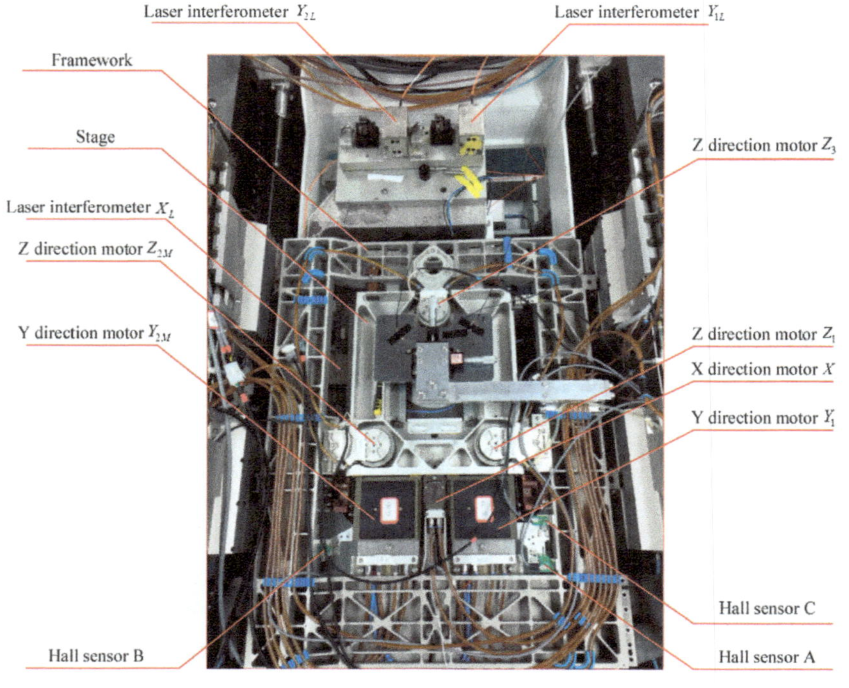

Figure 5. The precision stage.

MODEL ESTABLISHMENT

The Decoupling Model of the Hall Sensor Array

To establish the decoupling model of the Hall sensor array, the structure of the precision stage marked with geometrical information is shown in Figure 6. Assume that when the stage is at its initial position, the outputs of the three Hall sensors, A, B, C, are all zero. At this position, the coordinates of the probes of the three hall sensors are A(H1x,H1y,0), B(−H2x,H2y,0) and C(H3x,H3y,0) respectively. The movement of the stage can be decomposed into the following three scenarios: (1) firstly XY translation, then Rz rotation; (2) firstly Rz rotation, then XY translation; (3) firstly X (Y) translation, then Rz direction rotation, then Y (X) translation. The decoupling method of the sensor A under the three scenarios will be analyzed respectively. Assume that in the framework-fixed coordinate system, the stage is at its initial position at the initial moment, the movement of stage centroid in X, Y and Rz direction is Δx, Δy, ΔRzrespectively.

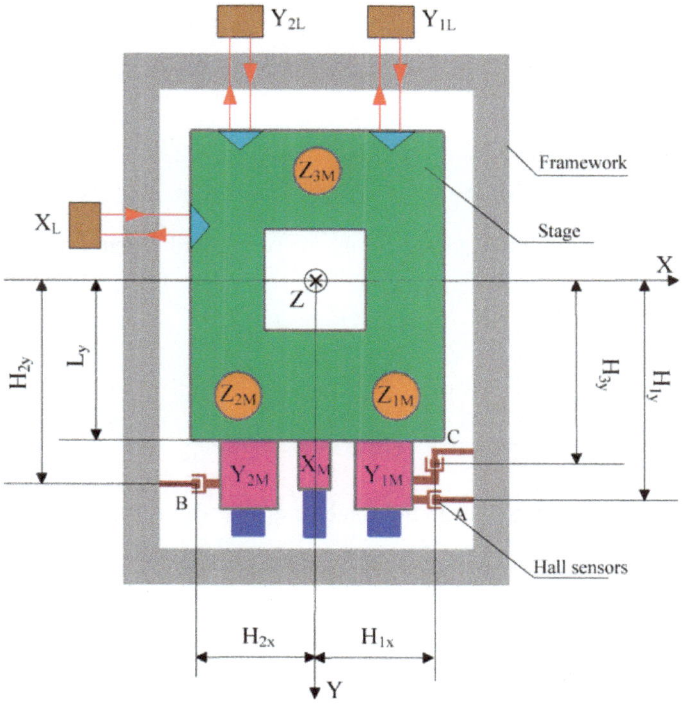

Figure 6. The dimension figure of the investigated stage.

Situation (1):

The stage moves Δx in X direction and then Δy in Y direction. Since the probe A is fixed on the framework, the relative coordinate of stage body centroid is

$$\begin{cases} x_A{}' = H_{1x} - \Delta x \\ y_A{}' = H_{1y} - \Delta y \end{cases} \tag{1}$$

Subsequently, the stage rotates along Rz direction, and we have

$$\begin{bmatrix} x_A{}'' \\ y_A{}'' \end{bmatrix} = \begin{bmatrix} \cos(-\Delta Rz) & -\sin(-\Delta Rz) \\ \sin(-\Delta Rz) & \cos(-\Delta Rz) \end{bmatrix} \begin{bmatrix} x_A{}' \\ y_A{}' \end{bmatrix} \tag{2}$$

where (xA″,yA″) is the coordinates of point A in the centroid coordinates. Equation (2) can be rewritten as

$$\begin{bmatrix} x_A{}'' \\ y_A{}'' \end{bmatrix} = \begin{bmatrix} \cos(\Delta Rz) & \sin(\Delta Rz) \\ -\sin(\Delta Rz) & \cos(\Delta Rz) \end{bmatrix} \begin{bmatrix} x_A{}' \\ y_A{}' \end{bmatrix} \tag{3}$$

The moving range of the stage is quite small, with Rz rotation within 1 mrad, and X, Y translation within 2 mm. According to the Maclaurin formula of sine and cosine function, within the moving range of the stage, it is reasonable to adopt approximations of sine and cosine functions to reduce the amount of calculation without affecting accuracy. As a result, the coordinate of point A in the centroid coordinates can be expressed as

$$\begin{bmatrix} x_A{}'' \\ y_A{}'' \end{bmatrix} = \begin{bmatrix} 1 & \Delta Rz \\ -\Delta Rz & 1 \end{bmatrix} \begin{bmatrix} H_{1x} - \Delta x \\ H_{1y} - \Delta y \end{bmatrix} \tag{4}$$

Similarly, the coordinates of B and C can be also obtained as

$$\begin{bmatrix} x_B'' \\ y_B'' \end{bmatrix} = \begin{bmatrix} 1 & \Delta Rz \\ -\Delta Rz & 1 \end{bmatrix} \begin{bmatrix} -H_{2x} - \Delta x \\ H_{2y} - \Delta y \end{bmatrix} \tag{5}$$

$$\begin{bmatrix} x_C'' \\ y_C'' \end{bmatrix} = \begin{bmatrix} 1 & \Delta Rz \\ -\Delta Rz & 1 \end{bmatrix} \begin{bmatrix} H_{3x} - \Delta x \\ H_{3y} - \Delta y \end{bmatrix} \tag{6}$$

Furthermore, the coordinates of the Hall sensor A, B, C could be written as

$$\begin{cases} x_A'' = H_{1x} - \Delta x + H_{1y} \cdot \Delta Rz - \Delta Rz \cdot \Delta y \\ y_A'' = -H_{1x} \cdot \Delta Rz + \Delta Rz \cdot \Delta x + H_{1y} - \Delta y \\ x_B'' = -H_{2x} - \Delta x + H_{2y} \cdot \Delta Rz - \Delta Rz \cdot \Delta y \\ y_B'' = H_{2x} \cdot \Delta Rz + \Delta Rz \cdot \Delta x + H_{2y} - \Delta y \\ x_C'' = H_{3x} - \Delta x + H_{3y} \cdot \Delta Rz - \Delta Rz \cdot \Delta y \\ y_C'' = -H_{3x} \cdot \Delta Rz + \Delta Rz \cdot \Delta x + H_{3y} - \Delta y \end{cases} \tag{7}$$

Given that the moving range is so small, the quadratic terms can be ignored. Then, Equation (7) is simplified as

$$\begin{bmatrix} x_A'' \\ y_A'' \\ x_B'' \\ y_B'' \\ x_C'' \\ y_C'' \end{bmatrix} = \begin{bmatrix} -1 & 0 & H_{1y} \\ 0 & -1 & -H_{1x} \\ -1 & 0 & H_{2y} \\ 0 & -1 & H_{2x} \\ -1 & 0 & H_{3y} \\ 0 & -1 & -H_{3x} \end{bmatrix} \begin{bmatrix} \Delta x \\ \Delta y \\ \Delta Rz \end{bmatrix} + \begin{bmatrix} H_{1x} \\ H_{1y} \\ H_{2x} \\ H_{2y} \\ H_{3x} \\ H_{3y} \end{bmatrix} \tag{8}$$

As a result, the coordinates of the sensing probes in the stage centroid-origined frame could be expressed as

$$\begin{bmatrix} \Delta x_A'' \\ \Delta y_A'' \\ \Delta x_B'' \\ \Delta y_B'' \\ \Delta x_C'' \\ \Delta y_C'' \end{bmatrix} = \begin{bmatrix} 1 & 0 & -H_{1y} \\ 0 & 1 & H_{1x} \\ 1 & 0 & -H_{2y} \\ 0 & 1 & -H_{2x} \\ 1 & 0 & -H_{3y} \\ 0 & 1 & H_{3x} \end{bmatrix} \begin{bmatrix} \Delta x \\ \Delta y \\ \Delta Rz \end{bmatrix} \tag{9}$$

where ΔxA″, ΔyA″, ΔxB″, ΔyB″, ΔxC″, ΔyC″ denote the translational displacement of Hall sensor in stage centroid-origin. However, the sensitive axis of a Hall sensor may not be aligned with the corresponding axis of the stage centroid-origin due to installation error. The unknown angular deviations are shown in Figure 7, where θA, θB, θC are the angles between the sensitive axis of the Hall sensors and the corresponding coordinate axes.

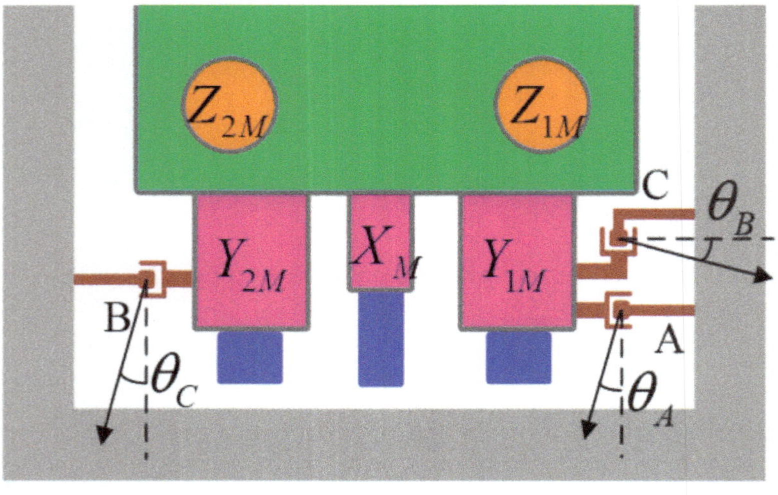

Figure 7. The angular assembly errors of the Hall sensor array.

By projecting the changes of coordinates of the points A, B, C in the stage centroid-origin frame onto the sensitive axis, the measurement values of the Hall sensors could be written as

$$\begin{bmatrix} \Delta y_A \\ \Delta y_B \\ \Delta x_C \end{bmatrix} = \begin{bmatrix} -\sin\theta_A & \cos\theta_A & 0 & 0 & 0 & 0 \\ 0 & 0 & -\sin\theta_B & \cos\theta_B & 0 & 0 \\ 0 & 0 & 0 & 0 & \cos\theta_C & \sin\theta_C \end{bmatrix} \begin{bmatrix} \Delta x_A'' \\ \Delta y_A'' \\ \Delta x_B'' \\ \Delta y_B'' \\ \Delta x_C'' \\ \Delta y_C'' \end{bmatrix} \tag{10}$$

Combining Equations (9) and (10) yields

$$\begin{bmatrix} \Delta y_A \\ \Delta y_B \\ \Delta x_C \end{bmatrix} = \begin{bmatrix} -\sin\theta_A & \cos\theta_A & H_{1x}\cdot\cos\theta_A + H_{1y}\cdot\sin\theta_A \\ -\sin\theta_B & \cos\theta_B & H_{2y}\cdot\sin\theta_B - H_{2x}\cdot\cos\theta_B \\ \cos\theta_C & \sin\theta_C & H_{3x}\cdot\sin\theta_C - H_{3y}\cdot\cos\theta_C \end{bmatrix} \begin{bmatrix} \Delta x \\ \Delta y \\ \Delta Rz \end{bmatrix} \tag{11}$$

Solving Equation (11), the decoupling model of the horizontal 3-DOF movement relative to the framework measured by the Hall sensor array could be expressed as

$$\begin{bmatrix} \Delta x \\ \Delta y \\ \Delta Rz \end{bmatrix} = \begin{bmatrix} -\sin\theta_A & \cos\theta_A & H_{1x}\cdot\cos\theta_A + H_{1y}\cdot\sin\theta_A \\ -\sin\theta_B & \cos\theta_B & H_{2y}\cdot\sin\theta_B - H_{2x}\cdot\cos\theta_B \\ \cos\theta_C & \sin\theta_C & H_{3x}\cdot\sin\theta_C - H_{3y}\cdot\cos\theta_C \end{bmatrix}^{-1} \begin{bmatrix} \Delta y_A \\ \Delta y_B \\ \Delta x_C \end{bmatrix} \tag{12}$$

Situations (2) and (3):

As can be seen by comparing Equations (7) and (8), the remaining terms are the linear superposition of the translational movement of the stage in X and Y direction after the quadratic terms such as ΔRz·Δx have been ignored. So, similarly, the decoupling models of the stage moment in situation (2) and (3) are the same as Equation (12).

In conclusion, the decoupling model of the stage movement relative to the framework could be written as Equation (12), which can guarantee the measurement accuracy and real-time performance of the decoupling process.

The Calibration Model of the Hall Sensor Array

According to the previous analysis, the actual distance between the Hall sensing probe and the permanent magnet is different from its designed value due to assembly error, which will change the output slope of the Hall sensor and therefore create a difference between the actual displacement and its computed value. Using Equation (12) directly as the decoupling model will introduce measurement errors and cause degradation of control precision. Therefore, the outputs of the Hall sensor array have to be calibrated.

In reference [37], by finite element simulation of a Hall sensor, Chen finds that the output in X direction of the Hall sensor is hardly influenced by the stage movement in the Y direction when the height of the sensing probe is fixed. So, in this paper, the influence of orthogonal movement on the output of the Hall sensor is ignored. Based on the previous analysis, considering the initial mechanical assembly error, the calibration model of the Hall sensor array is defined as

$$\begin{bmatrix} y_{Ae} \\ y_{Be} \\ x_{Ce} \end{bmatrix} = \begin{bmatrix} \varphi_{yA} & 0 & 0 \\ 0 & \varphi_{yB} & 0 \\ 0 & 0 & \varphi_x \end{bmatrix} \begin{bmatrix} y_A \\ y_B \\ x_C \end{bmatrix} \tag{13}$$

where yA, yB, xC denote the outputs of the Hall sensor array under designed parameters, and yAe, yBe, xCe denote the outputs of the Hall sensor array after calibration.

$$\varphi_{yA} = \frac{1}{\eta_{yA}} \qquad \varphi_{yB} = \frac{1}{\eta_{yB}} \qquad \varphi_x = \frac{1}{\eta_x} \tag{14}$$

where ηyA, ηyB, ηx are undetermined coefficients.

According to the analysis of Section 3.1 and Section 3.2, the measurement model of the Hall sensor array could be expressed as

$$\begin{bmatrix} \Delta x(t) \\ \Delta y(t) \\ \Delta Rz(t) \end{bmatrix} = \begin{bmatrix} -\sin\theta_A & \cos\theta_A & H_{1x}\cdot\cos\theta_A + H_{1y}\cdot\sin\theta_A \\ -\sin\theta_B & \cos\theta_B & H_{2y}\cdot\sin\theta_B - H_{2x}\cdot\cos\theta_B \\ \cos\theta_C & \sin\theta_C & H_{3x}\cdot\sin\theta_C - H_{3y}\cdot\cos\theta_C \end{bmatrix}^{-1} \begin{bmatrix} \varphi_{yA} & 0 & 0 \\ 0 & \varphi_{yB} & 0 \\ 0 & 0 & \varphi_x \end{bmatrix} \begin{bmatrix} \Delta y_A(t) \\ \Delta y_B(t) \\ \Delta x_C(t) \end{bmatrix} \tag{15}$$

SELF-ADAPTIVE HYBRID TLBO ALGORITHM

Teaching-learning-based optimization (TLBO) is a new meta-heuristic method, which is inspired by a teaching and learning process used in class. In this method, only population *Np* and iteration times *Gm* need to be selected before execution of the program. The whole process of TLBO can be decomposed into two phases: the teacher phase and the learner phase. In the teacher phase, the student who has the best grade (the best fitness value) is regarded as the teacher, and he will improve the grade of the whole class. The students can absorb the teacher's knowledge to improve quickly. In the learner phase, student can select a classmate randomly from the whole class, and then learn with him after comparing fitness with each other. This phase can further improve the performance of the whole class. At the end of the iterative process, the teacher is the output of the optimization.

Original TLBO

Teacher Phase

In the teacher phase, as the global best of the population, the teacher shares their knowledge to improve the performance of the whole class. However, due to the students' different learning capabilities, the improvement of each student is different. The updated formula of the *j*-th student St_{ij} is defined as

$$St_{ij}^{new} = St_{ij}^{old} + r_1 \cdot [T_i - T_F \cdot Mean_i] \tag{16}$$

where Stoldij and Stnewij denote the Stij before and after learning process respectively. ri is a random number in the range [0, 1]. Ti is the teacher and Meani is the mean at the *i*-th iteration. TFis a teaching factor set as either 1 or 2 presented as

$$T_F = round\,(1 + rand_2\,(0,1)) \tag{17}$$

The new learners will compete with their predecessors and replace them if a better fitness value is achieved.

Learner Phase

In the learner phase, a student will select a partner from class randomly, and then update their status according to the following formula,

$$St_{ij}^{new} = \begin{cases} St_{ij}^{old} + rand_3\left(St_{ik} - St_{ij}^{old}\right) if\ f\,(St_{ik}) < f\left(St_{ij}^{old}\right) \\ St_{ij}^{old} + rand_3\left(St_{ij}^{old} - St_{ik}\right) if\ f\left(St_{ij}^{old}\right) < f\,(St_{ik}) \end{cases} \tag{18}$$

where Stik is the randomly selected student whose serial number is k. The learning process will also improve the performance of the population. Similar to the teacher phase, if the new learner is better than the old one, the old one will be replaced.

Self-Adaptive Hybrid TLBO

Although the TLBO is effective in many practical problems, it could not guarantee an optimal solution, especially in high dimensional, non-linear problems. To improve the optimization results, the global exploitation and partial exploration ability of the TLBO should be balanced.

Based on bare-bones PSO proposed by Kennedy [38] and bare-bones DE proposed by Omran [39], a Gaussian mutation operation is introduced to TLBO by Ghasemi [36]. The updated formula is defined as

$$St_{ij}^{new} = N\left(\frac{St_{ij}^{old} + T_i}{2}, \left|T_i - St_{ij}^{old}\right|\right) \tag{19}$$

$$St_{ij}^{new} = N\left(\frac{St_{ij}^{old} + St_{ik}}{2}, \left|St_{ij}^{old} - St_{ik}\right|\right) k \neq j \tag{20}$$

where *N* represents a Gaussian distribution.

In GBTLBO, the population is generated by Gaussian distribution, which uses the mean and distance of Stoldij and Ti(Stik) as expectation and standard deviation, respectively. In order to improve the exploration ability and avoid prematurity, a novel mutation strategy is proposed in this paper, which is expressed as follows:

$$St_{ij}^{new} = St_{ij}^{old} + N\,(0,1) \cdot \left|T_i - St_{ij}^{old}\right| \tag{21}$$

$$St_{ij}^{new} = St_{ij}^{old} + N\,(0,1) \cdot \left|St_{ij}^{old} - St_{ik}\right| k \neq j \tag{22}$$

The Gaussian distribution generated by the modified GBTLBO still use Stoldij as the expectation. However, the population would explore the neighborhood of each individual instead of converging with the teacher or an excellent student, increasing the search range and boosting exploitation ability.

Combining the fast convergence feature of TLBO and the powerful exploitation ability of CBTLBO, a hybrid style of TLBO is proposed for balancing the capabilities of exploration and exploitation, which is expressed as

$$St_{ij}^{new} = \begin{cases} St_{ij}^{old} + r_1 \cdot [T_i - T_F \cdot Mean_i], & rand < p \\ St_{ij}^{old} + N(0,1) \cdot \left| T_i - St_{ij}^{old} \right|, & otherwise \end{cases} \tag{23}$$

where rand is a random number lies in [0, 1], p is a threshold to control the mutation probability. rand obeys the uniform distribution so that p could reflect the proportion of two kinds of mutation during the evolution.

Similarly, this hybrid mutation strategy is also introduced in Learner phase, with the corresponding updated formula written as

$$St_{ij}^{new} = \begin{cases} \begin{cases} St_{ij}^{old} + rand_3\left(St_{ik} - St_{ij}^{old}\right) & if\ f(St_{ik}) < f\left(St_{ij}^{old}\right) \\ St_{ij}^{old} + rand_3\left(St_{ij}^{old} - St_{ik}\right) & if\ f\left(St_{ij}^{old}\right) < f(St_{ik}) \end{cases}, & rand < p \\ St_{ij}^{old} + N(0,1) \cdot \left| St_{ij}^{old} - St_{ik} \right|, & otherwise \end{cases} \tag{24}$$

Obviously, the parameter p plays a critical role in the performance of the modified TLBO by affecting the converging speed and the exploration ability. So, the optimization results depend much on the selection of p. However, it is hard to determine the value of p without adequate prior knowledge. To tackle this intractable problem, a self-adaptive method is presented. Concerning self-tuning of parameters, Qin proposed a updated method for p [40], expressed as

$$p = \frac{ns_1 \cdot (ns_2 + nf_2)}{ns_2 \cdot (ns_1 + nf_1) + ns_1 \cdot (ns_2 + nf_2)} \tag{25}$$

where ns_1 and ns_2 denote the numbers of individuals successfully updated by Equations (1) and (2) in Equation (26), respectively, while nf_1 and nf_2 denote the numbers of those not. Using this strategy, a self-adaptive differential evolution (SaDE) with two mutation modes is presented as follows [40]:

$$St_{ij}^{new} = \begin{cases} Eq.(1), & rand < p \\ Eq.(2), & otherwise \end{cases} \tag{26}$$

The initial value of p is set to be 0.5. Once p updates, ns_1, ns_2, nf_1 and nf_2 would be reset.

However, the success ratio of mutation could not accurately reflect the performance of the updated strategy. In case of local convergence, for instance, although the individual is updated, the improvement in the cost-function is quite small as the algorithm has already lost its exploration ability. So, the successful update ratio is not a perfect criterion. To address this shortcoming, the declining ratio of the cost-function caused by the successfully mutating individuals is introduced to the updated formula. Then, p is updated as

$$p = \frac{vs_1 \cdot (ns_2 + nf_2)}{vs_2 \cdot (ns_1 + nf_1) + vs_1 \cdot (ns_2 + nf_2)} \tag{27}$$

where vs1, vs2 denote the declining ratio of cost-function caused by Equations (1) and (2) in Equation (26), respectively, calculated as

$$\begin{cases} vs_1 = \sum \Delta f\left(St_{ij}^{new}\right), if\ f\left(St_{ij}^{new}\right) < f\left(St_{ij}^{old}\right)\ and\ rand < p \\ vs_2 = \sum \Delta f\left(St_{ij}^{new}\right), if\ f\left(St_{ij}^{new}\right) < f\left(St_{ij}^{old}\right)\ and\ rand \geqslant p \end{cases} \tag{28}$$

$$\Delta f\left(St_{ij}^{new}\right) = \frac{f\left(St_{ij}^{old}\right) - f\left(St_{ij}^{new}\right)}{f\left(St_{ij}^{old}\right)} \tag{29}$$

This way, the updated strategy is directly related to the cost-function, and the preponderant mutation will have a greater opportunity in the next evolution iteration. The flow chart of the proposed Self-adaptive hybrid TLBO (SHTLBO) is shown in Figure 8.

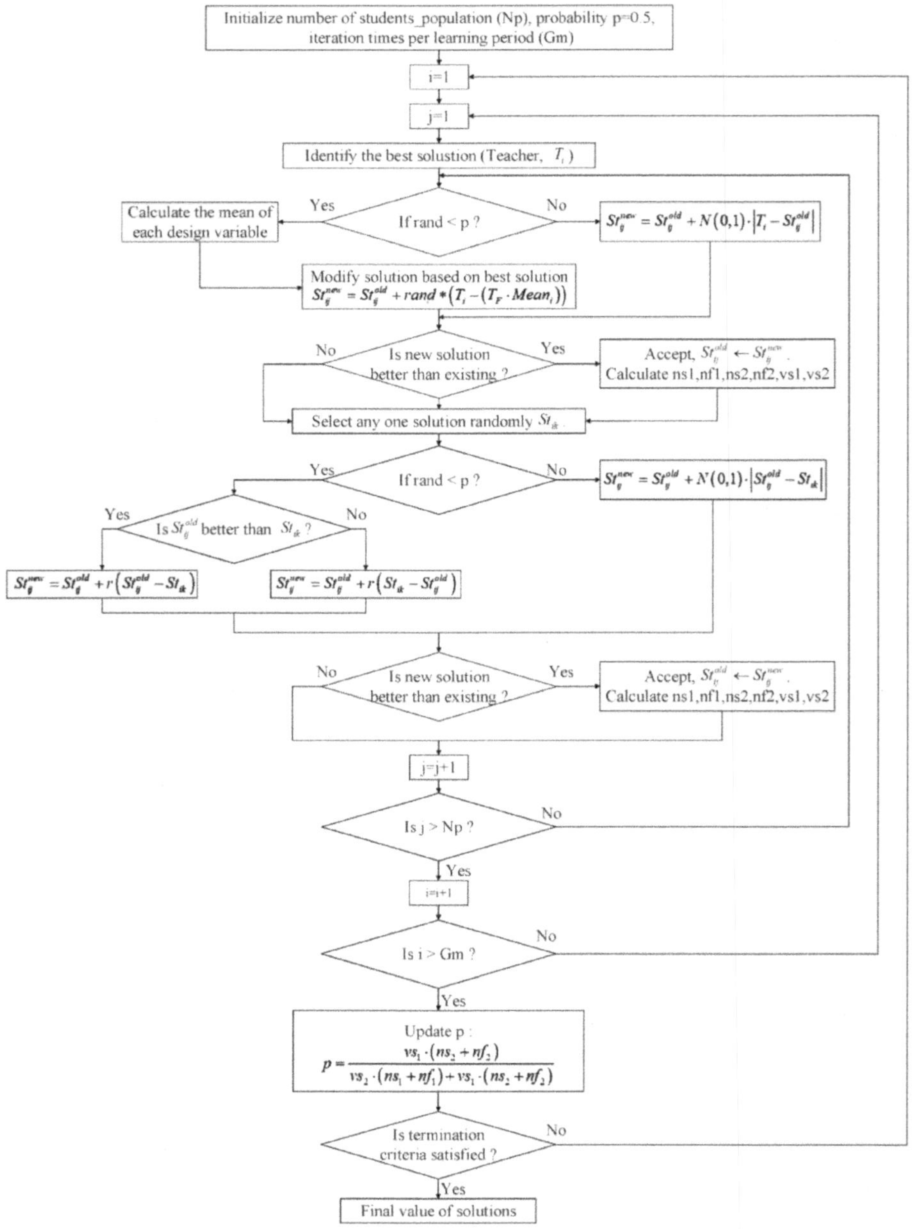

Figure 8. The flow chart of the self-adaptive hybrid TLBO.

EXPERIMENTS AND ANALYSIS

In this section, the proposed Self-adaptive hybrid TLBO method is applied to estimate the mechanical parameters of the designed Hall measurement system for the 6-DOF precision stage illustrated in Section 2. The desired accuracy of the Hall sensor array for each axis is 5 μm, and the vertical 3-DOF movement of the stage is obtained by capacitive sensors, CAPANCDT 6500 by Micro-Epsilon, with an accuracy of 12 nm.

To calibrate the Hall sensor array, a 3-DOF laser interferometer array is adopted to provide reference. Although environmental variations, such as temperature, humidity and vibration, limited the static accuracy of the interferometer to 500 nm, it is still adequate for calibration of the Hall sensor array. To synchronize different sensors, a clock with 2 ms is designed.

To conduct the calibration experiment, the designed values of parameters in the Hall sensor array are listed in Table 1.

Table 1. The designed values of parameters in Hall sensor array

Parameter	H_{1x}	H_{2x}	H_{3y}	H_{3x}	H_{1y}	H_{2y}
Value	0.1975 m	0.1975 m	0.36 m	0.1975 m	0.4 m	0.42 m
Parameter	η_{yA}	η_{yB}	η_x	θ_1	θ_2	θ_3
Value	1	1	1	0°	0°	0°

The precision stage is first controlled to stop at a fixed vertical level to avoid the disturbance induced by vertical moment. Considering the influence of vertical states on the measurements of Hall arrays, the stage is fixed at two heights: Z = −250 μm and Z = −250 μm. Then, the stage conducts the planar reset operation, providing the zero points for the 3-DOF interferometer array. After the reset, the planar motion of the stage could be measured by both the Hall sensors and the interferometers. Next, the stage is controlled to track certain trajectory with feedback from the interferometers. Then the output of the interferometers can be utilized to calibrate the Hall sensors. In this experiment, the trajectory along each axis is a sinusoidal wave: with amplitude of 300 μm and frequency of 1 Hz along *X* axis; 500 μm and 1 Hz along *Y*; 300 μrad and 1 Hz along *Z*. The corresponding outputs from the sensors are shown in Figure 9.

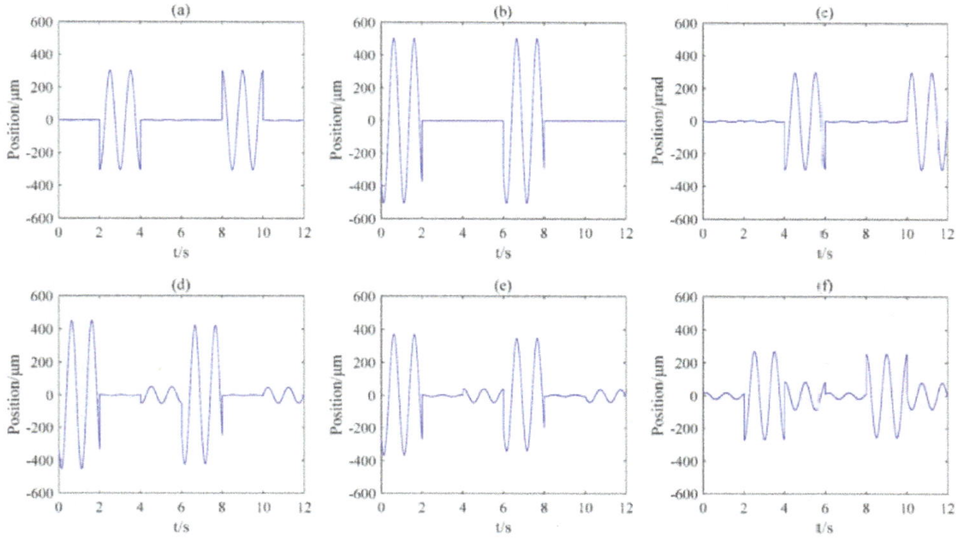

Figure 9. The displacements of the stage measured by the interferometers in X, Y, Rz directions (**a–c**) and the outputs of the Hall sensors A, B, C (**d–f**).

Using the designed parameters listed in Table 1 and the measurement model of the Hall sensor array built in Section 3, the output errors of Hall sensors with the designed parameters could be obtained, as shown in Figure 10.

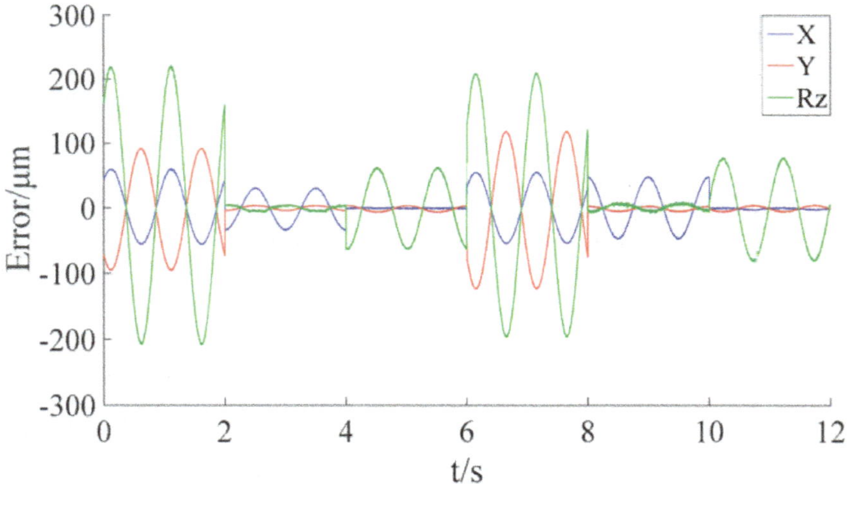

Figure 10. The output errors of the Hall sensor array with the designed parameters using Model (15) when the stage is on a different vertical level.

The peak errors and the standard deviations (RMSE) of each DOF with the designed parameters are listed in Table 2.

Table 2. The output errors of the 3-DOF with designed values of parameters using Model (15)

DOF	X	Y	Rz
RMSE	27.8795 μm	44.1243 μm	89.2927 μrad
Peak error	61.2092 μm	119.9458 μm	221.5275 μrad

As can be seen from Figure 10 and Table 2, the measurement accuracies of the Hall sensor array with the designed parameters fall far short of the requirements. The peak error in the translational direction is 119.9458 μm, while the peak error in the rotational direction is 221.5275 μrad. This is caused by the afore-mentioned mechanical assembly errors in the Hall sensor array. So, it is necessary to conduct the calibration. To set proper constraints on the optimization algorithm, the search ranges are given according to the mechanical design of the stage, as listed in Table 3.

Table 3. The search ranges of the parameters using Model (15)

Parameter	H_{1x}	H_{2x}	H_{3y}	H_{3x}	H_{1y}	H_{2y}
Lower range	0.1475 m	0.1475 m	0.26 m	0.1475 m	0.3 m	0.32 m
Upper range	0.2475 m	0.2475 m	0.46 m	0.2475 m	0.5 m	0.52 m
Parameter	η_{yA}	η_{yB}	η_x	θ_1	θ_2	θ_3
Lower range	0.78	0.65	0.72	−8°	−8°	−8°
Upper range	0.98	0.85	0.92	8°	8°	8°

In this experiment, the interferometers need to be reset using the feedback from the Hall measuring system, which contains about 5 μm measurement noise. As a result, the zero points of the interferometers vary between different rounds of movements, introducing uncertainty to the calibration system. To eliminate this offset, the initial aligning parameters should be considered in the calibration model. Therefore, Equation (15) is modified as

$$\begin{bmatrix} \Delta x(t) \\ \Delta y(t) \\ \Delta Rz(t) \end{bmatrix} = \begin{bmatrix} -\sin\theta_A & \cos\theta_A & H_{1x}\cdot\cos\theta_A + H_{1y}\cdot\sin\theta_A \\ -\sin\theta_B & \cos\theta_B & H_{2y}\cdot\sin\theta_B - H_{2x}\cdot\cos\theta_B \\ \cos\theta_C & \sin\theta_C & H_{3x}\cdot\sin\theta_C - H_{3y}\cdot\cos\theta_C \end{bmatrix}^{-1} \begin{bmatrix} \frac{1}{\eta_{yA}} & 0 & 0 \\ 0 & \frac{1}{\eta_{yB}} & 0 \\ 0 & 0 & \frac{1}{\eta_x} \end{bmatrix} \begin{bmatrix} \Delta y_A(t) - \delta_{Ai} \\ \Delta y_B(t) - \delta_{Bi} \\ \Delta x_C(t) - \delta_{Ci} \end{bmatrix} \quad (30)$$

where δAi, δBi, δCi denote the reset errors of HALL sensors A, B and C during the *i*-th movement, respectively. The range of these variables are all [−10 μm, 10 μm]. Then, parameter estimation for the measurement Model (30) was conducted, and the corresponding results are shown in Figure 11, while the RMSE and the peak errors are listed in Table 4. To estimate the parameters, the proposed self-adaptive hybrid TLBO is adopted. The training data is composed of 100 randomly selected data from the experimental results. The setting of the optimization algorithm is: $Np = Gm = 100$. The stop criterion is LP>100.

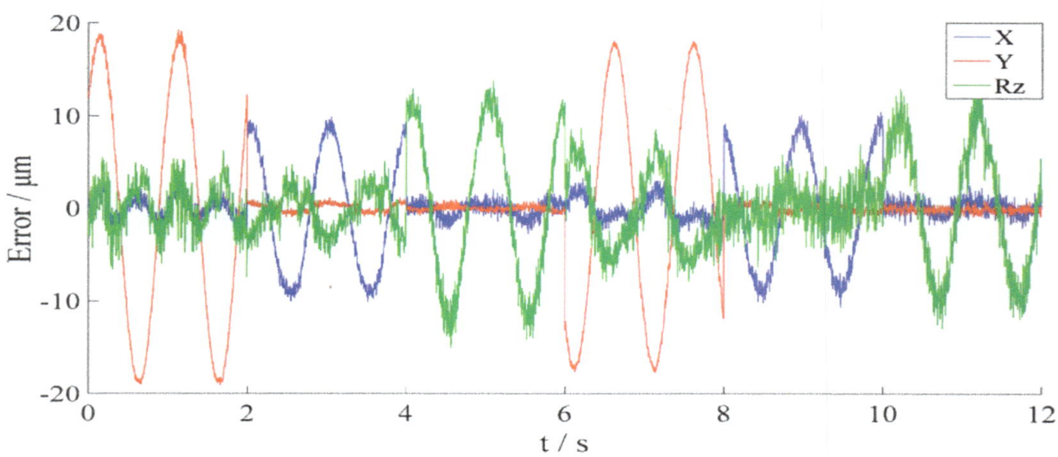

Figure 11. The output errors in horizontal 3-DOFs with the optimized parameters using Model (30).

Table 4. The statistics of the horizontal 3-DOFs' outputs with optimized parameters using Model (30)

DOF	X	Y	Rz
RMSE	2.1896 μm	4.2560 μm	2.9176 μrad
Peak error	10.6451 μm	19.2473 μm	15.0316 μrad

As can be seen from Figure 11 and Table 4, the measurement accuracy of the Hall sensor array is improved by the parameter estimation. The peak error in translational direction is reduced to 19.2473 μm, and that in rotational direction to 15.0316 μrad, approaching the desired accuracy. But this accuracy is still far from desired value, which means that the measurement Model (30) is not accurate enough for the purpose. Therefore, the influence of vertical moments should be incorporated in the model. According to the principle of Hall measurement, the slope of the magnetic flux density sensed by the Hall probe would change with the distance between the Hall sensor and the upper surface of the permanent magnet. To accommodate this feature, a new calibration model is constructed as in Equation (31), which considers the linear combinations of vertical movements.

$$\begin{bmatrix} \Delta x(t) \\ \Delta y(t) \\ \Delta Rz(t) \end{bmatrix} = PM^{-1} \cdot \left\{ \begin{bmatrix} \frac{1}{\varphi_A(t)} & 0 & 0 \\ 0 & \frac{1}{\varphi_B(t)} & 0 \\ 0 & 0 & \frac{1}{\varphi_C(t)} \end{bmatrix} \begin{bmatrix} \Delta y_A(t) \\ \Delta y_B(t) \\ \Delta x_C(t) \end{bmatrix} - \begin{bmatrix} \frac{1}{\varphi_A(0)} & 0 & 0 \\ 0 & \frac{1}{\varphi_B(0)} & 0 \\ 0 & 0 & \frac{1}{\varphi_C(0)} \end{bmatrix} \begin{bmatrix} \delta_{Ai} \\ \delta_{Bi} \\ \delta_{Ci} \end{bmatrix} \right\} \tag{31}$$

$$PM = \begin{bmatrix} -\sin\theta_A & \cos\theta_A & H_{1x} \cdot \cos\theta_A + H_{1y} \cdot \sin\theta_A \\ -\sin\theta_B & \cos\theta_B & H_{2y} \cdot \sin\theta_B - H_{2x} \cdot \cos\theta_B \\ \cos\theta_C & \sin\theta_C & H_{3x} \cdot \sin\theta_C - H_{3y} \cdot \cos\theta_C \end{bmatrix} \tag{32}$$

$$\begin{cases} \varphi_A(t) = a_A \cdot Z(t) + b_A \cdot Rx(t) + c_A \cdot Ry(t) + \eta_{yA} \\ \varphi_B(t) = a_B \cdot Z(t) + b_B \cdot Rx(t) + c_B \cdot Ry(t) + \eta_{yB} \\ \varphi_C(t) = a_C \cdot Z(t) + b_C \cdot Rx(t) + c_C \cdot Ry(t) + \eta_x \end{cases} \tag{33}$$

where *PM* is the matrix containing the error parameters, φA(t), φB(t), φC(t) denote respectively the calibration coefficients of the three Hall sensors' outputs which vary with the vertical moments of the stage. {aA,bA,…,cC} denote the coefficients of different DOFs, which are to be optimized. ηyA, ηyB, ηx denote the calibration coefficients of the vertical zero plane; δAi, δBi, δCi denote the zero deviations of the three Hall sensors at the *i*-th movement.

During the experiment, the vertical moving range is ±250 μm; the rotational ranges of *Rx* and *Ry*are both ±500 μrad. Considering the achievable assembly accuracy, the boundary of the TLBO searching is set to be 5000. Furthermore, the sign of the boundary could be determined according to reference [14]. Taking Hall sensor A as an example: the distance between the Hall probe and the upper surface of the permanent magnetic will increase if the stage moves in the positive vertical direction. As a result, the output slope of the Hall sensor A would decrease. Therefore, the coefficient aA should be negative. The signs of other coefficients can be determined likewise. Finally, the search boundary of the coefficients can be determined, as listed in Table 5.

Table 5. The search boundaries of the coefficients related with the vertical movements

Parameter	a_A	b_A	c_A	a_B	b_B
Lower range	−5000	−5000	0	−5000	−5000
Upper range	0	0	5000	0	0
Parameter	c_B	a_C	b_C	c_C	
Lower range	−5000	−5000	−5000	0	
Upper range	0	0	0	5000	

In the experiment, the stage is set at different vertical states as listed in Table 6. Under each vertical setting, the stage is controlled in the same manner as mentioned above. Alsom the sensors' data are sampled in the same way. Then, the parameters in Model (31) are optimized by the hybrid adaptive TLBO. The setting of the optimizing algorithm is the same as mentioned above. The training data is set to 200 points.

Table 6. Vertical movement setting

NO. \ DOF	Z/μm	Rx/μrad	Ry/μrad
1	0	0	0
2	250	0	0
3	−250	0	0
4	150	0	0
5	−150	0	0
6	50	0	0
7	−50	0	0
8	0	500	0
9	0	−500	0
10	0	250	0
11	0	−250	0
12	0	100	0
13	0	−100	0
14	0	0	500
15	0	0	−500
16	0	0	250
17	0	0	−250
18	0	0	100
19	0	0	−100

The estimated results of the measurement parameters are listed in Table 7, while the corresponding coefficients of zero deviations are shown in Table A1 (Appendix A).

Table 7. The estimated values of parameters using model (31)

Parameter	H_{1x}	H_{2x}	H_{3y}	H_{3x}	H_{1y}	H_{2y}
Value	0.1944 m	0.2173 m	0.3293m	0.1979 m	0.3381 m	0.3555 m
Parameter	a_A	b_A	c_A	η_{yA}		
Value	−188.1873	−69.7432	14.8664	0.8671		
Parameter	a_B	b_B	c_B	η_{yB}		
Value	−156.5783	−56.7511	−55.0764	0.7049		
Parameter	a_C	b_C	c_C	η_{yx}		
Value	−178.1902	−63.4133	12.7241	0.8632		
Parameter	θ_1	θ_2	θ_3			
Value	0.0145°	0.4143°	−0.0561°			

Apply the estimated results to Model (31), and then validate it using all the sampling data. The output errors of X, Y and Rz are shown in Figure 12 and Table 8.

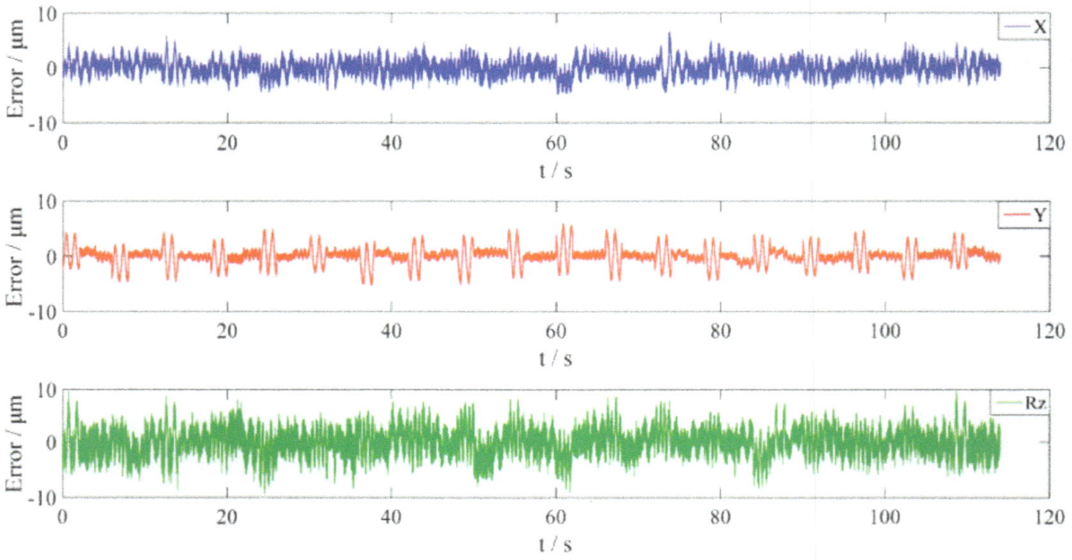

Figure 12. The output errors in horizontal 3-DOFs with the optimized parameters using Model (31).

Table 8. The statistics of the horizontal 3-DOFs' outputs with optimized parameters using Model (31)

DOF	X	Y	Rz
RMSE	1.3919 µm	1.5381 µm	2.3057 µrad
Peak error	6.6368 µm	5.9375 µm	9.6300 µrad

As seen from Figure 12 and Table 8, by calibration, the maximum translational and rotational errors of the Hall sensor array are reduced to 6.6368 µm and 9.63 µrad, respectively, which meets the measurement accuracy requirement.

CONCLUSIONS

In this paper, a 3-DOF measurement system based on the Hall sensor array is designed for a 6-DOF precision stage. Considering the mechanical assembly error of the Hall sensors, the measurement model and the corresponding calibration model of the Hall sensor array are established. Then, the influence of the vertical stage movements and the reset deviation of the interferometers are analyzed to ensure suitability for practical application. To estimate the related parameters in the measurement model, a self-adaptive hybrid TLBO method is proposed. The experimental results show that the measurement accuracy of the Hall sensor array could reach 6 micrometers using the proposed measurement parameter estimation algorithm.

ACKNOWLEDGMENTS

This work is supported by Chinese National Science Foundation (51405097), the State Key Program of National Natural Science of China under Grant 51537002 and Natural Scientific Research Innovation Foundation in Harbin Institute of Technology (HIT.NSRIF.2015031).

AUTHOR CONTRIBUTIONS

Yang Liu and Shenmin Song conceived and designed the experiments; Zhenyu Chen performed the experiments; Zhenyu Chen and Yang Liu analyzed the data; Jiubin Tan contributed the investigated stage; Yang Liu wrote the paper; Zhenxian Fu polished the paper.

REFERENCES

1. Lee, C.B.; Lee, S.K. Multi-degree-of-freedom motion error measurement in an ultraprecision machine using laser encoder—Review. *J. Mech. Sci. Technol.* 2013, *27*, 141–152.
2. Mura, A. Multi-dofs MEMS displacement sensors based on the Stewart platform theory.*Microsyst. Technol.* 2012, *18*, 575–579.
3. Mura, A. Sensitivity analysis of a six degrees of freedom displacement measuring device. *J. Mech. Eng. Sci. Proc. Inst. Mech. Eng. Part C* 2014, *228*, 158–168.
4. Allred, C.J.; Jolly, M.R.; Buckner, G.D. Real-time estimation of helicopter blade kinematics using integrated linear displacement sensors. *Aerosp. Sci. Technol.* 2015, *42*, 274–286.
5. Kim, J.A.; Bae, E.W.; Kim, S.H.; Kwak, Y.K. Design methods for six-degree-of-freedom displacement measurement systems using cooperative targets. *Precis. Eng.* 2002, *26*, 99–104.
6. Rhyu, S.H.; Jung, I.S.; Kwon, B.I. 2-D modeling and characteristic analysis of a magnetic position sensor. *IEEE Trans. Magn.* 2005, *41*, 1828–1831.
7. Schott, C.; Racz, R.; Betschart, F.; Popovic, R.S. A new two-axis magnetic position sensor. In Proceedings of IEEE Sensors, Orlando, FL, USA, 12–14 June 2002; pp. 911–915.
8. Han, X.T.; Cao, Q.L.; Wang, M. A linear Hall Effect displacement sensor using a stationary two-pair coil system. In Proceedings of the IEEE International Instrumentation and Measurement Technology Conference, Hangzhou, China, 10–12 May 2011; pp. 1342–1345.
9. Manzin, A.; Nabaei, V. Modelling of micro-Hall sensors for magnetization imaging. *J. Appl. Phys.* 2014, *115*.
10. Rajkumar, R.K.; Asenjo, A.; Panchal, V.; Manzin, A.; Iglesias-Freire, Ó.; Kazakova, O. Magnetic scanning gate microscopy of graphene Hall devices. *J. Appl. Phys.* 2014, *111*, 172606-1–172606-6.
11. Xu, Y.; Pan, H.B.; He, S.Z.; Li, L. A highly sensitive CMOS digital Hall Sensor for low magnetic field applications. *Sensors* 2012, *12*, 2162–2174.
12. Hyeonh, J.A.; Kyoung, R.K. 2D Hall sensor array for measuring the position of a magnet matrix. *Int. J. Precis. Eng. Manuf. Green Technol.* 2014, *1*, 125–129.
13. Kim, S.Y.; Choi, C.; Lee, K.; Lee, W. An improved rotor position estimation with vector—Tracking observer in PMSM drives with low-resolution Hall-effect sensors. *IEEE Trans. Ind. Electron.* 2011, *58*, 4078–4086.
14. Zhao, B.; Wang, L.; Tan, J.B. Design and Realization of a Three Degrees of Freedom Displacement Measurement System Composed of Hall Sensors Based on Magnetic Field Fitting by an Elliptic Function. *Sensors* 2015, *15*, 22530–22546.
15. Kawato, Y.; Kim, W.J. Multi-degree-of-freedom precision position sensing and motion control using two-axis Hall-effect sensors. *J. Dyn. Syst. Meas. Control* 2006, *128*, 980–988.
16. Kawato, Y.; Kim, W.J. A novel multi-DOF precision positioning methodology using two-axis Hall-effect sensors. In Proceedings of the American Control Conference, Portland, OR, USA, 8–10 June 2005.
17. Sun, Y.T.; Hu, J.C.; Zhu, Y.; Chen, L.M. Study on fast and precise measurement of three-dimensional displacement using hall sensors. *Adv. Mater. Res.* 2013, *694–697*, 1034–1038.
18. Brosillow, C. Inferential control of process. *AIChE J.* 1978, *24*, 485–509.
19. Prasad, V.; Schley, M.; Russo, L.P.; Bequetta, W. Product property and production rate control of styrene polymerization. *J. Process Control* 2002, *12*, 353–372.
20. Zamprogna, E.; Barolo, M.; Seborg, D.E. Optimal selection of soft sensor inputs for batch distillation columns using principal component analysis. *J. Process Control* 2005, *15*, 39–52.
21. Gonzaga, J.C.B.; Meleiro, L.A.C.; Kiang, C.; Filho, R.M. ANN-based soft-sensor for real-time process monitoring and control of an industrial polymerization process. *Comput. Chem. Eng.*2009, *33*, 43–49.
22. Liu, X.Q.; Li, K.; Mcafee, M.; Deng, J. 'Soft-sensor' for Real-time Monitoring of Melt Viscosity in Polymer Extrusion Process. In Proceedings of the 49th IEEE Conference on IEEE Decision and Control (CDC), Atlanta, GA, USA, 15–17 December 2010.
23. Elmas, C.; Zelaya-De, L.P.H. Applilcation of a full-oirder extended Luenberger observer for a position sensorless operation of a switched reluctance motor drive. *IEEE Proc. Control Theory Appl.* 1996, *143*, 401–408.
24. Szabat, K.; Tran-Van, T.; Kaminski, M. A Modified Fuzzy Luenberger Observer for a Two-Mass Drive System. *IEEE Trans. Ind. Inform.* 2015, *11*, 531–539.
25. Shirai, H.; Kageyama, Y.; Ohuchi, A.; Takaya, T.; Nishida, M. On-line Parameter Estimation of Interior Permanent Magnet Synchronous Motor using an Extended Kalman Filter. *J. Electr. Eng. Technol.* 2014, *9*, 600–608.
26. Rajamani, R.; Hedrick, J.K. Adaptive Observer for Active Automotive Suspensions. In Proceedings of the American Control Conference, San Francisco, CA, USA, 2–4 June 1993; pp. 706–710.

27. Lim, T.M.; Cheng, S. Parameter estimation and statistical analysis on frequency-dependent active control forces. *Mech. Syst. Signal Process.* 2007, *21*, 2112–2124.
28. Holland, J.H. *Adaptation in Natural and Artificial Systems*; MIT Press: Cambridge, MA, USA, 1992.
29. Das, S.; Suganthan, P.N. Differential Evolution: A Survey of the State-of-the-Art. *IEEE Trans. Evolut. Comput.* 2011, *15*, 4–31.
30. Kennedy, J.; Eberhart, R. Particle swarm optimization. In Proceedings of the IEEE International Conference on Neural Networks, Perth, WA, USA, 27 Novemebr–1 December 1995; pp. 1942–1948.
31. Rao, R.V.; Savsani, V.J.; Vakharia, D.P. Teaching-Learning-Based Optimization: A novel method for constrained mechanical design optimization problems. *Comput.-Aided Des.* 2011,*43*, 303–315.
32. Rao, R.V.; Savsani, V.J.; Vakharia, D.P. Teaching-Learning-Based Optimization: An optimization method for continuous non-linear large scale problems. *Inf. Sci.* 2012, *183*, 1–15.
33. Neri, F.; Mininno, E.; Iacca, G. Compact Particle Swarm Optimization. *Inf. Sci.* 2013, *239*, 96–121.
34. Niu, Q.; Zhang, H.Y.; Li, K. An improved TLBO with elite strategy for parameters identification of PEM fuel cell and solar cell models. *Int. J. Hydrog. Energy* 2014, *39*, 3837–3857.
35. Niknam, T.; Azizipanah-Abarghooee, R.; Narimani, M.R. A new multi objective optimization approach based on TLBO for location of automatic voltage regulators in distribution systems. *Eng. Appl. Artif. Intell.* 2012, *25*, 1577–1588.
36. Ghasemi, M.; Taghizadeh, M.; Ghavidel, S.; Aghaei, J.; Abbasian, A. Solving optimal reactive power dispatch problem using a novel teaching-learning-based optimization algorithm. *Eng. Appl. Artif. Intell.* 2015, *39*, 100–108.
37. Chen, Y.M. A 3-DOF Precision Position Sensing Using Linear Hall Sensors. Master's Thesis, Harbin Institute of Technology, Harbin, China, June 2014.
38. Kennedy, J. Bare bones particle swarms. Swarm Intelligence Symposium. In Proceedings of the IEEE Swarm Intelligence Symposium, Indianapolis, IN, USA, 24–26 April 2003; pp. 80–87.
39. Omran, M.G.H.; Engelbrecht, A.P.; Salman, A. Bare bones differential evolution. *Eur. J. Oper. Res.* 2009, *196*, 128–139.
40. Qin, A.K.; Suganthan, P.N. Self-adaptive differential evolution algorithm for numerical optimization. *IEEE Congr. Evolut. Comput. Proc.* 2005, *2*, 1785–1791.

CHAPTER 5

DYNAMIC RESPONSE ANALYSIS OF MICROFLOW ELECTROCHEMICAL SENSORS WITH TWO TYPES OF ELASTIC MEMBRANE

Qiuzhan Zhou[1,2,3], Chunhui Wang[2], Yongzhi Chen[2], Shuozhang Chen[2] and Jun Lin[3]

[1]State Key Laboratory of Automotive Simulation and Control, Jilin University, Changchun 130022, China

[2]College of Communication Engineering, Jilin University, Changchun 130022, China

[3]College of Instrumentation and Electrical Engineering, Jilin University, Changchun 130026, China

ABSTRACT

The Molecular Electric Transducer (MET), widely applied for vibration measurement, has excellent sensitivity and dynamic response at low frequencies. The elastic membrane in the MET is a significant factor with an obvious effect on the performance of the MET in the low frequency domain and is the focus of this paper. In simulation experiments, the elastic membrane and the reaction cavity of the MET were analysed in a model based on the multiphysics finite element method. Meanwhile, the effects caused by the elastic membrane elements are verified in this paper. With the numerical simulation and practical experiments, a suitable elastic membrane can be designed for different cavity structures. Thus, the MET can exhibit the best dynamic response characteristics to measure the vibration signals. With the new method presented in this paper, it is possible to develop and optimize the characteristics of the MET effectively, and the dynamic characteristics of the MET can be improved in a thorough and systematic manner.

INTRODUCTION

As an important type of sensor widely used in the fields of seismological measurement and geography, acceleration sensors hold a significant position in the marketplace. Among all types of acceleration sensors, acceleration sensors equipped with a solid-mass inertial element account for a high proportion. However, for low or ultralow frequency signal measurement applications, acceleration sensors with a solid-mass inertial element are limited by having a more complex structure, low resistance to interference, higher cost and other shortcomings [1]. Compared to the acceleration sensors equipped with a solid-mass inertial element, a novel microchannel electrochemical sensor Molecular Electric Transducer (MET) with an electrolytic solution as the liquid-mass inertial element achieves a better performance in the low frequency range field. Electrochemical sensors are used in seismology, seismic exploration, perimeter security systems, structural monitoring [2], among other applications. In the low frequency domain, a MET offers advantages of higher sensitivity, lower cost, high shock resistance and a wider bandwidth. An in-depth study on the use of METs to measure low frequency vibration signals is

Citation: Qiuzhan Zhou, Chunhui Wang, Yongzhi Chen, Shuozhang Chen and Jun Lin, Dynamic Response Analysis of Microflow Electrochemical Sensors with Two Types of Elastic Membrane, doi:10.3390/s16050657.

necessary. Generally, a MET consists of a reaction cavity, an external conditioning circuit and a protective shell. As the core element in the MET, the reaction cavity, filled with electrolyte, with the reversible reaction $I_3^- + 2e^- \rightarrow 3I^-$, ensures the sensing of a vibration signal [3]. In addition, both sides of the reaction cavity are encapsulated by two elastic membranes that influence the dynamic response of the MET. Figure 1 presents the basic concept of the MET and consists of two anode-cathode pairs [1]. The multichannel structure immersed in the electrolyte allows the liquid flow to move inertially, with four electrodes arranged in the order of anode-cathode-cathode-anode (ACCA), which are separated by dielectric layers. The electrolyte is capable of flowing in and out of the multichannel through the electrodes. The usual method to bond the membrane to the reaction cavity is to apply the four-bolt structure shown in Figure 1 (right), thus preventing liquid leakage.

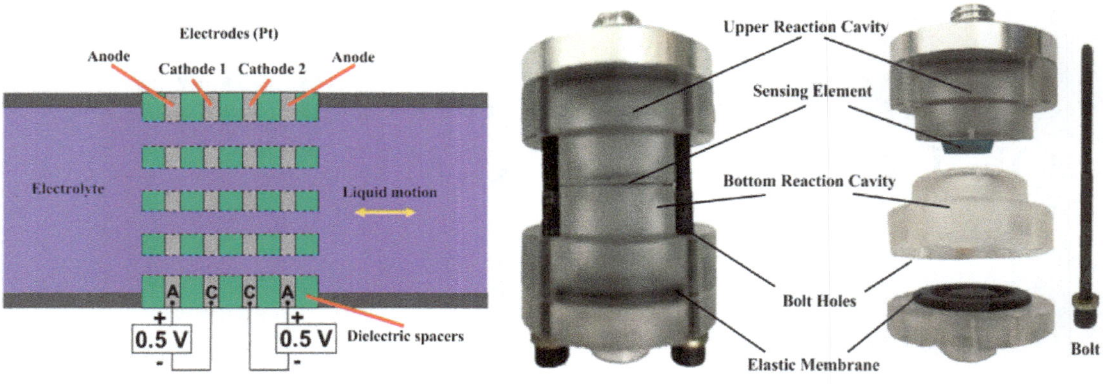

Figure 1. Schematics of the MET (**left**), real product graph and decomposing graph (**right**) structures.

A sensing element with microflow channels (Figure 2) is found in the middle of the reaction cavity. Alternately overlapping platinum layers (four) and insulating layers (five) are obtained by a procedure of bonding, pressuring and puncturing by Low Temperature Cofired Ceramic (LTCC) technology (Figure 3).

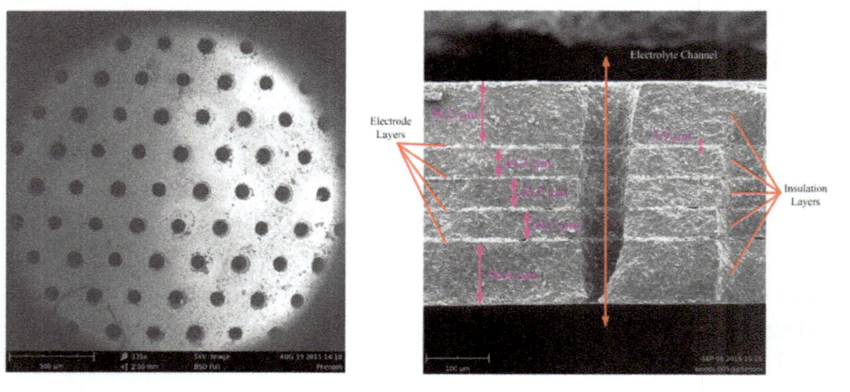

Figure 2. Sensing element with microflow channels (**left**) and dimensions of the intersection of the surface of a single channel (**right**).

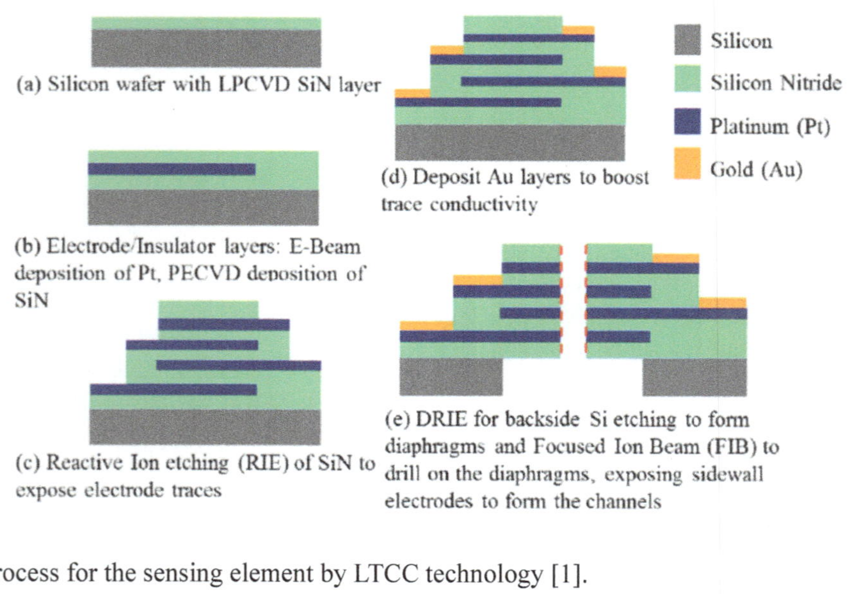

Figure 3. Fabrication process for the sensing element by LTCC technology [1].

In the sensing element, due to the fact the external vibration signal and the multichannel are oriented in the same direction, the electrolyte flow changes with the measured vibration signal when the MET is in the working condition. Because the MET is an active sensor, a fixed electric potential difference imposed alternately on the platinum layers is arranged as anode-cathode-cathode-anode (ACCA). With the reversible reaction $I_3^- + 2e^- \rightarrow 3I^-$ on the interface of the cathodes, a miniature electrochemical system is created in the cavity of the MET [1]. When the electrolyte shows ion movement caused by external vibration excitation, a concentration gradient of the electrolyte is generated in the whole reaction cavity, especially on the electrodes in the sensing element. Meanwhile, the enhanced effect of electrode kinetics controls the rate of the electrochemical reaction. Thus, the changing current density derived from the reversible electrochemical reaction on the surface of the electrodes is proportional to the signal from the external vibration. Thus, the differential current generated by two cathodes, which is the source of the MET output, can be the output from the cathodes after amplification by the conditioning circuit because the interface charge transfer is associated with the generation and absorption of the tri-iodide ions on the electrode surface. Therefore, the electrical current through any electrode can be related to the flux of active ions towards or away from the electrode, according to the following equation:

$$I = \mathrm{Dq}\left(\oint_S (\nabla c, \mathbf{n})\right) dS \quad (1)$$

where D is the diffusion coefficient, c is the concentration of the active charge carriers, q is the charge transferred across the interface in single electrochemical reaction, **n** is a unit vector normal to the surface of the electrode, and integration is performed over S, the electrode surface area.

The operating principle of MET can be described as follows: when an electric voltage is applied to the electrodes, an electrochemical (background) current appears. When an external vibration appears, the electrolyte starts moving due to inertia, and the convective transport of ions changes the electrode current according to the mechanism described above. The difference in the cathode currents in the two anode-cathode pairs is used as the MET output signal. Although each electrode current is non-linear with respect to the fluid velocity, the combined output of both cathodes is linear for a very wide range of fluid velocities. The MET output current is given by [3]:

$$I_{out} = I_2 - I_1 = \mathrm{Dq}\left(\oint_{S2} (\nabla c, \mathbf{n})\right) dS2 - \mathrm{Dq}\left(\oint_{S1} (\nabla c, \mathbf{n})\right) dS1 \quad (2)$$

where I_2 and I_1 are the currents through the surface of the corresponding cathodes, and S_2 and S_1 are the surface areas of the corresponding cathodes.

The optimization of the design of the MET requires too many physical parameters of the complex electrochemical process to be obtained directly from practical measurement. Under practical conditions, due to the different materials, manufacturing processes, experimental conditions and other issues, many variables such as flow field distribution in the reaction cavity, a variety of charging ion concentration gradients and the electrode surface ion concentration, etc., cannot always be measured directly by practical experiments, so a few problems of the MET cannot be solved by traditional methods.

Recently, with the development of numerical simulation technology, multiphysics finite element analysis methods can process the joint simulation of a complex electrochemical and microfluidic model. Combined with the traditional methods, the use of numerical simulation aids in the development of a comprehensive analysis of the study of METs. Since 2010, many numerical simulation models have provided effective solutions in the relevant MET study fields. Because the study of the MET combines electrochemical, electrical, microfluidic and other complex multiphysics aspects, this study takes advantage of the current new simulation technology methods to accomplish the modelling of the study object. With the numerical simulation, physical characteristics that could not be measured and observed before in the practical experiments can be studied effectively. Beginning in 2012, based on numerical simulation and practical experiments, Sun [4,5], Vadim [6,7,8] and Huang [1,9,10] have studied the theory and the complex electrochemical changes in the MET reaction cavity. By combining the results of practical experiments with numerical simulations, a number of performance optimizations can be accomplished by analysing the actual results measured by data acquisition equipment and the simulations that involve effective concentrations of charged ions and comprehensive velocity-field distributions.

Compared with traditional acceleration sensors, its special structure and measurement principles make the MET an effective transition from a solid inertial element to a liquid inertial element due to a great innovation. With its special inertial element, unique advantages for vibration signal measurement in the low frequency domain can be achieved. For the MET, new methods have become available to guarantee its measurement accuracy.

First, the elastic membrane element in the MET causes the electrolyte to be encapsulated in the reaction cavity. When an external excitation signal arrives, the liquid inertial element not only generates a movement in the same direction as the vibration direction but also generates a movement in the perpendicular direction, and a trend in the flow is followed. Under normal conditions, the electrolyte should remain in a state of standard laminar flow in the reaction cavity. However, due to the kinetic energy caused by the high frequency or large amplitude of the external vibration signal, a slip happens between the layers of the electrolyte that is exposed to laminar flow. When the electrolyte is moving, to prevent the electrolyte from generating a series of slips between the laminar layers, the elastic membranes that are positioned on both ends of the reaction cavity are capable of maintaining the laminar flow of the electrolyte under various forms of external vibration motion. Only if the laminar flow state is maintained, will the MET be operated under the best performance conditions.

According to the theory mentioned above, the elastic force generated by the elastic membrane can control the movement of the electrolyte in the reaction cavity. When an external vibration occurs, the electrolyte in the reaction cavity of the MET moves in the same way as the vibration signal. Thus, the electrochemical current signal that acts as the source of the output of the MET is proportional to the vibration intensity on the cathodes. The elastic membrane element has the main function of adjusting the electrolyte motion. However, for different measurement fields, the design of the elastic membrane influences the dynamic response of the MET. To compensate and decrease the negative impact caused by improper design, in this paper, a novel elastic membrane design method that can adapt to the different structures of the MET is proposed. Combining the practical size of the MET with numerical modelling, the state of movement of the electrolyte in the reaction cavity and elastic membranes with different thicknesses can be observed by the multiphysics finite element method. With the same pulse signal imposed on two METs with different elastic membranes, a series of evaluations can be accomplished. The optimal design of the elastic membrane can be calculated by comparing variables such as the response time. In addition, in this paper, relevant practical experiments are also applied to verify the correctness of the numerical simulation. Finally, amplitude-frequency and phase-frequency characteristics are discussed to describe the influence of different membranes on the MET.

MATHEMATICAL MODELLING

The geometric structure of the MET is built as shown in Figure 4 (left). The elastic membrane element (yellow) is placed on the top of the cavity. The reaction cavity is marked as blue. The inside structure is presented in Figure 4 (right). The membrane element is on the top of the numerical model. In the simulation, a Gaussian pulse response signal is used as the external vibration excitation. By observing the movement state of the electrolyte in the reaction cavity and the elastic membrane, the dynamic response characteristics of the membrane can be calculated by the numerical simulation.

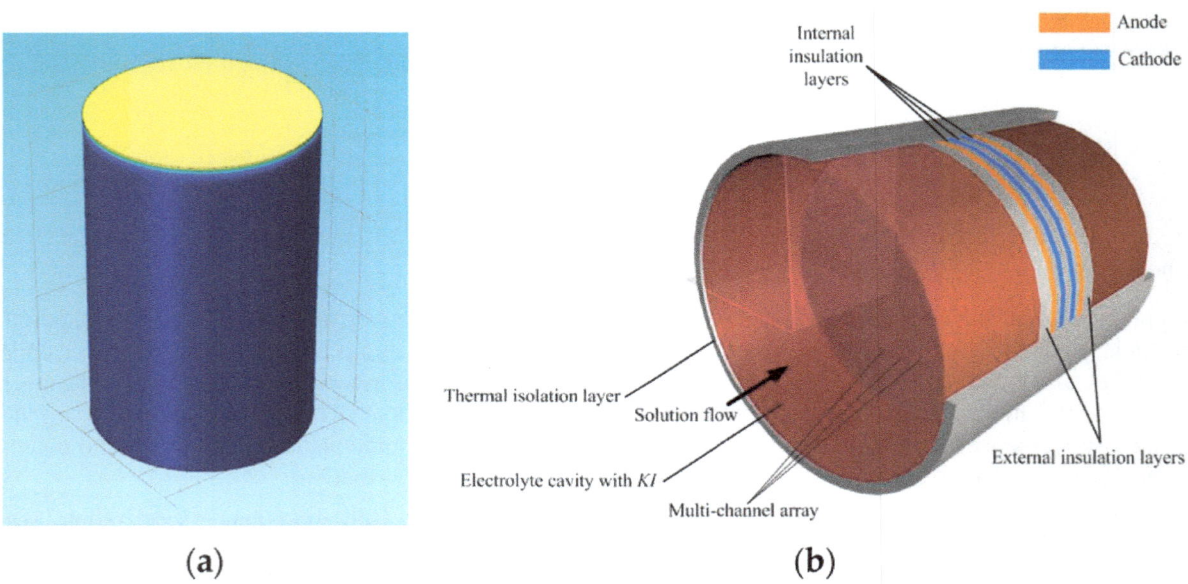

Figure 4. The structure of the reaction cavity (**a**) and the elastic membranes in the numerical simulation and the structure inside the reaction cavity (**b**).

The numerical simulation parameters of the dimensions and materials are shown in Table 1.

Table 1. Numerical simulation parameters

Membrane Material	**Low Density Polyethylene (LDPE)**
Membrane Radius	r = 1.25 cm
Membrane Thickness	h_1 = 1.0 mm, h_2 = 1.5 mm
Young's Modulus	E = 0.2 GPa
Poisson's Ratio	V = 0.4
Mass Density	ρ = 900 kg/m^3
Cavity Height	H = 4.0 cm
Cavity Material	Polymeric Methyl Methacrylate (PMMA)
Cavity Radius	R = 1.2 cm
Multi-channel Centre Distribution Radius	R_{centre} = 0.13 cm

Navier-Stokes equations are applied to build the fluid dynamic model of the electrolyte as follows:

$$\nabla v=0 \quad (3)$$

$$\frac{\partial v}{\partial t} = -\frac{\nabla P}{\rho} + \frac{\mu}{\rho}\nabla^2 \mathbf{v} + a \quad (4)$$

where v and t denote the velocity of the electrolyte flow and time. The electrolyte density is ρ=1.473×103kg/m. P, a scalar, represents the pressure, and $\mu = 1.4 \times 10^{-3}$ Pa is the dynamic viscosity. Parameter $\boldsymbol{a}$ is the acceleration of the external signal.

In this study, Gauss's law is applied to demonstrate charge conservation:

$$-\nabla \cdot (\varepsilon \nabla \varphi) = \rho' \quad (5)$$

where ε, φ and ρ′ are the permittivity ($F{\cdot}m^{-1}$), potential (V) and charge density ($C{\cdot}m^{-3}$), respectively.

In addition, the Nernst-Plank-Poisson equation is used to describe the mass transport (Equations (6) and (7)) subject to mass continuity (Equation (8)). Because of the high degree of nonlinearity, multiple length and time scales are observed, and the applicable approximation, which applies to scales larger than nanometres, is the assumption of electroneutrality (Equation (7)) [11]:

$$\vec{N_i} = -D_i \nabla c_i - z_i m_i c_i \nabla \varphi + c_i \vec{u} \quad (6)$$

$$c_{K+} = c_{I_3^-} + c_{I^-} \quad (7)$$

$$\frac{\partial c_i}{\partial t} + \nabla \cdot \vec{N_i} = R_i = 0 \quad (8)$$

where $\vec{N}$ i is the flux of species i ($mol{\cdot}m^{-2}{\cdot}s^{-1}$) and D_i is the diffusion coefficient of species i($m^2{\cdot}s^{-1}$). Here we set $D_{I}^{-} = D_{K}^{+} = 2.8 \times 10^{-9}\ m^2{\cdot}s^{-1}$ and $D_{I3}^{-} = 2.0 \times 10^{-9}\ m^2{\cdot}s^{-1}$. c_i is the concentration of species i ($mol{\cdot}m^{-3}$), z_i is the charge number of species i; m_i is the mobility of species i ($m^2{\cdot}V^{-1}{\cdot}s^{-1}$), $\vec{u}$ =[u,v,w] is the velocity vector ($m{\cdot}s^{-1}$), and R_i is the mass source of species i ($mol{\cdot}m^{-3}{\cdot}s^{-1}$).

Assuming electroneutrality and small absolute concentration gradients of charge-carrying electrolyte species, the electrolyte current obeys Ohm's law (Equation (9)) subject to a constant conductivity (Equation (10)). Meanwhile, the concentration of the electrolyte is effectively infinite, Fick's 2nd law for an incompressible flow is applied (Equation (11)) as follows:

$$-\nabla \cdot (\sigma \nabla \varphi) = Q \quad (9)$$

$$\sigma \approx F \sum_i z_i^2 m_i c_i \quad (10)$$

$$\frac{\partial c_i}{\partial t} = \nabla^2 c_i - \vec{u} \cdot \nabla c_i \quad (11)$$

where σ is the conductivity ($S{\cdot}m^{-1}$), Q is the charge source ($A{\cdot}m^{-3}$·), and F is the Faraday constant.

Combining the equations from Equation (6) to Equation (11), the partial differential equations (PDEs) are applied in the Comsol Multiphysics software as shown below [3]:

$$\frac{\partial c_{I^-}}{\partial t} + \frac{\partial c_{I^-}u}{\partial x} + \frac{\partial c_{I^-}v}{\partial y} + \frac{\partial c_{I^-}w}{\partial z} = -m_{I^-}F\left[\nabla c_{I^-} \cdot \nabla \varphi + c_{I^-}\nabla^2\varphi\right] + D_{I^-}\nabla^2 c_{I^-} \quad (12)$$

$$\frac{\partial c_{I_3^-}}{\partial t} + \frac{\partial c_{I_3^-}u}{\partial x} + \frac{\partial c_{I_3^-}v}{\partial y} + \frac{\partial c_{I_3^-}w}{\partial z} = -m_{I_3^-}F\left[\nabla c_{I_3^-} \cdot \nabla \varphi + c_{I_3^-}\nabla^2\varphi\right] + D_{I_3^-}\nabla^2 c_{I_3^-} \quad (13)$$

$$\frac{\partial c_{K+}}{\partial t} + \frac{\partial c_{K+}u}{\partial x} + \frac{\partial c_{K+}v}{\partial y} + \frac{\partial c_{K+}w}{\partial z} = m_{K+}F\left[\nabla c_{K+} \cdot \nabla \varphi + c_{K+}\nabla^2\varphi\right] + D_{K+}\nabla^2 c_{K+} \quad (14)$$

In addition, the Butler-Volmer condition is set up to convey the procedure of electrode kinetics:

$$2\vec{n} \cdot \vec{N}_{I_3^-} = -\frac{2\vec{n} \cdot \vec{N}_{I^-}}{3} = -2k_a c_{I_3^-} e^{(-\alpha F/RT)(U-\varphi-E_0)} + k_c c_{I^-} e^{(1-\alpha)(F/RT)(U-\varphi-E_0)}\alpha \quad (15)$$

where n is the out-normal of the electrode surface, and ka=kc=4×10−9 m2/s are the anodic and cathodic reaction constants, respectively. The parameter α is the charge transfer coefficient for the cathodic reaction and is set to 0.5 in this study. U = 0.8 V is the imposed electric potential at the electrodes, and E_0 = 0.54 V is the equilibrium potential.

Instead of applying the shell interface, which is barely coupled with the hydrodynamic model of the electrolyte in the cavity, extensive efforts were made to build a suitable solid-liquid interface, which was inspired by tracking the immiscible fluid-fluid interface. The membrane model offered by the structural mechanics of Comsol Multiphysics can be coupled with the modified Navier-Stokes equations, including the surface tension force [11,12]. To be the input to the stress load interface of the membrane model, the surface tension force is imposed on the membrane in the vertical direction. According to the relationship between stress and train yields to the constraints such as linear elastic condition and solid material setting, the force distribution on the membrane in the traverse direction is also capable of being calculated. As a feedback effect from the membrane to the electrolyte, the movement state of the electrolyte in the reaction cavity can be influenced, corresponding to the movement state of the membrane. Therefore, variables such as velocity and force distribution for both electrolyte and membrane can be solved by the finite element analysis method.

The topology of the solid-liquid interface changes with time for the contact interface of the solid elastic membrane and the electrolyte. The level set method as well as the phase field method are both well suited for modelling moving boundaries where topology changes occur [13]. Both methods are available in the CFD module and the structure mechanics model. In this solid-liquid two phase model, the level set function determines the solid-fluid interface by tracing the isolines of the level set function, Φ. The equation governing the transport and initialization of Φ is:

$$\frac{\partial \Phi}{\partial t} + \mathbf{u} \cdot \nabla \Phi = \gamma \nabla \cdot \left(\varepsilon \nabla \Phi - \Phi (1 - \Phi) \frac{\nabla \Phi}{|\nabla \Phi|} \right) \tag{16}$$

where **u** is the fluid velocity, and γ (m/s) and ε (m) are the initialization parameters. The ε parameter determines the thickness of the membrane around the interface where Φ goes from zero to one. When stabilization is used for the level set equation, we can typically use an interface with a thickness of $\varepsilon = h_c/2$ to describe the initial membrane, where h_c is the characteristic mesh size in the region encompassed by the solid-liquid interface. A suitable value for γ is the magnitude of the maximum velocity occurring in the model. Because the level set function is a smooth step function, it is also used to determine the density and dynamic viscosity globally by:

$$\rho = \rho_l + (\rho_m - \rho_l)\Phi$$
$$\mu = \mu_l + (\mu_m - \mu_l)\Phi \tag{17}$$

Here ρ_l, μ_l, ρ_m and μ_m denote the constant density and viscosity of the electrolyte and the membrane, respectively [11]:

$$\rho \left(\frac{\partial \mathbf{u}}{\partial t} + \mathbf{u} \cdot \nabla \mathbf{u} \right) = -\nabla p + \nabla \cdot \mu \left(\nabla \mathbf{u} + \nabla \mathbf{u}^T \right) + \rho g + \mathbf{F}_{st}$$
$$\nabla \cdot \mathbf{u} = 0 \tag{18}$$

In Equation (18), ρ represents the density (kg/m³) of the electrolyte, and **u** is the velocity (m/s). The parameters t and p are time (s) and pressure (Pa), respectively. The momentum equations contain gravity, ρg, and the surface tension force is also denoted by F_{st}.

The surface tension force in Equation (16) F_{st} is defined by:

$$\mathbf{F}_{st} = \nabla \cdot \mathbf{T} = \nabla \cdot \left\{ \sigma \left[\mathbf{I} + \left(-\mathbf{n}\mathbf{n}^T \right) \right] \delta \right\} \tag{19}$$

where σ is the surface tension coefficient, **I** is the identity matrix, **n** is the interface unit normal, and δ is a Dirac delta function. The interface normal is calculated from:

$$\mathbf{n} = \frac{\nabla \Phi}{|\nabla \Phi|} \tag{20}$$

The level set parameter Φ is also used to approximate the delta function by a smooth function defined by [13]:

$$\delta = 6 \left| \Phi (1 - \Phi) \right| \left| \nabla \Phi \right| \tag{21}$$

when the surface tension force F_{st} is confirmed by using the stress and strain interface of the membrane model, the surface tension force is proportional to the stress in the vertical direction. Constrained by the material setting and the relationship between stress and strain in the membrane model, the elastic force generated in the transverse direction is also calculated.

The constraints and boundary conditions are described as follows: the bottom of the cavity uses no-slip conditions, **u** = 0. The sidewall of the cavity surrounding the electrolyte uses the wetted wall as its boundary condition. The centre axis of the cylinder cavity corresponds to the symmetry boundary condition. For the membrane constraints, the outer edge of the membrane is supported in the transverse direction and the edge connecting the membrane with the sidewall is set as a fixed constraint. To summize the modelling process, a modified Navier-Stokes equation based on the principles of a two phase flow is presented to describe the state of the movement of the electrolyte. Meanwhile, the surface tension force at the solid-liquid surface is derived by the modified Navier-Stokes equation. To strongly couple the liquid phase with the solid phase, the surface tension force acts as the input of the stress force in the membrane model. Due to the elastic material setting and the stress-strain relationship of a membrane structure, the force distribution on the membrane will be solved integrally. As the feedback from the solid phase to liquid phase, the force and velocity change would correspondingly affect the movement state of the electrolyte. With the constraints and the boundary conditions mentioned above, the level set function is applied to describe the solid-liquid interface. Coupling the membrane with the electrolyte tightly also makes effective communication between the electrolyte and the membrane by the variable velocity.

ANALYSIS AND DISCUSSION

To compare the effect of using different elastic membrane elements, the same parameters of the MET are adopted in this paper. Two thickness types of elastic membrane with the same reaction cavity are simulated in the numerical experiments. In the numerical simulation experiments, the same Gaussian pulse signals of acceleration y = exp(–((x − 0.3)/0.2)^2)) are applied in the models as the external excitation signals. By observing the movement of the elastic membranes, elastic membranes that use different thicknesses can be evaluated. The result of the numerical simulation is shown in Figure 5. The same pulse signals are imposed on the reaction cavities that only have different elastic membranes.

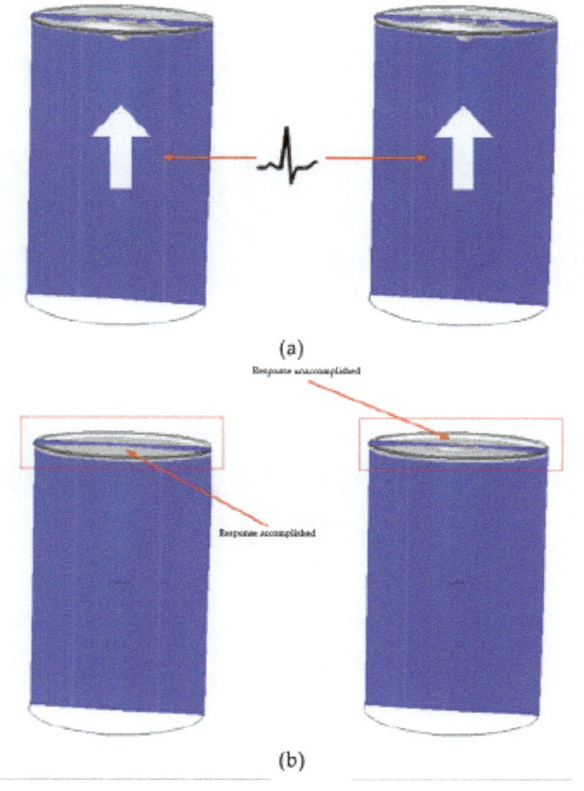

Figure 5. The performance of the different membrane thicknesses under the same Gaussian pulse signal excitation. (**a**) 0–0.3 s, the same pulse signals are imposed on the 1.5 mm thick membrane (**left**) and on the 1.0 mm thick membrane (**right**); (**b**) 0.3–0.5 s, the response of the reaction cavity with a membrane thickness of 1.0 mm is not determined.

In the beginning, the electrolyte is accelerated by the same pulse signal. Due to the energy loss, the amplitude of the moving membranes is reduced, and the state of the electrolyte tends to be stable. After 0.5 s, the MET with the membrane thickness of 1.5 mm regresses to the stable state and waits for the arrival of the next signal. However, because of the 1.0 mm membrane thickness, the other MET is still responding to the pulse signal at 0.5 s. From the result of the numerical simulation, the MET with 1.0 mm thick membrane is not capable of reducing the energy in a short time. To observe the result of the numerical simulation, a probe is inserted on the centre of the membrane to acquire the magnitude of the moving velocity of the membrane at all times. From the probe's observations, the movement state of the membrane is shown with the specific curves in Figure 6.

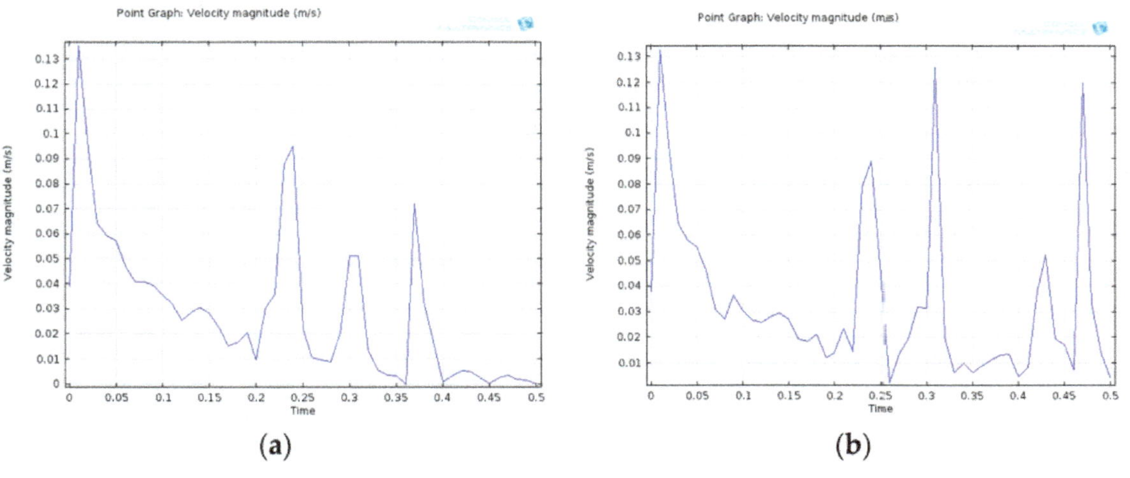

Figure 6. Time (s) vs. velocity magnitude (m/s) of the MET with a membrane thickness of 1.5 mm (**a**) and a membrane thickness of 1.0 mm (**b**).

In Figure 6, at 0 s, the two METs give similar pulse signals with the same magnitude. From 0 to 0.5 s, the MET equipped with a 1.5 mm membrane effectively restrains the electrolyte and helps the liquid inertia regress to the stable state. Meanwhile, the 1.0 mm membrane affects the strength of the constraint. The MET with a 1.0 mm thick membrane is more suitable for the measurement of long-periodic vibration signals. Above all, by using numerical simulation, different elastic membrane models can be built by the multiphysics field element method. The performance and the effect of the membranes can also be evaluated in this way.

RESULTS AND DISCUSSION

To support the numerical simulation results, practical experiments comparing different membrane structures were also applied in this paper. The experimental parameters of the materials and the structure and dimensions for membrane are listed in Table 2. Figure 7 shows the experimental setup and the recorded signals.

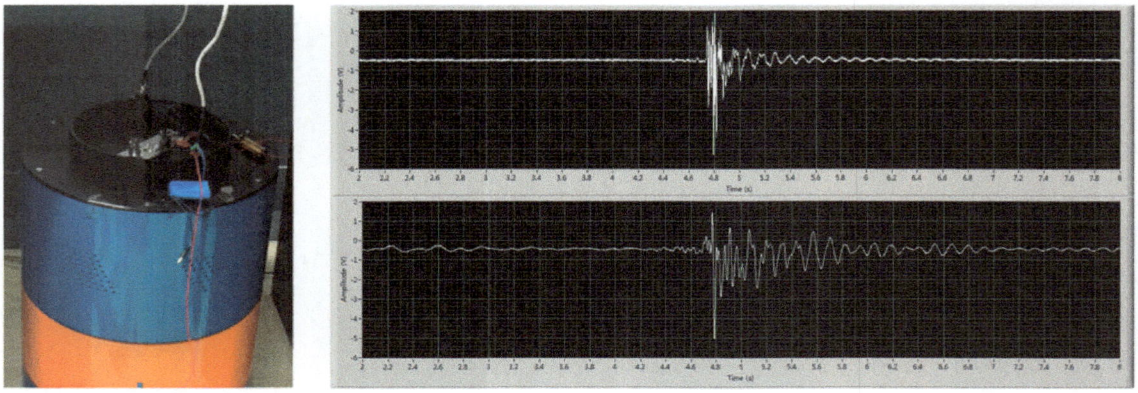

(a) (b)

Figure 7. Experimental environment of the vibration table and the data acquired from NI PXI (a product of National Instrument Corporation).

Table 2. Experimental parameters

Membrane Material	Low Density Polyethylene (LDPE)
Membrane Radius	r = 1.25 cm
Membrane Thickness	h_1 = 1.0 mm, h_2 = 1.5 mm
Membrane Density	ρ = 900 kg/m³
Cavity Height	H = 4.0 cm
Cavity Material	Polymeric methyl methacrylate (PMMA)
Cavity Radius	R = 1.2 cm
Shape of the Membrane	Circular

By using the NI PXI data acquisition equipment the results of the practical experiment for two METs with different membrane thicknesses are presented in Figure 8.

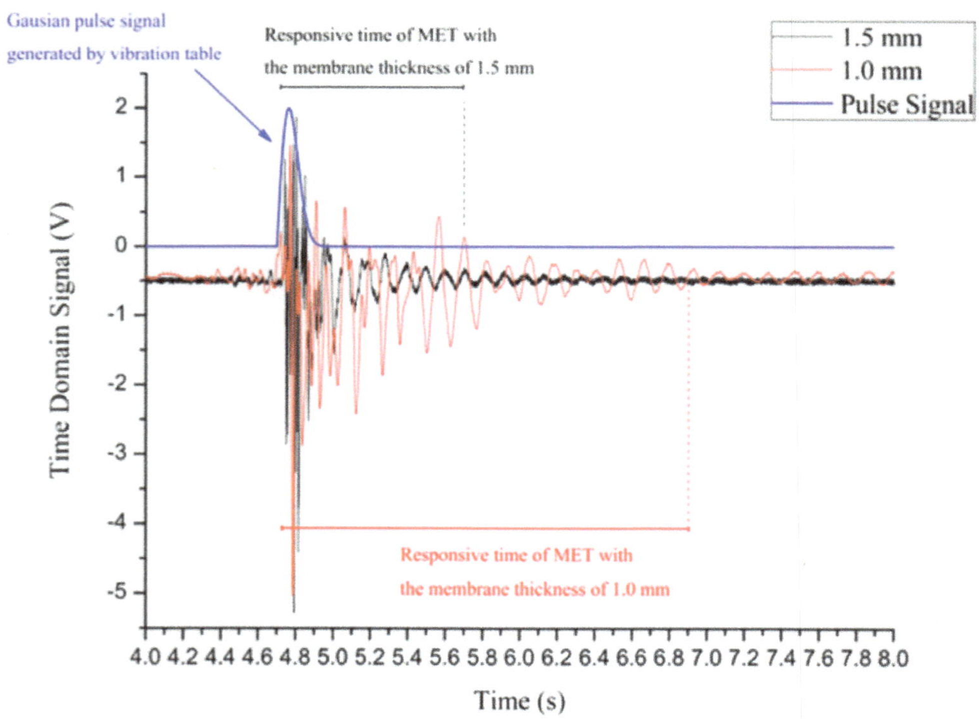

Figure 8. Practical experiment for the MET with different elastic membranes.

By analysing the results shown in Figure 8, at the beginning, the Gaussian pulse signals marked in blue produce an effective external vibration excitation with the same amplitude for both METs. The MET with the 1.0 mm membrane a longer response time. The MET with a 1.5 mm membrane can regress to the stable state in a shorter time.

To discuss the correspondence between modelling and the experimental data, a segment captured from the simulation results in Figure 6 is compared with the experimental result in Figure 9. First, comparing both the left figures in Figure 6 and Figure 9, the MET with the 1.5 mm membrane is activated by an external pulse signal, and the first peak appears in both left figures simultaneously. The peaks following this highest peak dampen gradually, and the curves in the practical experiment convey the same trend. The curves of the right figures in both Figure 6 andFigure 9 share the same trend and shape. Therefore, this practical experiment consisting of a pulse signal response test conveys that the results meet the expectations of the numerical simulation inFigure 6, and the result of the practical experiment in Figure 9 is presented to verify the numerical simulation.

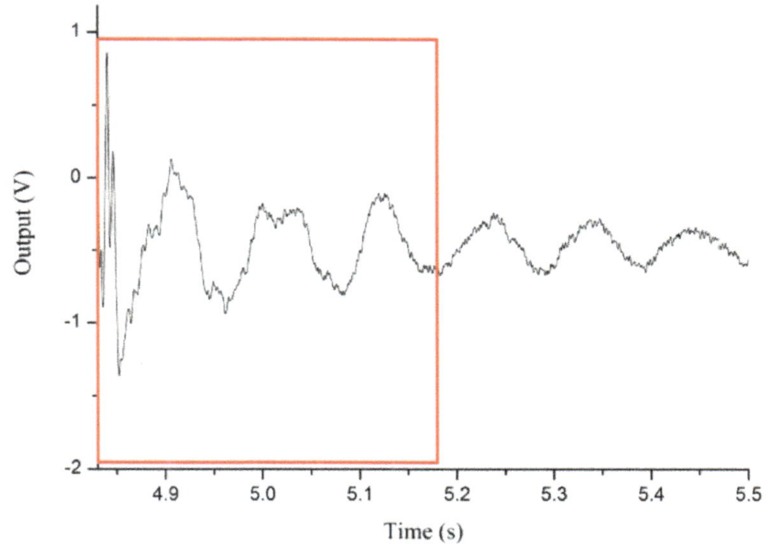

(a)

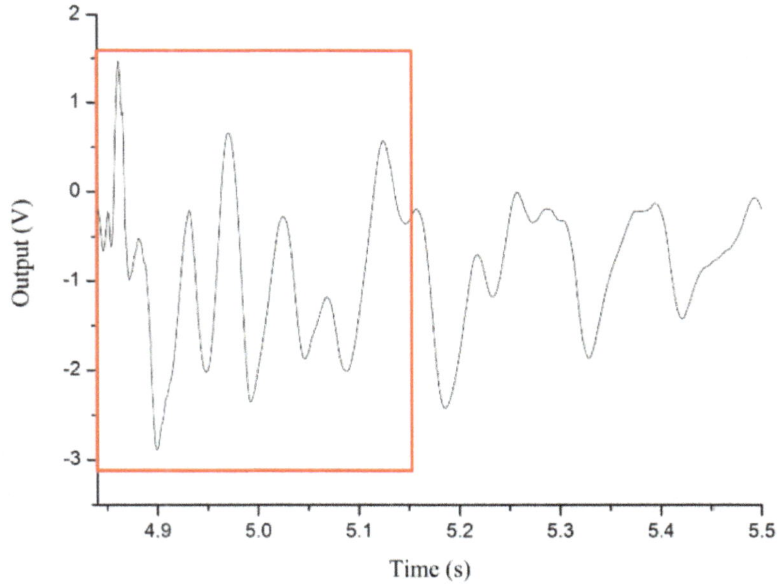

(b)

Figure 9. Segments captured from Figure 8 corresponding to the simulation results in Figure 6, and (**a**) is the result of practical experiment for MET with 1.5mm membrane thickness corresponding to the simulation in Figure 6a; (**b**) is the result of practical experiment for MET with 1.0mm membrane thickness corresponding to the simulation in Figure 6b.

However, for the dynamic response characteristics, the pulse signal is insufficient to completely reveal the dynamic response characteristics of the MET with different types of membranes. Another practical experiment was studied to investigate the influence of different elastic membranes.

The sensitivity-frequency curves for two types of MET equipped with both membranes are shown in Figure 10. From the distribution of the points, the dynamic response for the MET can be represented by the result of the practical experiments. In Figure 10, the MET with the 1.5 mm thick membrane has a higher natural frequency. The natural frequency as a significant parameter affects the sensitivity in the low frequency domain. For a MET that has a low natural frequency, the capacity to detect the low-frequency signal will be much better. The MET with a 1.0 mm thick membrane has a lower natural frequency, so this type of membrane is more suitable for a MET that requires better measurement performance in the low frequency range. However, from the bandwidth point of view, the MET with the 1.0 mm membrane covers a shorter range in the frequency domain. Compared with the 1.5 mm membrane, the MET with the 1.0 membrane always has a narrow band range but a stronger capacity to measure low frequency signals. From the practical experiment data in Figure 10, the amplitude-frequency characteristics are revealed to describe the dynamic response of the MET.

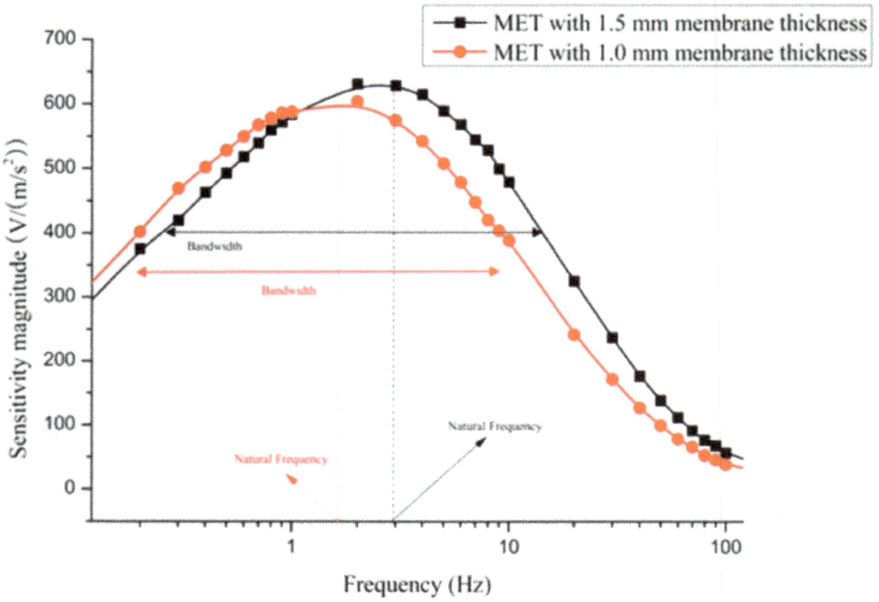

Figure 10. Frequency vs. sensitivity magnitude of the MET that is equipped with different membrane thicknesses.

Moreover, to completely study the dynamic response of the MET affected by different types of membranes, another practical experiment involving the phase-frequency characteristics is necessary, and the results are shown in Figure 11. By comparing two curves from the aspect of phase-frequency, the MET with the 1.5 mm membrane can maintain a little phase shift in the frequency domain. Thus, analysis of the dynamic response for the METs with two types of membrane is explained, and the influence of the elastic membrane is also presented above.

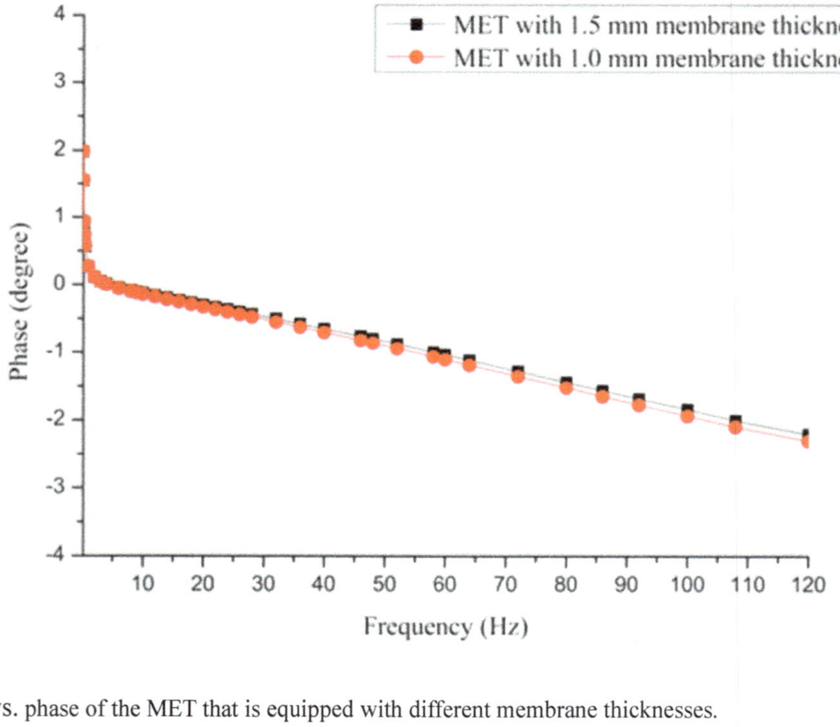

Figure 11. Frequency vs. phase of the MET that is equipped with different membrane thicknesses.

CONCLUSIONS

The elastic membrane element, a significant element in METs, is the main focus of this paper. From the numerical simulation experiments based on the multiphysical field finite method the elastic membrane and the cavity of the MET are studied integrally. Meanwhile, the relationship between different elastic membrane elements and the dynamic response of the MET are derived in this paper. Combining the numerical simulation with practical experiments, the elastic membrane can be customized for several of vibration measurement fields. Thus, the adaptation mentioned above can ensure the best dynamic response characteristics for the MET.

ACKNOWLEDGMENTS

This work is supported by China Postdoctoral Science Foundation (No.: 801151014537), and Department of science and technology of Jilin Province (No.: 20150203018GX).

AUTHOR CONTRIBUTIONS

Chunhui Wang and Yongzhi Chen conceived and designed the experiments; Qiuzhan Zhou performed the experiments; Chunhui Wang and Shuozhang Chen analyzed the data; Jun Lin contributed experiment and analysis tools; Qiuzhan Zhou wrote the paper.

REFERENCES

1. Huang, H.; Agafonov, V.M.; Yu, H.Y. Molecular electric transducers as motion sensors: A review. Sensors **2013**, 13, 4581–4597.
2. Agafonov, V.M.; Egorov, E.V.; Zaitsev, D.L. Molecular electronic linear accelerometers. Preliminary test results. Gyrosc. Navig. **2010**, 1, 246–251.
3. Krishtop, V.G.; Agafonov, V.M.; Bugaev, A.S. Technological principles of motion parameter transducers based on mass and charge transport in electrochemical Microsystems. Russ. J. Electrochem. **2012**, 48, 746–755.
4. Sun, Z.; Agafonov, V.M. Computational study of the pressure-driven flow in a four-electroderectangular micro-electrochemical accelerometer with an infinite aspect ratio.Electrochim. Acta **2010**, 55, 2036–2043.
5. Sun, Z.; Agafonov, V.M. 3D numerical simulation of the pressure-driven flow in a four-electrode rectangular micro-electrochemical accelerometer. Sens. Actuat. B Chem. **2010**, 146, 231–238.
6. Zaitsev, D.; Agafonov, V.; Egorov, E.; Antonov, A.; Shabalina, A. Molecular Electronic Angular Motion Transducer Broad Band Self-Noise. Sensors **2015**, 15, 29378–29392.
7. Neeshpapa, A.; Antonov, A.; Agafonov, V. A Low-Noise DC Seismic Accelerometer Based on a Combination of MET/MEMS Sensors. Sensors **2014**, 14, 365–381.
8. Agafonov, V.; Egorov, E. Electrochemical accelerometer with DC response, experimental andtheoretical study. J. Electroanal. Chem. **2016**, 761, 8–13.
9. Huang, H.; Carande, B.; Tang, R.; Oiler, J.; Zaitsev, D.; Agafonov, V.; Yu, H.Y. A micro seismometer based on molecular electronic transducer technology for planetary exploration.Appl. Phys. Lett. **2013**, 102, 193512.
10. Huang, H.; Carande, B.; Tang, R.; Oiler, J.; Zaitsev, D.; Agafonov, V.; Yu, H.Y. Development of a micro seismometer based on molecular electronic transducer technology for planetary exploration. In Proceedings of the IEEE 26th International Conference on Micro Electro Mechanical Systems, MEMS 2013, Taipei, Taiwan, 20–24 January 2013; pp. 629–632.
11. Gross, S.; Reusken, A. Numerical Methods for Two-Phase Incompressible Flows; Springer-Verlag: Berlin/ Heidelberg, Germany, 2011; Volume 40, p. 482.
12. Hysing, S. A new implicit surface tension implementation for interfacial flows. Int. J. Numer. Meth. Fluids **2006**, 51, 659–672.
13. Engquist, B.; Tornberg, A.K.; Tsai, R. Discretization of Dirac delta functions in level set methods. J. Comput. Phys. **2005**, 207, 28–51.

CHAPTER 6

AUTOMATED KICK CONTROL PROCEDURE FOR AN INFLUX IN MANAGED PRESSURE DRILLING OPERATIONS

Jing Zhou[1], Jan Einar Gravdal[2], Per Strand[3], Svein Hovland[4]

[1]Department of Engineering Sciences, University of Agder, N-4898 Grimstad, Norway.

[2]International Research Institute of Stavanger, N-5008 Bergen, Norway

[3]Statoil ASA, Bergen, Norway

[4]National Oilwell Varco, Stavanger, Norway

ABSTRACT

Within drilling of oil and gas wells, the Managed Pressure Drilling (MPD) method with active control of wellbore pressure during drilling has partly evolved from conventional well control procedures. However, for MPD operations the instrumentation is typically more extensive compared to conventional drilling. Despite this, any influx of formation fluids (commonly known as a kick) during MPD operations is typically handled by conventional well control methods, at least if the kick is estimated to be larger than a threshold value. Conventional well control procedures rely on manual control of the blow out preventer, pumps, and choke valves and do not capitalize on the benefits from the instrumentation level associated with MPD. This paper investigates two alternative well control procedures specially adapted to backpressure MPD: the dynamic shut-in (DSI) procedure and the automatic kick control (AKC) procedure. Both methods capitalize on improvements in Pressure While Drilling (PWD) technology. A commercially available PWD tool buffers high-resolution pressure measurements, which can be used in an automated well control procedure. By using backpressure MPD, the choke valve opening is tuned automatically using a feedback-feedforward control method. The two procedures are evaluated using a high fidelity well flow model and cases from a North Sea drilling operation are simulated. The results show that using AKC procedure reduces the time needed to establish control of the well compared to DSI procedure. It also indicates that the AKC procedure reduces the total kick size compared to the DSI procedure, and thereby reduces the risk of lost circulation.

INTRODUCTION

During drilling operations, there are many events that can evolve rapidly and have large consequences related to the flow of drilling fluid in the well. Fluid influx is one of these, which if not properly handled in due time, may lead to very dangerous and very expensive situations. Eliminating or minimizing drilling fluid in- flux reduces costs, improves safety, increases well bore stability, and decreases formation damage. Therefore, it is of great importance to control it in its early phase, so that it can be attenuated in a controlled manner. Recent experience indicates that in order to optimize the drilling operation the entire drilling system, not just the mechanics or software, needs to be designed from a control system point of view. Automatic

Citation: Jing Zhou, Jan Einar Gravdal, Per Strand, Svein Hovland. Automated Kick Control Procedure for an Influx in Managed Pressure Drilling Operations, DOI 10.4173/mic.2016.1.3

control of drilling operations in a well can be a challenging task, due to the very complex dynamics of the multiphase flow potentially consisting of drilling mud, oil, gas and cuttings. Managed Pressure Drilling (MPD) is a technology that enables precisely control of the annular pressure during drilling and aims to prevent drilling related problems in Hannegan et al. (2004). Constant bottomhole pressure control, one of the methods within MPD, utilizes a pressurized drilling system with a rotating control head and a surface choke to precisely control the wellbore pressure. Using this method, the bottomhole pressure can be more easily maintained within the bounds of a narrow porepressure to fracture-pressure gradient window. Manual control of the choke valve in well control is commonly applied in today's drilling operations. The control systems are operated by the drilling crew, where the various inputs to the drilling system are adjusted independently. Therefore, it is low reaction to changes in set-point and disturbances. Proportional-IntegralDerivative (PID) control is a powerful control method because of the simple parameter tuning and limited requirements for knowledge about the process. PID control in the drilling process has been studied in Godhavn (2009), Zhou and Nygaard (2010) and Siahaan et al. (2014). Based on the developed drilling model, the model-based control for drilling operation has been studied in Nygaard et al. (2007), Carlsen et al. (2008), Breyholtz et al. (2009), Zhou et al. (2009), Zhou et al. (2010), Zhou et al. (2011), Zhou and Nygaard (2011), Kaasa et al. (2013). However, this method depends on the accuracy of the model and the complexity is increased by the fact that a large set of parameters in such models are uncertain or unknown and possibly changing.

The concept of MPD using annulus backpressure has partly evolved from conventional well control procedures. However, for MPD operations the instrumentation is typically more extensive compared to conventional drilling. Despite this, a kick incident during MPD operations is typically handled by conventional well control methods, at least if the kick is estimated to be larger than a threshold value. Conventional well control procedures rely on manual control of the blow out preventer, pumps, and choke valves and do not capitalize on the benefits from the instrumentation level associated with MPD. As a result of the ability to manipulate backpressure, MPD offers the capability to control a kick dynamically without shutting in. Once an influx is detected during MPD, an important decision is whether to control the well dynamically or to revert to conventional well control techniques. Automatic well control without shutting in is only applicable up to a certain size, rate, and duration of influx. The influx must be sufficiently small that it is within the limits of MPD surface equipment to circulate the influx out safely while maintaining well control. Dynamic well control during MPD has been of interest in the literature for many years. Santos et al. (2003) and Santos et al. (2007) presented a Micro-flux control method, which is the monitoring and adjustment of return flow to control influx during drilling operations. The alternative initial responses to kicks during MPD were explored in Bacon et al. (2012); Davoudi et al. (2011); Smith and Patel (2012), by increasing casing pressure until flow out equals flow in. There is a significant potential to improve existing methods, and in what follows we propose a controller that automatically attenuates a kick without shutting down the pumps. In this paper, we investigate two alternative well control procedures specially adapted to backpressure MPD: The dynamic shut-in procedure and the automatic kick control procedure. Both methods capitalize on improvements in Pressure While Drilling (PWD) technology. A new commercially available PWD tool buffers high-resolution pressure measurements, which can be used in an automated well control procedure. Typically, only minimum, maximum and average downhole pressure is transmitted to surface after circulation is regained. Using new PWD technology, recorded measurements at 0.5 Hz can be transmitted after regaining circulation. By using backpressure MPD, the choke opening is automatically controlled using a feedback-feedforward control method. The rest of the paper is organized as follows. The second section presents the two kick control procedures. In the third section, the control theory is proposed. Results from case study are presented in the fifth section and finally the conclusions are drawn in the sixth section.

KICK CONTROL PROCERURES

This section presents two well control procedures used in this paper, namely, the dynamic shut-in (DSI) procedure and the automatic kick control (AKC) procedure. In the DSI procedure, both the mud pump and the backpressure pump are shut down automatically and the well is shut in by closing the MPD choke valve. In the AKC procedure, circulation is maintained, resulting in less pressure variations in the annulus and the choke line system.

Dynamic Shut-in Procedure

The DSI procedure is a method for attenuating a kick and controlling bottomhole pressure by adjusting the mud pump, backpressure pump and MPD choke valve. When using an automatic control system during the drilling operation, the bottomhole pressure is kept constant at a set-point defined by the driller. An indicator that a kick has occurred is an increase in mudflow out of the well. The proposed system monitors the flow out vs. flow in and activates the kick control procedure. Immediately after the kick detection, the system starts the DSI procedure by closing the mud pump and backpressure pump while the choke valve is controlled by keeping the bottomhole pressure at a set-point. When the mud pump flow rate is less than a threshold value, the choke valve is closed completely. During the time when the pumps and the choke are fully closed, the well pressure will increase and balance the pore pressure of the influx zone. During this sequence the standpipe pressure, choke pressure and PWD measurements are recorded for later use as part of the pore pressure estimation of the reservoir kick zone.

After shut-in the circulation procedure starts by increasing the backpressure pump rate to the setting, which is the same rate before the kick situation began, and increasing the mud pump to a pre-determined slow circulation rate, 600 l/min while the choke valve is controlled by keeping the choke pressure at a calculated value. The PWD tool has an activation flow rate, which we have set to 600 l/min. The mud pump rate is kept at a constant value for 5 minutes to transmit the recorded PWD measurements. After that, the mud pump rate is increased to the setting, which is the same rate before the kick situation began, and the choke valve is controlled by keeping the bottomhole pressure at a new set-point, which corresponds to the estimated pore pressure of the reservoir zone. Since the well is completely shut in during the pressure buildup, the new set point can be found using e.g. the pore pressure estimation method proposed by Gravdal et al. (2010).

Automatic Kick Control Procedure

Immediately after the kick detection, AKC automatically starts to increase the BHP by using the MPD choke control system. The choke valve is automatically controlled by balancing the flow-in and flow-out, resulting in reduced influx from the reservoir. The proposed response involves reducing flow out until equal to flow in, followed by maintaining this condition of fluid continuity for some time. The duration for this condition will be based on the confirmation that the influx has ceased to flow and experience. This phase is similar to the Micro-flux control method by Santos et al. (2003) and Santos et al. (2007). At the same time, the flow rates through the mud pump and backpressure pump are maintained at the initial values. When the pressures have stabilized and the flow-out vs. flowin balance is re-established, the bottomhole pressure is recorded for calculation of pore pressure of the reservoir zone. Then the choke valve is controlled by keeping the bottomhole pressure at a new set-point, which is higher than the estimated pore pressure of the reservoir zone.

FEEDBACK-FEEDFORWARD CONTROLS

In this section, the automatic control method is described, where a feedback Proportional-Integral control and a feedforward control are applied. By using backpressure MPD, the choke valve opening is controlled automatically using a feedback-feedforward control method while the main pump flow and the backpressure pump flow are manually operated. The proposed control scheme can be described by the structure in Figure 1, where the feedback controller calculates an error value as the difference between a measured process variable and a desired set-point. The controller attempts to minimize the error by adjusting the process by use of a manipulated variable and compensate the effects of the disturbance. The disturbance consists of known and unknown disturbances, where the flows through the main pump and the backpressure pump are regarded as measured disturbances to the process.

The proposed feedback-feedforward control is expressed as

$$u(k) = u_{PI}(k) + u_{ff}(k) \tag{1}$$

$$u_{PI}(k) = u_{PI}(k-1) - K_P(e(k) - e(k-1)) - K_I e(k) \tag{2}$$

$$u_{ff}(k) = K_{ff}\, d(k) \tag{3}$$

$$e(k) = y(k) - y_{sp} \tag{4}$$

where the control input u(k) is the choke valve opening designed by combining a feedback control u_{PI} (k) with a feedforward control u_{ff} (k), e(k) is the error value as the difference between a measured process variable y(k) and a desired set-point y_{sp}, d(k) is the disturbance, the proportional gain K_P , integral gain K_I, and feedforward gain K_{ff} are three positive constants chosen by the designer. By tuning the three parameters in the controller algorithm, the controller can provide control action designed for specific process requirements.

For our case, the disturbance variable d(k) is considered as the change of flowrates, such as

$$d(k) = q_{pump}(k) - q_{pump}(k-1) + q_{back}(k) - q_{back}(k-1)) \tag{5}$$

In the AKC procedure, the process variable y(k) is set to be the downhole pressure BHP during the constant bottomhole pressure control, such as

$$y(k) = BHP(k) \tag{6}$$

$$y_{set} = BHP_{sp} \tag{7}$$

During the kick handling in the AKC procedure, we set the process variable y(k) to be the flowrate, such as

$$y(k) = q_{pump}(k) + q_{back}(k) \tag{8}$$

$$y_{set} = q_{choke}(k) \tag{9}$$

The objective is to rapidly balance the flow in and out of the well by automatically controlling the MPD choke, resulting in reduced influx from the reservoir. In order that the AKC procedure is effective and the flow control does not lead to too high BHP, the time keeping flow-in equals flow-out should be limited. Therefore, we have a pre-defined time window when using flow control. In the DSI procedure, the process variable y(k) can be set as the measured downhole pressure BHP during constant bottomhole pressure control and mud mump pressure P_{pump} or the choke pressure Pc during times with the low flowrate circulation, such as starting-up of the pumps.

In practice, the PID controller will be reduced to a PI controller in a field implementation since the derivative term is sensitive to measurement noise. The PI controller is a closed-loop feedback controller widely used in industrial control systems. One advantage of PI control is that it requires little knowledge about the process such that a process model is not necessary. Another advantage of PI control is its simplicity of structure and ease of implementation. In the absence of knowledge of the underlying process, a PI controller has historically been considered the best controller. By tuning the two parameters in the PI controller algorithm, the controller can provide control action designed for specific process requirements. The response of the controller can be described in terms of the responsiveness of the controller to an error, the degree to which the controller overshoots the set point, and the degree of system oscillation. PI controllers, when used alone, can give poor performance when the PI loop gains must be reduced so that the control system does not overshoot or oscillate around the control set-point value. They also have limitations in the presence of nonlinearities, such as have longer response time and have lag in responding to large disturbances.

The control system performance can be improved by combining the feedback control of a PI controller with feed-forward control. Knowledge about the system (such as the impact from changes of pump rate) can be utilized and combined with the PI output to improve the overall system performance. The feed-forward value alone can often provide the major portion of the controller output. The PI controller primarily has to compensate whatever difference or error remains between the set-point and the system response to the open loop control. Since the feed-forward output is not affected by the process feedback, it can never cause the control system to oscillate, thus improving the system response without affecting stability. Feed-forward can be based on the setpoint and on extra measured disturbances.

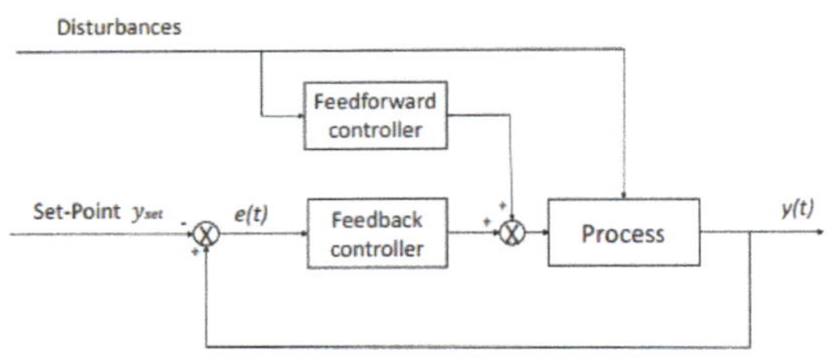

Figure 1: Feedback-Feedforward Control structure.

CASE STUDIES

In this section, the two alternative well control procedures, DSI and AKC, are evaluated using a high fidelity well flow model WeMod Lage et al. (2000), which is a high fidelity drilling simulator developed by the International Research Institute of Stavanger (IRIS). WeMod solves a set of partial differential equations for transient two-phase flow referred to as the drift-flux formulation Lorentzen and Fjelde (2005). To illustrate the use of the proposed well control procedures, a drilled gas-kick situation is considered. MPD is simulated when drilling into a permeable formation containing gas and with higher pore pressure than expected. An indicator that a kick has occurred is an increase in mud flow out. The influx is assumed to be detected by some kick-detection method. The detection methodology is not the focus of this work. The MPD setup for this work is a traditional automatic annular backpressure control system with backpressure pump, as shown in Figure 2. The drilling equipment consists of an MPD choke manifold including a high accuracy flow meter, main mud pump, and annulus backpressure pump. The annulus is pressurized by a rotating control device and the pressure can be controlled by adjusting the opening of the surface choke valve. A surface backpressure pump gives additional flow rate through the choke. By utilizing a PWD tool, the downhole measurement is transmitted every 20 seconds when the pump flow rate is above 600 l/min. There is no transmitted data during times with low flow rate (less than 600 l/min) and no circulation. However, the PWD tool will automatically record pressure every 2 seconds when the flow rate is lower than 600 l/min. Using the PWD technology, recorded pressure measurements at 0.5 Hz can be transmitted after regaining circulation. Thus, the PWD tool records and transmits full pressure data during that period. The transmission time for the recorded data after resuming circulation above 600 l/min is 5 minutes.

For the entire simulation, the choke valve is controlled automatically by the feedback-feedforward control algorithm described above. Two example wells are considered in this paper.

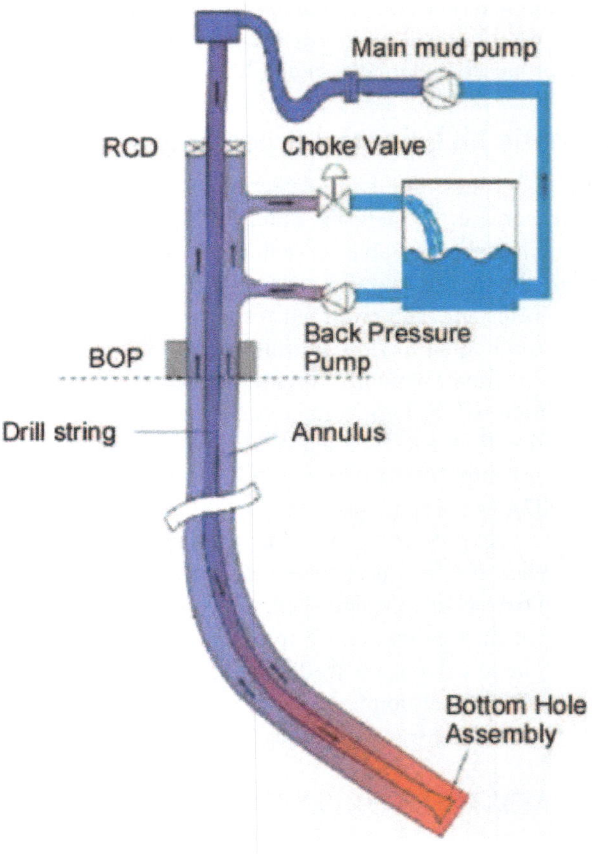

Figure 2: Schematic layout of MPD.

Well 1: short well with water-based Mud

Well 1 is a vertical well with a total depth of 2300 meter and the water-based drilling fluid density is 1.475 s.g. Further details of the well geometry, the drilling fluid and the reservoir conditions of Well 1 are given in Table 1. The well is initially in MPD mode, giving a bottomhole pressure of 261 bar. Prior to the kick incident, the main pump is running at a fixed rate of 2000 l/min and the backpressure pump is running at a fixed rate of 400 l/min. At t_{kick}=300 seconds a kick incident occurs, resulting in gas influx from the reservoir into the well.

Well 2: Long well with oil-based mud

Well 2 is a longer deviated well with oil-based mud, where the total depth is 6700 meter and the oil-based drilling fluid density is 1.670 s.g. Further details of the well geometry, the drilling fluid and the reservoir conditions in Well 2 are given in Table 2. The well is in MPD mode, giving a bottomhole pressure of 667 bar. The gas dissolution is ignored.

Throughout the simulations, we assume that a set of good control parameters is already provided. In this work, the good control parameters are obtained by exhaustive simulations. Hence, for each control configuration, we use the parameters, which give the best results among all the exhaustive simulations using the configuration. It should be noted that choosing parameters in the AKC procedure is much easier than that in DSI. In our cases, the control parameters used in AKC for both wells are set as $K_P = 0.01$, $K_I = 0.05$, $K_f = 0$ during the normal drilling operation with constant bottomhole pressure control and $K_P = 50$, $K_I = 0.5$, $K_f = 0$ during the kick handling. In DSI, the PI gains are tuned for different conditions, including normal constant bottomhole pressure control, shutin pumps, start-up pumps, constant backpressure control, and constant pump pressure control. In general, the proposed control can be robust against some disturbances if it is properly tuned.

Initially, an MPD operation is in progress in drilling mode. Prior to the kick incident, the main mud pump is running at a fixed rate of 2000 l/min and the backpressure pump is running at a fixed rate of 400 l/min. At $t_{kick} = 300$ seconds for Well 1 and 600 seconds for Well 2 a kick incident occurs, resulting in gas in- flux from the reservoir into the well. The gas influx is then handled using the two different handling procedures, namely DSI procedure and AKC procedure. During the DSI procedure, the main mud pump and the backpressure pump are shut down and the choke is closed. After the kick is attenuated, the main pump is gradually increased to 600 l/min and kept at this rate for five minutes, allowing the recorded PWD measurements to be transmitted. In the AKC procedure, the pumps are running with the same speed and the choke valve is automatically controlled. After the kick is attenuated, the choke valve is automatically controlled by the feedback-forward control algorithm, keeping the well pressure constant and above the pore pressure in the kick zone.

Table 1: Well geometry and fluid data in Well 1

Parameter	Value
Well length	2300 m
True vertical depth	1720 m
Annulus inner diameter	0.2445 m
Drillpipe outer diameter	0.1397 m
Drillpipe inner diameter	0.1183 m
Open hole length	291 m
Open hole diameter	0.2159 m
Mud density	1.475 sg
Mud system	Water based
Reservoir start TD	2290 m
Reservoir height	2 m
Reservoir pore pressure	263 bar
Reservoir kick-off time	300 sec

Table 2: Well geometry and fluid data in Well 2

Parameter	Value
Well length	6700 m
True vertical depth	4120 m
Annulus inner diameter	0.4508 m
Drillpipe outer diameter	0.1390 m
Drillpipe inner diameter	0.1183 m
Open hole length	200 m
Open hole diameter	0.2159 m
Oil density	0.805 sg
High gravity density	4.2 sg
Oil-water ratio	3.160
Mud system	Oil based
Reservoir start TD	6695 m
Reservoir height	2 m
Reservoir pore pressure	670 bar
Reservoir kick-off time	600 sec

Influx Control Procedure

The two well control procedures are described below. Dynamic shut-in prcedure (DSI):

- Start time T0: Choke control for BHP to make BHP = SetPoint
- Time point T1: influx initiated
- Time point T2: influx detection
  - Close BackPressure Pump to 0 in 2 mins ($q_{bpp} = 0$)
  - Close MudPump to 0 in 2 mins ($q_{pump} = 0$)
  - Choke control of BHP = SetPoint
- Time point T3: qpump < 600 l/min
  - Close choke to zero in 45 seconds
- Time point T4: Pumps and choke are fully closed
  - Wait several minutes for pressure stabilization and kill influx
- Time point T5:
  - Start BackPressure Pump to initial speed in 1 min
  - Choke control of Pc = Constant (consider friction pressure loss)
- Time point T6: qback = initial speed
  - Start Mud Pump to 600 l/min in 1 min
  - Choke control of Pc = Constant (consider annulus friction pressure loss)
- Time point T7: Keep qpump = 600 l/min for 5 mins
  - Choke control of Ppump = Constant (consider friction pressure loss)
  - 5 minutes is for transmission of recorded PWD data
- Time point T8:
  - Start MudPump to initial speed in 1 min
  - Choke control of BHP = New SetPoint (Estimated pore pressure+safety margin)

Automatic kick control procedure (AKC):

- Start time T0:
  - Choke control of BHP = SetPoint
- Time point T1: influx initiated
- Time point T2: influx detected
  - Keep constant MudPump and BackPressurePump as initial speeds
  - Choke control of flowrate to make flowrate out = flowrate in
  - Choke control of flowrate out = flowrate in for pressure stabilization and influx killed
- Time point T3: Choke control of BHP = New SetPoint (Estimated pore pressure + safety margin)

RESULTS

This section presents the results from the simulations of a kick scenario. Four cases have been run. In Case 1 and 2, we have used Well 1 to simulate the effect of the DSI and the AKC, respectively. In Case 3 and 4, we have shown the performance in Well 2 from using the DSI and AKC, respectively. When discussing the results, several variables from the various locations of the well should be evaluated together. The results are shown in Figures 3-6, their contents will be presented firstly, and then the results will be discussed further. Figure 3 and Figure 4 show the results from Case 1 and 2 in a short well with water based mud. Figure 5 and Figure 6 show the results from Case 3 and 4 in a long well with oil based mud. Each figure includes the flow rates through the main pump, backpressure pump and the MPD choke, the corresponding choke valve, the mud pump pressure, the backpressure, the bottomhole pressure for the whole simulation, and the reservoir influx.

Prior to the kick incident, the main mud pump is running at a fixed rate of 2000 l/min and the backpressure pump is running at a fixed rate of 400 l/min. The kick situation initiates at time T1= 300 sec. Since the flow rate out of the well can be measured relatively accurately e.g. using a Coriolis flow meter, the increase in flow can be monitored for detection of a possible kick. When the flow out of the well increases, the flow through the choke also increases, and results in an increased choke pressure between time T1 and T2. This also causes both the bottomhole pressure and the standpipe pressure to decrease since the automatic control system is designed to maintain a constant bottomhole pressure. At time T2, this situation is diagnosed as a possible kick situation. When the possible kick situation is detected, the two different kick control procedures are initiated. The DSI procedure starts at time T2 by shutting down the backpressure pump and the mud pump in 2 minutes.

The choke is then closed completely during 48 seconds at time T3 when the mud pump flow rate is less than 600 l/min, since there is no downhole pressure measurement available. The well is fully shutin at time T4. The bottomhole pressure will then increase and stabilize when the bottomhole pressure equals pore pressure. The choke pressure and the mud pump pressure are recorded. After time T5, the backpressure pump is started gradually to the rate of 400 l/min in 1 minute, which is the same value as before the shut-in procedure. The main pump is then started to the rate of 600 l/min at time T6 in 1 minute and kept at this speed for 5 minutes. From T5 to T8, the automatic control system is designed to maintain constant choke pressure, which depends on the previous recorded choke pressure and the calculated friction pressure loss in the annulus. The time between T7 and T8 is used for buffering the high-resolution PWD data.

The pore pressure of the reservoir kick zone is evaluated based on the recorded bottomhole pressure measurement. After time T8, the main pump rate is increased to the rate of 2000 l/min, which was the speed before the shut-in procedure. The new set-point reference for bottomhole pressure is calculated based on the estimated pore pressure and a pre-defined safety margin. The influx is sufficiently small that it is within the limits of MPD surface equipment to circulate the influx out safely while maintaining well control. The circulation of gas is not the focus of this work. When comparing the procedures, it can be seen that the influx is larger when using the DSI compared to the AKC. In Well 1, the total kick size is approximately 800 liters when using DSI and 450 liters when using AKC. In Well 2, the total kick size is approximately 2000 liters when using DSI and 500 liters when using AKC. The reason is that AKC increases pressure more rapidly than DSI and also results in less pressure variations downhole. In addition, the tuning of control parameters in DSI is more complicated compared to the tuning in AKC due to more disturbances and uncertainties in the DSI procedure, such as the disturbance when changing the main pump flow rate, the large sampling time and the delay for the transmission of data is also a factor.

CONCLUSION

This paper evaluates two well control procedures adapted to backpressure MPD: The dynamic shutin (DSI) procedure and the automatic kick control (AKC) procedure. Both methods capitalize on improvements in Pressure While Drilling (PWD) technology. A new commercially available PWD tool buffers high-resolution pressure measurements, which can be used in an automated well control procedure. Typically, only minimum, maximum and average downhole pressure is transmitted to surface after circulation is regained. Using new PWD technology, recorded measurements at 0.5 Hz can be transmitted after regaining circulation. By using backpressure MPD, the choke opening is tuned automatically using a feedbackfeedforward control method.

A feedback-feedforward control algorithm is developed to maintain BHP at the desired set-point and attenuate an influx in MPD operations. The two procedures, DSI and AKC, are evaluated using a high fi- delity well flow simulator and cases from a North Sea drilling operation are simulated. The results show that using AKC procedure reduces the time needed to establish control of the well, the total kick size, and the risk of lost circulation compared to DSI procedure in our cases.

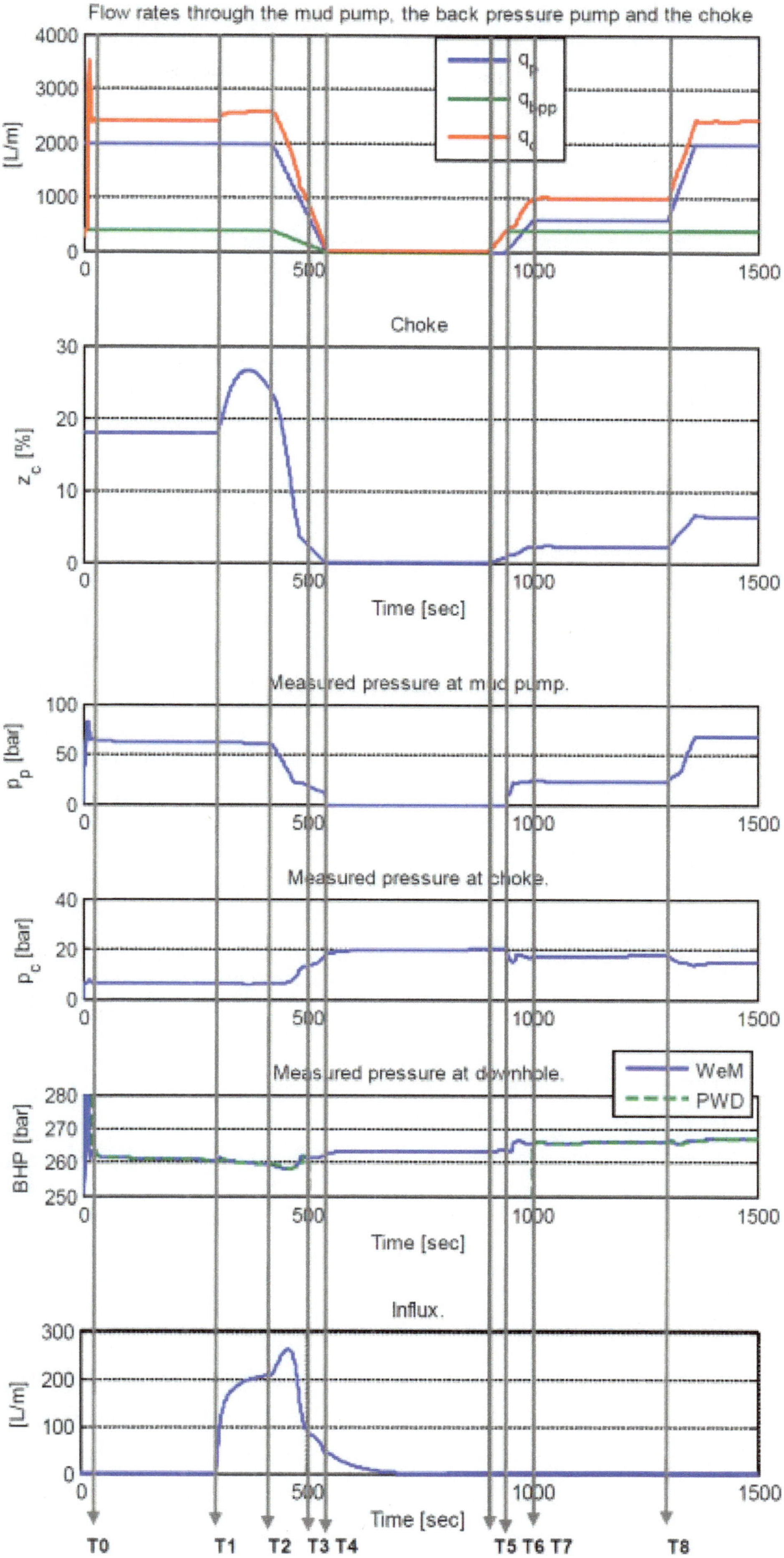

Figure 3: Case 1: Flowrates, choke opening, measured pressure and Influx flow rate using DSI on Well 1.

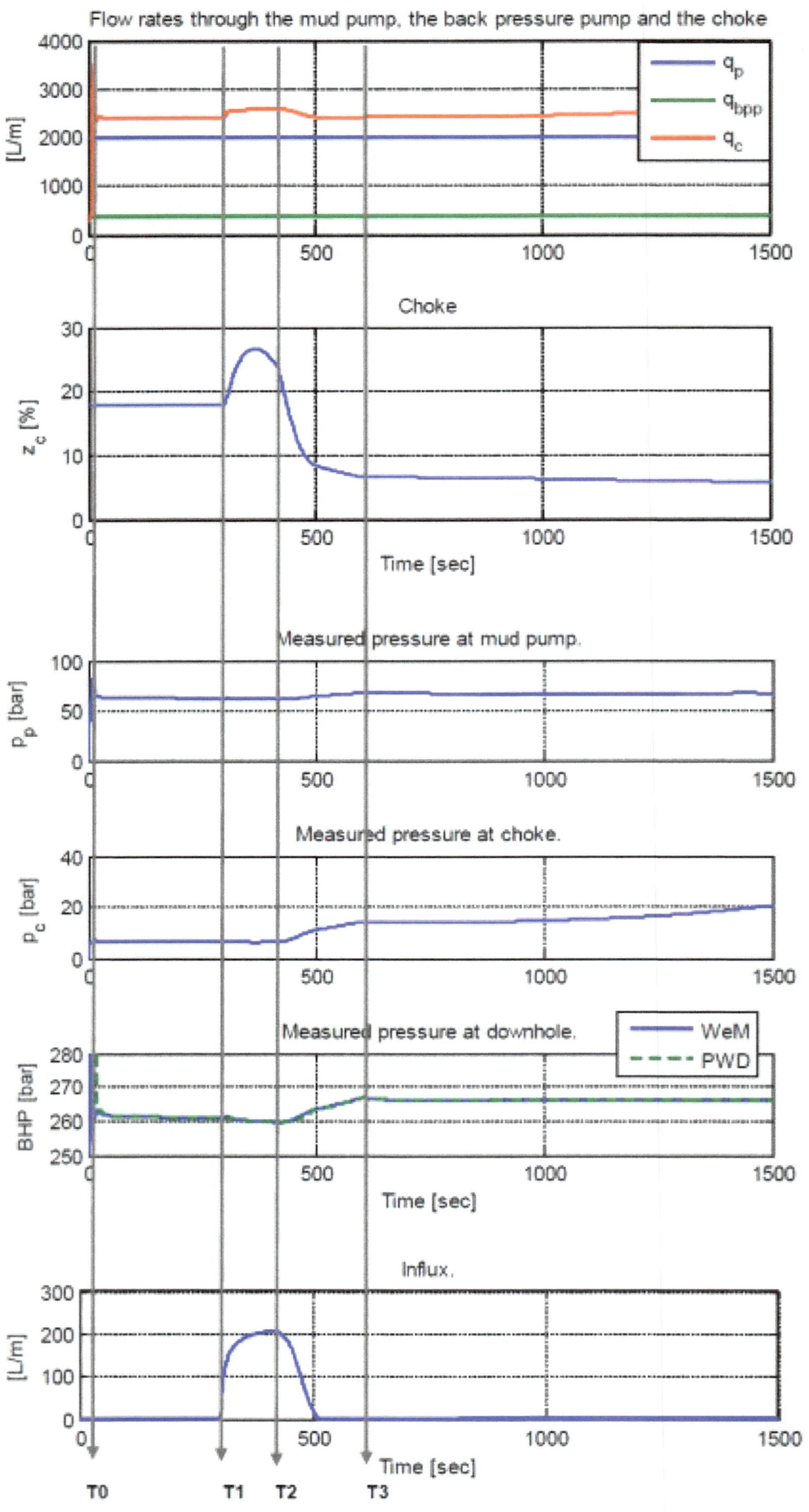

Figure 4: Case 2: Flowrates, choke opening, measured pressure and Influx flow rate using AKC on Well 1.

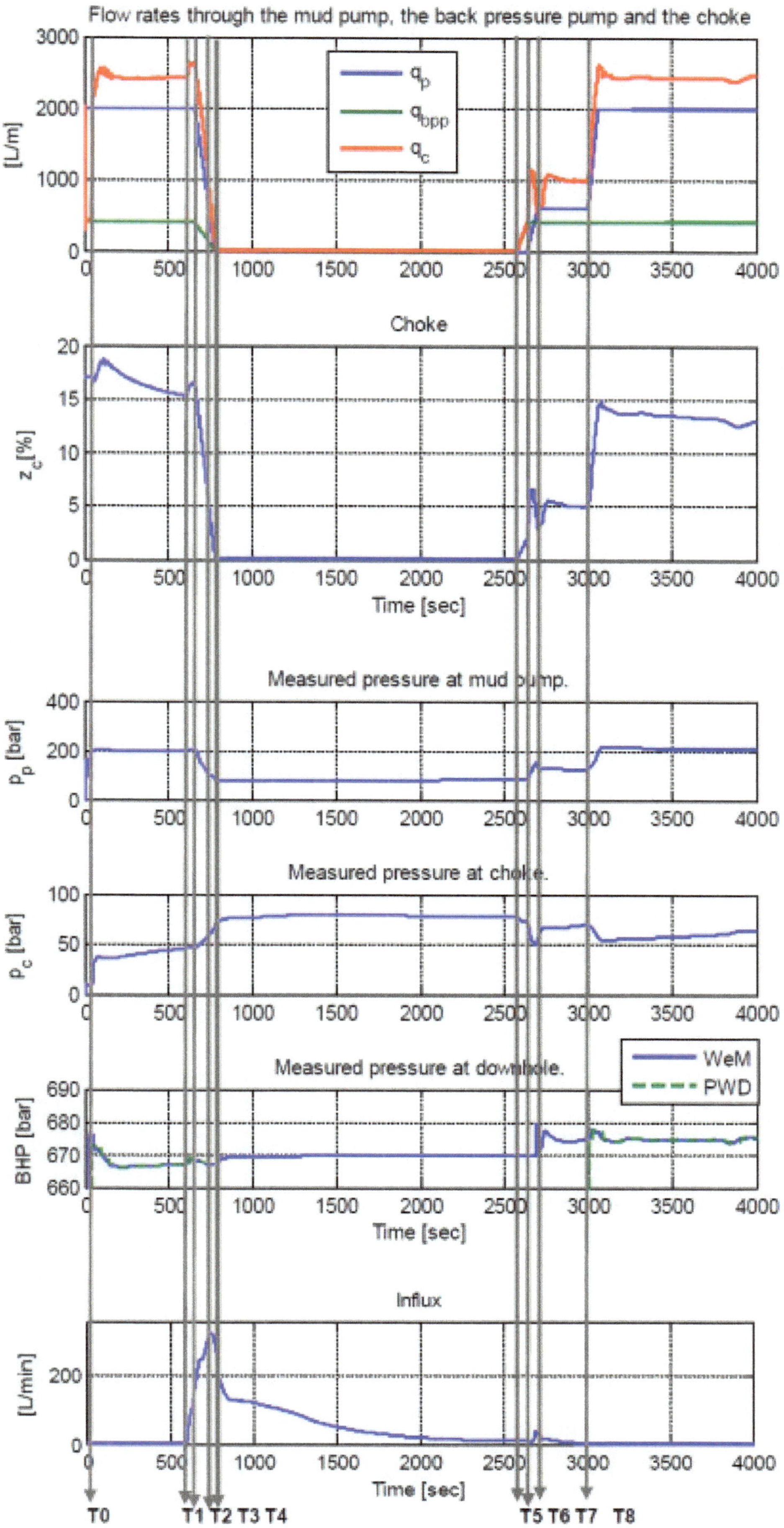

Figure 5: Case 3: Flowrates, choke opening, measured pressure and Influx flow rate using DSI on Well 2.

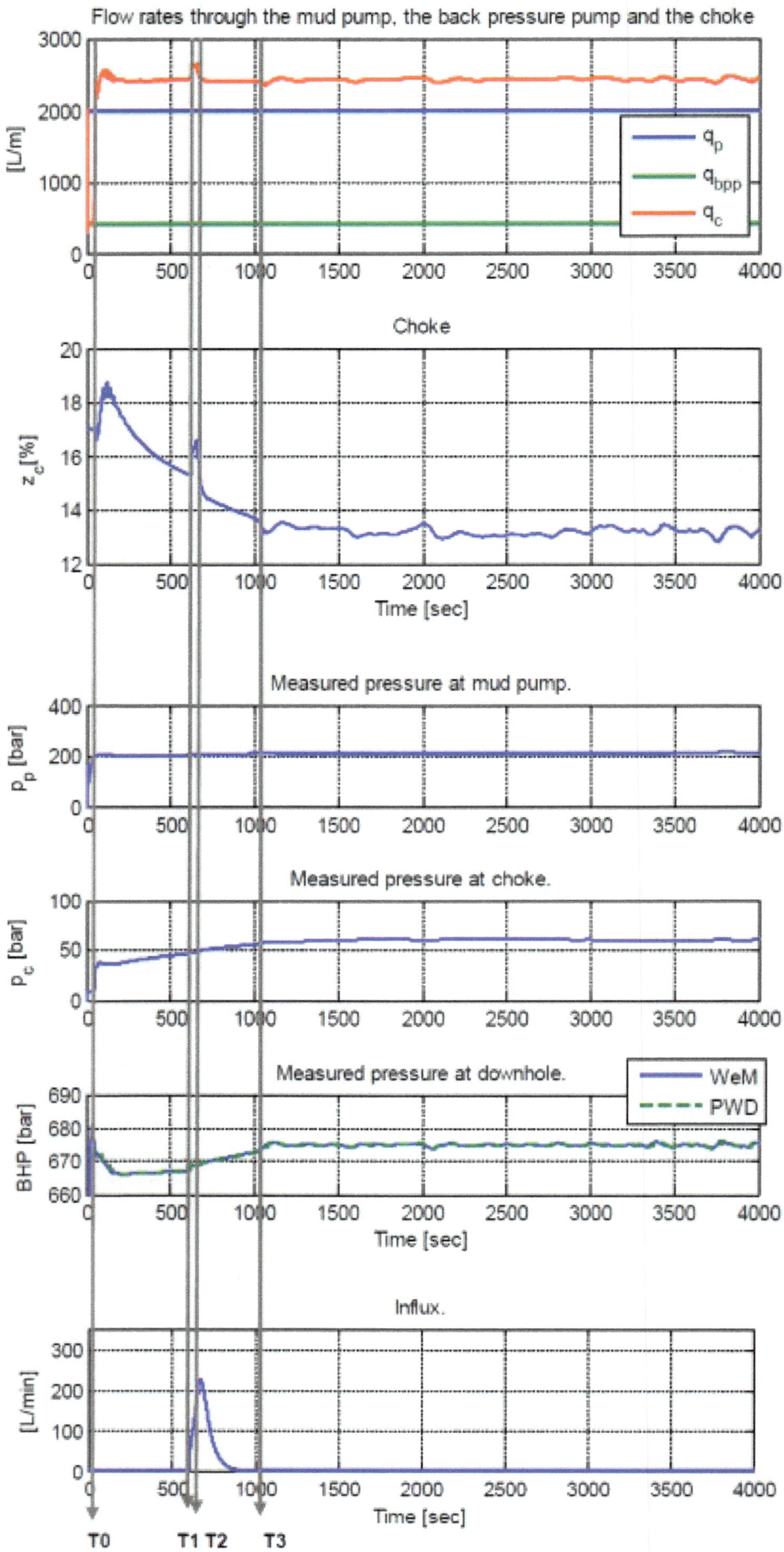

Figure 6: Case 4: Flowrates, choke opening, measured pressure and Influx flow rate using AKC on Well 2.

ACKNOWLEDGMENTS

The authors acknowledge the Research Council of Norway, ConocoPhillips, Det norske oljeselskap, Lundin Norway, Statoil and Wintershall for financing the work through the research centre DrillWell - Drilling and Well Centre for Improved Recovery, a research cooperation between IRIS, Norwegian Univereity of Science and Technology (NTNU), SIINTEF and University of Stavanner.

REFERENCES

1. Bacon, W., Tong, A. Y., Gabaldon, O. R., Sugden, C., and Suryanarayana, P. An improved dynamic well control response to a gas influx in managed pressure drilling operations. In IADC/SPE Drilling Conference and Exhibition, SPE-151392. San Diego, California, USA, 2012. doi:10.2118/151392-ms.
2. Breyholtz, Ø., Nygaard, G., Godhavn, J.-M., and Vefring, E. H. Evaluating control designs for co-ordinating pump rates and choke valve during managed pressure drilling operations. In IEEE International Conference on Control and Applications. Russia, pages 731 – 738, 2009. doi:10.1109/CCA.2009.5281013.
3. Carlsen, L. A., Nygaard, G., Gravdal, J. E., Nikolaou, M., and Schubert, J. Performing the dynamic shutin procedure becasue of a kick incident when using automatic coordinated control of pump rates and chokevalve opening. In SPE/IADC managed Pressure Drilling and Underbalanced Operations Conference and Exhibition, SPE 113693. Abu Dhabi, UAE, 2008.
4. Davoudi, M., Smith, J., and Chirinos, J. Evaluation of alternative initial responses to kicks taken during managed-pressure drilling. SPE Drilling & Completion, SPE-128424-MS, 2011. 26(2):194–203.
5. Godhavn, J.-M. Control requirements for high-end automatic MPD operations. In SPE/IADC Drilling Conference and Exihibition, SPE 119442. Amsterdam, Neitherlands, 2009. doi:10.2118/119442-ms.
6. Gravdal, J. E., Lorentzen, R., Fjelde, K., and Vefring, E. Improved kick management during MPD by realtime pore-pressure estimation. SPE Drilling & Completion, SPE 124054, 2010. 25:577–584.
7. Hannegan, D., Todd, R. J., Pritchard, D. M., and Jonasson, B. MPD uniquely applicable to methane hydrate drilling. In SPE/IADC Underbalanced Technology Conference and Exhibition, SPE 91560. Houston, Texas, U.S.A., 2004. doi:10.2118/91560- ms.
8. Kaasa, G. O., Stamnes, Ø. N., Imsland, L., and Aamo, O. Simplified hydraulics model used for intelligent estimation of downhole pressure for a managedpressure-drilling control system. SPE Drilling & Completion, SPE 143097, 2013. 27:127–138.
9. Lage, A., Frøyen, J., and Saevareid, O. Underbalanced drilling dynamics: Two-phase flow modeling and experiments. In IADC/SPE Asia Pacific Drilling Technology, SPE -62743-MS. Kuala Lumpur, Malaysia, 2000.
10. Lorentzen, R. and Fjelde, K. Use of slopelimiter techniques in traditional numerical methods for multiphase flow in pipelines and wells. International Journal for Numerical Methods in Fluids, 2005. 48:723–745. doi:10.1002/fld.952.
11. Nygaard, G., Vefring, E. H., Fjelde, K.-K., Nævdal, G., Lorentzen, R. J., and Mylvaganam, S. Bottomhole pressure control during drilling operations in gas-dominant wells. SPE Journal, SPE-91578-PA, 2007. 12:49–61.
12. Santos, H., Catak, E., Kinder, J., and Sonnemann, P. Kick detection and control in oil-based mud: Real well-test results using micro-flux control equipment. In SPE/IADC Drilling Conference, SPE 105454. Amsterdam, Netherlands, 2007.
13. Santos, H., Leuchtenberg, C., and S.Shayegi. Micro-flux control:the next generation in drilling process for ultra-deepwater. In Offshore Technology Conference. Houston, Texas, USA, 2003. doi:10.4043/15062-MS.
14. Siahaan, H. B., Bjrkevoll, K. S., and Gravdal, J. E. Possibilities of using wired drill pipe telemetry during managed pressure drilling in extended reach wells. In SPE Intelligent Energy Conference and Exhibition, SPE 167856. Utrecht, Netherlands, 2014.
15. Smith, J. R. and Patel, B. M. A proposed method for planning the best initial response to kicks taken during managed-pressure-drilling operations. SPE Drilling & Completion, SPE-143101-PA., 2012. 27(2):194–203.
16. Zhou, J. and Nygaard, G. Control and estimation of downhole pressure in managed pressure drilling operations. In Proceedings of the 4th International Symposium on Communications, Control and Signal Processing. Limassal, Cyrpus, 2010. doi:10.1109/isccsp.2010.5463474.
17. Zhou, J. and Nygaard, G. Automatic model-based control scheme for stabilizing pressure during dual gradient drilling. Journal of Process Control, 2011. 21:1138–1147. doi:10.1016/j.jprocont.2011.06.022.
18. Zhou, J., Nygaard, G., Godhavn, J.-M., Breyholtz, Ø., and Vefring, E. H. Adaptive observer for kick detection and switched control for bottomhole pressure regulation and kick attenuation during managed pressure drilling. In Proceedings of American Control Conference. USA, 2010. doi:10.1109/ACC.2010.5531551.
19. Zhou, J., Stamnes, Ø. N., Aamo, O. M., and Kaasa, G.-O. Pressure regulation with kick attenuation in a managed pressure drilling system. In Proceedings of the 48th IEEE Conference on Decision and Control. Shanghai, China, 2009. doi:10.1109/CDC.2009.5400792.
20. Zhou, J., Stamnes, Ø. N., Aamo, O. M., and Kaasa, G.-O. Switched control for pressure regulation and kick attenuation in a managed pressure drilling system. IEEE Transactions on Control Systems Technology, 2011. 19:337–350. doi:10.1109/TCST.2010.2046517.

CHAPTER 7

SPEED CONTROL BASED ON ESO FOR THE PITCHING AXIS OF SATELLITE CAMERAS

BingYou Liu[1,2], Yi Jin[1], ChangZheng Chen[3], and HuiCheng Yang[2]

[1]Department of Precision Machinery and Precision Instrumentation, University of Science and Technology of China, Hefei 230027, China

[2]Key Lab of Electric and Control of Anhui Province, Anhui Polytechnic University, Wuhu 241000, China

[3]Changchun Institute of Optics, Fine Mechanics and Physics, Chinese Academy of Sciences, Changchun 130033, China

ABSTRACT

The pitching axis is the main axis of a satellite camera and is used to control the pitch posture of satellite cameras. A control strategy based on extended state observer (ESO) is designed to obtain a fast response speed and highly accurate pitching axis control system and eliminate disturbances during the adjustment of pitch posture. First, a sufficient condition of stabilization for ESO is obtained by analyzing the steady-state error of the system under step input. Parameter tuning and disturbance compensation are performed by ESO. Second, the ESO of speed loop is designed by the speed equation of the pitching axis of satellite cameras. The ESO parameters are obtained by pole assignment. By ESO, the original state variable observes the motor angular speed and the extended state variable observes the load torque. Therefore, the external load disturbances of the control system are estimated in real time. Finally, simulation experiments are performed for the system on the cases of nonload starting, adding external disturbances on the system suddenly, and the load of system changing suddenly. Simulation results show that the control strategy based on ESO has better stability, adaptability, and robustness than the PI control strategy.

INTRODUCTION

Pitching axis is an important axis of the posture control system of a satellite camera. The control system of pitching axis adopts a three-loop servo control system: the current loop, speed loop, and position loop. The disturbances of the speed loop are greater than those in the other two loops. Therefore, the speed loop should be controlled in real time to achieve good performances for the control system of the pitching axis. This paper uses an extended state observer (ESO) in the speed loop control system to obtain good performances.

Citation: BingYou Liu, Yi Jin, ChangZheng Chen, and HuiCheng Yang, "Speed Control Based on ESO for the Pitching Axis of Satellite Cameras," Mathematical Problems in Engineering, vol. 2016, Article ID 2138190, 9 pages, 2016. doi:10.1155/2016/2138190

A control strategy based on ESO was first presented in [1]. The main idea of the control strategy is to extend the disturbance that influences the output of the controlled object into a new state and establish an observer to observe this state by using a special feedback method. The control strategy is a good control method for solving the problem of uncertain nonlinear system. The control strategy has various features, such as high-control accuracy, fast response speed, and good robust performance, and does not rely on the model of a controlled object. Therefore, this control strategy is widely used in the field of aeronautics and astronautics, precision machinery, modern weapon systems, and so on.

Extensive research on the application of ESO has been conducted locally and internationally and has shown some achievements. For example, Yang and Huang [2] analyzed the capability of ESO in estimating uncertainty. Wang et al. [3] proved the second-order nonlinear stability of ESO by the Lyapunov function. Shao et al. [4] not only extended the ESO from continuous space to discrete space but also presented the discrete linear specific form and stability analysis of ESO. Wang and Deng [5] proposed an operator-based robust nonlinear multivariable tracking control for a manipulator with uncertainties by using a robust right coprime factorization approach and discussed the robust stability in the presence of model uncertainties. Deng and Bu [6, 7] discussed the robust control for nonlinear feedback system with unknown perturbations based on the robust right coprime factorization. These research results have been widely used in engineering practice. For example, Hasanzadeh et al. [8] used ESO in the frequency control of the multiregion load and obtained better anti-interference ability than the traditional method. Lin et al. [9] designed a rotor flux observer based on ESO to solve the problem of rotor resistance uncertainty. Shi and Chang [10] designed a time-optimal control strategy based on ESO and integrated it into an accurate point-to-point movement system. Chan et al. [11] used ESO into two-degree-of-freedom control system of mechanical arms. Compared with a traditional controller, the unique advantages of ESO are embodied in the following aspects. ESO was compared with a sliding mode observer in [12], and the results showed that the upper bound of the interference does not need to be known. However, chattering phenomenon exists in the output of the sliding mode observer. ESO was compared with a high-gain observer in [13] and was used to estimate the unobservable and extended states of the system according to the system input and output. A set of mature experience formulas is available for reference during the parameter setting of ESO. A setting method for ESO parameters was presented in [14].

Research on the posture control of satellite cameras has shown some achievements in recent years. Guan and Zheng [15] introduced an integrated design method of posture control in satellite cameras. Xiubin et al. [16] introduced a control method for the side swing of satellite cameras. However, few reports have focused on the use of ESO in the speed loop control of the pitching axis of satellite cameras.

The adjustment of the pitching posture of satellite camera is presented under the environment of weightlessness. Therefore, a small interference will cause a serious error. Hence, the control system needs a control strategy with stronger anti-interference ability. The control strategy based on ESO does not depend on the precise model of a control object and can estimate the disturbances of a system in real time. This control strategy has a stronger anti-interference ability and is used in this study.

This study analyzed the mathematical models for the actuator permanent magnet synchronous motor (PMSM) of the pitching axis. General-type ESO principles are presented first. The features of ESO nonlinear functions were then analyzed. Given that the disturbances in the speed of satellite cameras need to be eliminated, an ESO was designed and used in the speed loop control of the pitching axis of satellite cameras. The advantages of this type of control strategy were verified by comparing it with a PI control strategy.

PRINCIPLE AND MODELS

Principle of Pitching Axis

The pitching posture of satellite cameras needs adjusting when tracking objects. The pitching axis of satellite cameras can achieve this function. The control schematic diagram of pitching posture of satellite camera is shown in Figure 1. The adjustment of pitching posture is realized on the camera posture rotating platform by using the control board. The PMSM is used as the actuator in the control system of the pitching axis, where R_a is the resistance of stator winding. u_d, u_q are the equivalent voltage of axes $d$and q, respectively. i_d, i_q are the equivalent current of axes$d$ and q, respectively. Ld, L_q are the equivalent inductance of axes d and q, respectively. A three-loop AC servo control system has a current loop, a speed loop, and a position loop used as the control strategy [17]. The speed loop plays an important role in this system. A high-precision speed loop can restrain the uncertain disturbances of the current loop and position loop.

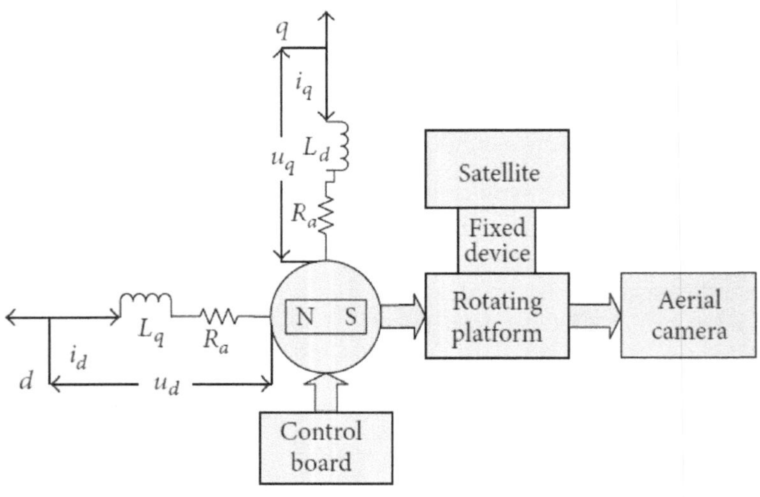

Figure 1: The control schematic diagram of pitching posture of satellite camera.

PMSM Models

The control schematic diagram of pitching posture of satellite camera shows that the control board drives the PMSM. The PMSM drives the rotating platform which was fixed on satellite by fixed device. The rotating platform drives the satellite camera to control the pitching posture. This control function is realized by the control method of the motor speed. Therefore, the models of the actuator PMSM are established in this study. The PMSM models include voltage equation, flux equation, torque equation, and kinematic equation. The following assumptions are presented. First, the saturation effects of iron core are ignored. Second, the air-gap magnetic field shows a normal distribution. Third, the wastage of the eddy current and magnetic field is ignored. Fourth, no damper winding on the motor rotor exists. The kinematic equation of PMSM on the two-phase coordinate system (i.e., the dq coordinate system) is shown as follows:

$$\dot{\omega} = \frac{1}{J}\left(T_e - T_L - B\omega\right), \tag{1}$$

where B is the friction coefficient, J is the rotational inertia of the motor, ω is the angular speed of the motor rotor, T

e is the electromagnetic torque of the motor, and T L is the load torque of the motor.

GENERAL FORM OF ESO

The classic PI control strategy focuses on the elimination of errors but does not predict and estimate the variable quantities of the system. Hence, this strategy cannot control the magnitude of the integral. The robust performance of the system can be predicted and estimated in real time. Therefore, a new control strategy based on ESO is presented. The nonlinear dynamics, model uncertainty, and external disturbances of the system are considered as extended states that need to be observed.

A nonlinear single-input single-output system with unknown disturbances is expressed as follows:

$$x^{(n)} = f\left(x, \ldots, x^{(n-1)}, t\right) + w(t) + bu(t), \tag{2}$$

where $f(x, \ldots, x(n-1), t)$ is an unknown function constituted by the system state variables, $w(t)$ is the unknown external disturbances of the system, $u(t)$ is the input, $x, \ldots, x(n-1)$ are the state variables of the system, and $x(t)$ can be measured.

$$\begin{aligned} \dot{x}_1 &= x_2 \\ \dot{x}_2 &= x_3 \\ &\vdots \\ \dot{x}_n &= f\left(x_1, x_2, \ldots, x_n\right) + w(t) + bu(t). \end{aligned} \tag{3}$$

Assume that $x_1 = x(t), x_2 = x'(t), \ldots, x_n = x^{(n-1)}(t)$, state space of the system can be expressed as follows:

$$\begin{aligned} \dot{z}_1 &= z_2 - g_1\left(z_1 - x(t)\right) \\ \dot{z}_2 &= z_3 - g_2\left(z_1 - x(t)\right) \\ &\vdots \\ \dot{z}_n &= z_{n+i} - g_n\left(z_1 - x(t)\right) + bu(t) \\ \dot{z}_{n+1} &= -g_{n+1}\left(z_1 - x(t)\right), \end{aligned} \tag{4}$$

Where $z_1 \to x(t), \ldots, z_n \to x^{n-1}(t), z_{n+1} \to a(t)$ are the estimations of unknown functions $f(x, x^{(1)}, \ldots, x^{(n-1)}, t)$ and disturbances (t). A nonlinear system was constructed by treating (t) as an input. If nonlinear functions $g_1(z), g_2(z), \ldots, g_{n+1}(z)$ were selected appropriately, the state variables of the nonlinear system could track the state variables of the original system and (t). Hence, the ESO of the system can be obtained. Furthermore, the ESO can compensate the uncertain objects and external disturbances in the original system. A nonlinear function is themathematical fitting of big error and small gain or small error and big gain. This function is generally selected as a nonlinear combination power function and can be shown as follows [18]:

$$\mathrm{fal}(\varepsilon, \alpha, \delta) = \begin{cases} |x|^{\alpha}\,\mathrm{sign}(x), & |x| > \delta \\ \dfrac{x}{\delta^{1-\alpha}}, & |x| \le \delta, \end{cases} \quad \delta > 0. \tag{5}$$

Simulations were carried out based on the fal function to evaluate the function of performances. The curve of the fal

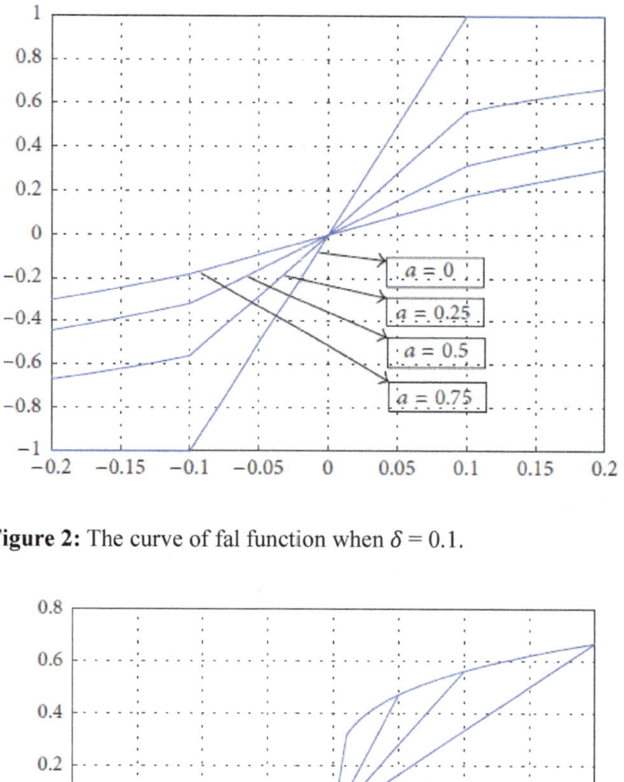

Figure 2: The curve of fal function when $\delta = 0.1$.

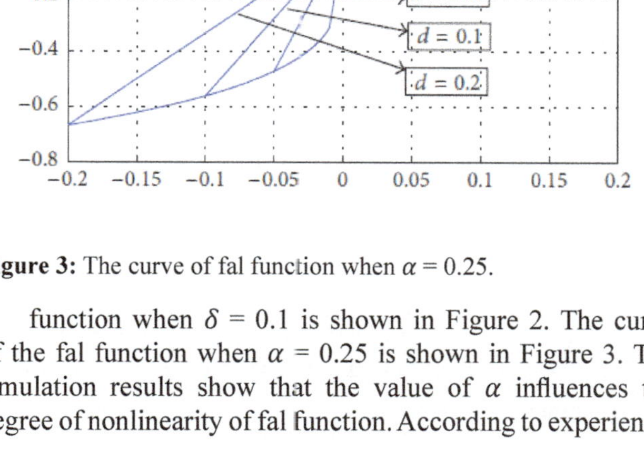

Figure 3: The curve of fal function when $\alpha = 0.25$.

function when $\delta = 0.1$ is shown in Figure 2. The curve of the fal function when $\alpha = 0.25$ is shown in Figure 3. The simulation results show that the value of α influences the degree of nonlinearity of fal function. According to experience, the value of α is chosen between zero and one. When $\alpha = 0$, the degree of nonlinearity of fal function is most serious. When $\alpha = 0.75$, the degree of linearity of fal function is best. The value of δ is the linear interval width of the fal function and is related to the error range of the system. In practical applications, the value of δ is generally selected as $\delta = 0.01$. So when the input is an error signal, the systemcould achieve the stable status rapidly by adjusting the parameter values of fal function.

Design of Speed Loop ESO

The design process of ESO is based on the following steps. First, the ESO parameters are designed, and the stability of ESO is ensured. Second, the disturbance observation bandwidth of ESO can be set according to the characteristics of object disturbance. Third, the ESO parameters could be set by using the parameter tuning method. Finally, the output signal of ESO is amended according to the case of the noise interference of object output signal. The schematic of the control system of pitching axis is shown in Figure 4.

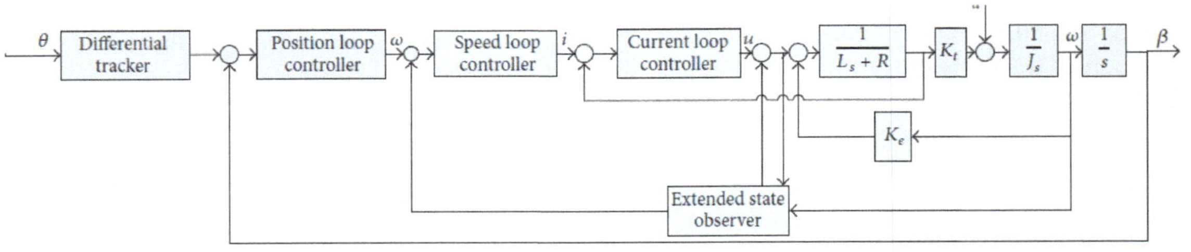

Figure 4: Schematic of the control system of pitching axis.

Design of the Speed Loop ESO for the Pitching Axis of Satellite

Cameras

The key of the speed loop control technique is the control of electromagnetic torque. The feedback control strategy of the speed loop can be designed on the basis of the speed error and the given speeds, and the test speeds are obtained. Then effective torque control can be achieved. The speed loop is also influenced by uncertain disturbances such as the rotational inertia and load torque.

Therefore, the main purpose of the speed loop controller is to eliminate these influences. The disturbances and error of model linearization are eliminated by ESO. Hence, the purpose of eliminating uncertain factors can be achieved. Considering the unknown external disturbances, the input is selected as $u = i_q$, and the output is selected as $y = \omega$. The PMSM kinematic equation can also be expressed as follows:

$$\dot{\omega} = \frac{3p\phi_f}{2J} i_q - \frac{B}{J}\omega - \frac{T_L}{J}. \tag{6}$$

The ESO of the speed loop can be designed by the following steps. Some definitions are given as follows:

$$\begin{aligned} x &= \omega, \\ a &= -\frac{B}{J}, \\ f(x,t) &= ax, \\ b &= 1.5\frac{p\phi_f}{J}, \\ w(t) &= -\frac{T_L}{J}, \\ u &= i_q, \\ y &= \omega. \end{aligned} \tag{7}$$

The expression of kinematic equation with extended states can be shown as follows:

$$\begin{aligned} \dot{x} &= f(x,t) + w(t) + bu, \\ y &= x. \end{aligned} \tag{8}$$

The ESO of the system is designed as follows:

$$\begin{aligned} \dot{z}_1 &= z_2 - g_1 \mathrm{fal}(e, \alpha_1, \delta_1) + f_0(z_1) + b_0 u, \\ \dot{z}_2 &= -g_2 \mathrm{fal}(e, \alpha_2, \delta_1), \end{aligned} \tag{9}$$

where $z1$ is the tracking of the motor speed signal ω, z_2 is the tracking of the motor load torque disturbance T, and $\alpha 1, \alpha_2$ are nonlinear factors and satisfy the condition $0 < \alpha 2 < \alpha_1 \leq 1$.This study selected the numerical values of α_1, α_2 as$\alpha_1 = 1$, $\alpha_2 = 0.5$. δ_1 is a filtering factor with a numerical value of $\delta_1 = 0.01$. The two expressions include $e = z_1 - \omega$ and$f_0(z_1) = -Bz_1/J$. Therefore, load torque can be estimated by these methods.

Stability Analysis of the Speed Loop ESO

The speed loop ESO is selected to perform stability analysis. An uncertain nonlinear single-input single-output system that applies unknown external disturbances is given as follows [19]:

$$\dot{y} = f + b_0 u, \tag{10}$$

where $f = g + (b - b0)u$ is an extended state, a differential function, and the sum of disturbances that need to be estimated. b is the high-frequency gain whose precise numerical value is difficult to obtain. b_0 is the estimation value of b.

Assume that $f = x_2$, $\dot{f} = h$. Then the following equation can be derived:

$$\begin{aligned} y &= x_1, \\ \dot{x}_1 &= x_2 + b_0 u, \\ \dot{x}_2 &= h. \end{aligned} \tag{11}$$

The above equation can be simplified as follows:

$$\begin{bmatrix} \dot{x}_1 \\ \dot{x}_2 \end{bmatrix} = \begin{bmatrix} 0 & 1 \\ 0 & 0 \end{bmatrix}\begin{bmatrix} x_1 \\ x_2 \end{bmatrix} + \begin{bmatrix} b_0 \\ 0 \end{bmatrix} u + \begin{bmatrix} 0 \\ 1 \end{bmatrix} h$$

$$y = \begin{bmatrix} 1 & 0 \end{bmatrix}\begin{bmatrix} x_1 \\ x_2 \end{bmatrix}. \quad (12)$$

The ESO of the above system can be designed as follows:

$$\dot{z}_1 = z_2 + \beta_1 (y - z_1) + b_0 u$$

$$\dot{z}_2 = \beta_2 (y - z_1); \quad (13)$$

β_1, β_2 are the parameters of ESO. The state-space expression of the above equation can be shown as follows:

$$\begin{bmatrix} \dot{z}_1 \\ \dot{z}_2 \end{bmatrix} = \begin{bmatrix} -\beta_1 & 1 \\ -\beta_2 & 0 \end{bmatrix}\begin{bmatrix} z_1 \\ z_2 \end{bmatrix} + \begin{bmatrix} b_0 & \beta_1 \\ 0 & \beta_2 \end{bmatrix}\begin{bmatrix} u \\ y \end{bmatrix}. \quad (14)$$

The following equation can be obtained by performing Laplace transformation on the above equation:

$$z_1(s) = \frac{b_0 s}{s^2 + \beta_1 s + \beta_2} u(s) + \frac{\beta_1 s + \beta_2}{s^2 + \beta_1 s + \beta_2} y(s),$$

$$z_2(s) = \frac{-\beta_2 b_0}{s^2 + \beta_1 s + \beta_2} u(s) + \frac{\beta_2 s}{s^2 + \beta_1 s + \beta_2} y(s). \quad (15)$$

The state-space expression of the above equation can be shown as follows:

$$\begin{bmatrix} z_1(s) \\ z_2(s) \end{bmatrix} = \begin{bmatrix} \dfrac{b_0 s}{s^2 + \beta_1 s + \beta_2} & \dfrac{\beta_1 s + \beta_2}{s^2 + \beta_1 s + \beta_2} \\ \dfrac{-\beta_2 b_0}{s^2 + \beta_1 s + \beta_2} & \dfrac{\beta_2 s}{s^2 + \beta_1 s + \beta_2} \end{bmatrix}\begin{bmatrix} u(s) \\ y(s) \end{bmatrix}. \quad (16)$$

The following equation can be obtained by performing Laplace transformation on formula (10):

$$sy(s) = f(s) + b_0 u(s). \quad (17)$$

The following equation can be obtained by combining formula (16) with (17):

$$z_2(s) = \frac{\beta_2}{s^2 + \beta_1 s + \beta_2} f(s). \quad (18)$$

If input (t) is a step input, the steady-state error can be calculated as follows:

$$\lim_{t\to\infty} e(t) = \lim_{t\to\infty} [z_2(t) - f(t)] = \lim_{s\to 0} s \cdot \frac{1}{s} \cdot e(s)$$

$$= \lim_{s\to 0}\left[\frac{-s(s+\beta_1)}{s^2 + \beta_1 s + \beta_2}\right] = 0. \quad (19)$$

Therefore, the ESO can achieve robust stability when its state matrix meets the condition of Hurwitz stability.

Parameter Design of the Speed Loop ESO

To eliminate observation errors, β_1 must meet the condition $|\beta_1| \gg 0$ at the high-frequency stage, and β_2 must meet the condition$|\beta_2| \gg 0$ at a low-frequency stage. However, if β_1 and β_2 are oversized, the sensitivity of ESO to noise will increase and the stability of the closed-loop system will be negatively affected.

Therefore, the observation accuracy and sensitivity of ESO to noise should be considered when valuating parameters $\beta_1$and β_2.The system poles are assigned in the point with a value of h. If the case of the above conditions is met and the stability of ESO is guaranteed, parameters $\beta 1$, β_2 and poles h satisfy the following relation [14]:

$$s^2 + \beta_1 s + \beta_2 = (s + h)^2. \quad (20)$$

Hence, the following equation can be obtained:

$$\beta_1 = 2h,$$

$$\beta_2 = h^2. \quad (21)$$

Therefore, only two ESO parameters, namely, β_1, $b0$, need to be adjusted. A large number of simulation experiments give the following rules during the adjustment of the two parameters. On one hand, a larger β_1 leads to the faster observation and control compensation of errors by ESO. However, the sensitivity of ESO to noise is increased and the sample frequency of the system becomes restricted. During the adjustment of β_1, the numerical value of β_1 should gradually increase until it meets the performance requirements. On the other hand, the numerical value of b_0 should be close to the actual high-frequency gain of the system as possible. According to the initial-value theorem, the numerical value can be determined by calculating the system's rising rate of the initial time for step response. If the numerical value of$b0$ is difficult to calculate, it can be debugged rom big too small .A stronger control action corresponds to a smaller stability margin.

The numerical values of β_1, b_0 can be chosen as $\beta_1 = 1$, $b_0 = 1$. If the output of ESO appears to vibrate, the numerical value of β_1 must be decreased until the output of ESO appears smooth. Otherwise, the numerical value of β_1 should be increased until the output of ESO appears to vibrate. If the adjustment of β_1 cannot meet the requirements, the numerical value of b_0 should be adjusted to achieve good ESO performance with good estimated accuracy on the premise of low noise sensitivity.

By considering the assigned system poles on the position of −10, the numerical values of β_1 and β_2 can be calculated as$\beta_1 = 10$ and $\beta_2 = 25$. The numerical value of b_0 is selected as$b_0 = 1$, and the ESO of the system is designed.

Disturbance Compensation

The input of the system can be compensated if ESO can estimate the state variables of the system in real time. The system can approximately simplify an integral series connection system after compensation. Therefore, the nonlinear state error feedback control strategy can use a linear proportional function instead of a nonlinear function. The disturbance outputs are observed by ESO to control the input. Then a controlling regularity can be designed as follows:

$$u = \frac{u_0 - z_2}{b_0}. \tag{22}$$

By combining formula (10) with (22), the following equation can be obtained:

$$\dot{y} = f + b_0 \left(\frac{u_0 - \tilde{f}}{b_0} \right) = u_0. \tag{23}$$

The open-loop system can be approximated into an integration element. The pole assignment can be realized by a proportionality coefficient:

$$u_0 = k_p \left(r - z_1 \right). \tag{24}$$

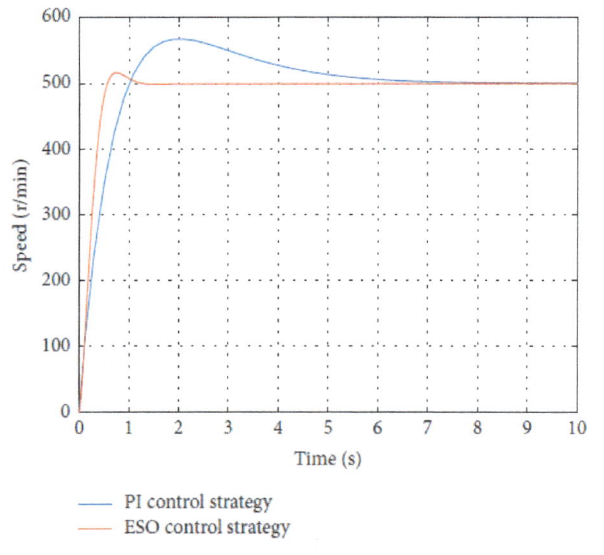

Figure 5: Motor speed response curve of nonload starting.

Therefore, the expected dynamic characteristics of the closed loop system can be expressed as follows:

$$G_c(s) = \frac{y(s)}{r(s)} = \frac{1}{s + k_p}. \tag{25}$$

SIMULATION EXPERIMENTS

The control strategy is taken to compare with the PI control strategy to accurately judge the performances of the control strategy based on ESO. A PMSM produced by Kollmorgen (KBM-43X01) was selected as the executing agency of the system. Table 1 shows the motor parameters. Simulations were carried out in the following four cases for two types of control strategies.

First, nonload starting. A given speed of 500 r/min is suddenly applied on the rotor for motor nonload starting at the time 0s. Figure 5 shows the motor speed response curve under the two types of control strategies. This figure shows that the control strategy based on ESO has a shorter adjusting time and smaller overshoot amount than the PI control strategy. From the simulation results it can be seen that the overshoot amount of the control strategy based on ESO is 10.098% less than that of the PI control strategy and the response time of the control strategy based on ESO is 5.885 s less than that of the PI control strategy. The speed loop parameters of the overshoot amount and the response time for the PI control strategy and ESO control strategy for nonload starting are shown in Table 2.

Second, the external load disturbance is applying on the rotor suddenly.The initial motor speed is given as 500 r/min. An external load disturbance of 300 r/min was suddenly applied on the rotor at the time 5 s. Figure 6 shows the motor speed response curve under the two types of control strategies. Fromthe simulation results it can be seen that after the external load disturbance applied on the rotor suddenly, the control strategy based on ESO has a smaller response curve fluctuation, shorter recovery time, and smaller load disturbance influence than that of the PI control strategy.

The simulation results show that the overshoot amount of the control strategy based on ESO is 6.068% less than that of the PI control strategy and the response time of the control strategy based on ESOis 2.341 s less than that of the PI control strategy.The speed loop parameters of the overshoot amount and recovery time for the PI control strategy and ESO control strategy on the case of the external load disturbance applying on the rotor suddenly are shown in Table 3.

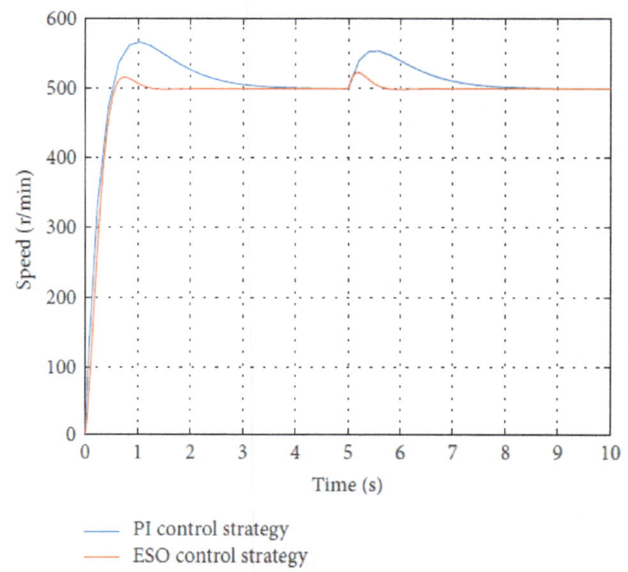

Figure 6: Motor speed response curve of external load disturbance suddenly applied.

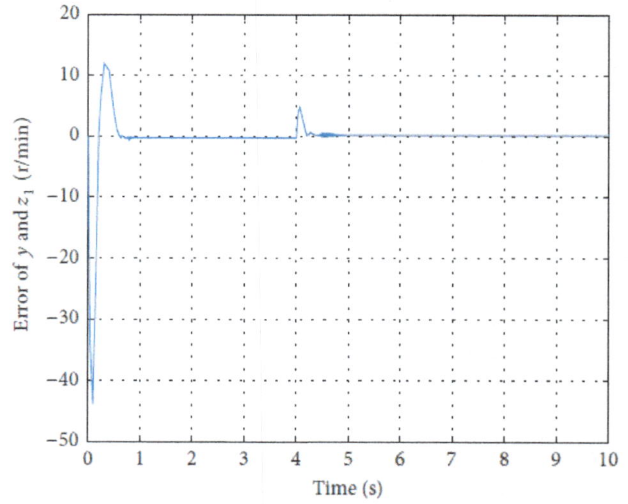

Figure 7: Error of y and z_1.

Simulation results of the error of y and z_1 are shown in Figure 7 and the errors of dy/dt and z_2 are shown in Figure 8. The simulation results show that the compensation time of disturbance is about 0.673 s when the bandwidth of regulating error is 5%.That is to say, the state variables z_1 and z_2 can indicate y and dy/dt timely, respectively. Therefore the control strategy based on ESO can eliminate the influence of the disturbance rapidly and can achieve a result that the steady-state error is zero.

Table 1: Parameters of the experiment motor

Type	Power supply voltage	Rated power	Maximum nonload speed	Continuous current	Continuous torque	Peak current	Peak torque	Motor pole logarithmic	Resistance	Inductance
	VAC	W	r/min	A	N·m	A	N·m		Ω	mH
KBM-43X01	400	2400	2750	5.1	6.11	18	18	16	2.9	6.8

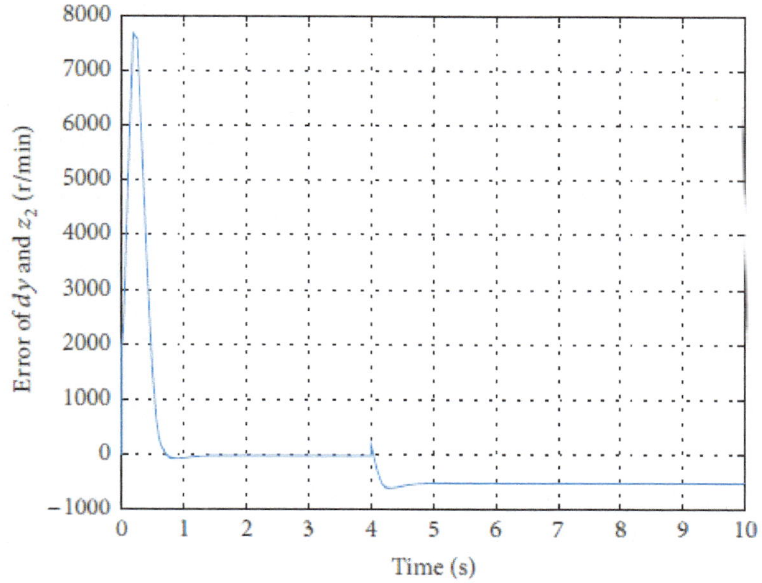

Figure 8: Error of dy/dt and $z2$.

Table 2: Response parameters of the system for nonload starting

Strategy	Overshoot amount	Response time
PI	13.461%	7.098 s
ESO	3.363%	1.213 s

Table 3: Response parameters of external load disturbance applied on rotor suddenly.

Strategy	Overshoot amount	Response time
PI	10.813%	2.973 s
ESO	4.745%	0.632 s

Third, in the case of the speed of motor changing suddenly, the initial motor speed is 500 r/min. The motor speed decreases from 500 r/min to 200 r/min at the time 4 s. The motor speed response curve under the two types of control strategies is shown in Figure 9. The figure shows that the control strategy based on ESO has smaller response curve fluctuations and better recovery capability than the PI control strategy when the speed of motor changes suddenly. From the simulation results it can be seen that the overshoot amount of the control strategy based on ESO is 3.034% less than that of the PI control strategy and the response time of the control strategy based on ESO is 2.688 s less than that of the PI control strategy. The speed loop parameters of the overshoot amount and recovery time for the PI control strategy and ESO control strategy when the speed of motor changes suddenly are shown in Table 4.

Table 4: System response parameters when the speed of motor changes suddenly

Strategy	Overshoot amount	Response time
PI	5.055%	3.367 s
ESO	2.021%	0.679 s

Table 5: Parameters of motor speed error

Strategy	Speed maximum error	Error vanishing time
PI	69.797 r/min	7.596 s
ESO	19.703 r/min	1.433 s

Fourth, response analysis is conducted on the motor speed error. The speed error response curve can be obtained in Figure 10 by comparing the actual motor speed output with the given speed. The figure shows that the speed error of the control strategy based on ESO decreases to 0 r/min faster than that of the PI control strategy. From the simulation results it can be seen that the speed maximum error of the control strategy based on ESO is 50.094 r/min less than that of the PI control strategy and the error vanishing time of the control strategy based on ESO is 6.163 s less than that of the PI control strategy. The parameters of motor speed error are shown in Table 5.

CONCLUSIONS

A speed control system for the pitching axis of satellite cameras based on an ESO was designed. The unknown states and uncertain factors of the system were estimated by the ESO. This control strategy compensates the influences produced by disturbances, and some good control performances are obtained. The simulation results show that the control strategy based on ESO has better control effects than the PI control strategy. This research provides a theoretical basis for future research. The application of ESO to the drift axis control system of satellite cameras, as well as the study of high-order ESO, will be conducted in future works.

ACKNOWLEDGMENTS

This work was supported by the Key Project of Natural Science by Education Department of Anhui Province (no. KJ2015A316), the Major Project of Natural Science by Education Department of Anhui Province (no. KJ2014ZD004), Outstanding Young Talents at Home Visit the School Training Project (no. gxfxZD2016101), Natural Science Foundation of Anhui Province of China (no. 1508085QE83), and the Key Science and Technologies Program of Anhui Province (no. 1301021005).

REFERENCES

1. J.-Q. Han, Active Disturbance Rejection Control Technique, National Defense Industry Press, Beijing, China, 2008.
2. X.-X. Yang and Y. Huang, "Capabilities of extended state observer for estimating uncertainties," inProceedings of the American Control Conference (ACC '09), pp. 3700–3705, IEEE, St. Louis, Mo, USA, June 2009.
3. Y.-H. Wang, Y. Yao, and K.-M. Ma, "Error estimation of second order extended state observer," Journal of Jilin University (Engineering and Technology Edition), vol. 40, no. 1, pp. 143–147, 2010.
4. L.-W. Shao, X.-Z. Liao, Y.-Q. Xia, et al., "Stability analysis and synthesis of third order discrete extended state observer," Information and Control, vol. 37, no. 2, pp. 135–139, 2008.
5. A. Wang and M. Deng, "Robust nonlinear multivariable tracking control design to a manipulator with unknown uncertainties using operator-based robust right coprime factorization," Transactions of the Institute of Measurement and Control, vol. 35, no. 6, pp. 788–797, 2013.
6. M. Deng and N. Bu, "Robust control for nonlinear systems with unknown perturbations using simplified robust right coprime factorisation," International Journal of Control, vol. 85, no. 7, pp. 842–850, 2012.
7. M. Deng and N. Bu, "Robust control for nonlinear systems using passivity-based robust right coprime factorization," IEEE Transactions on Automatic Control, vol. 57, no. 10, pp. 2599–2604, 2012.
8. M. Hasanzadeh, M. R. Jahed-Motlagh, and M. H. Kazemi, "Decentralized disturbance rejection control of multi-area power systems using generalized extended state observer," in Proceedings of the 20th Iranian Conference on Electrical Engineering (ICEE '12), pp. 940–945, IEEE, Tehran, Iran, May 2012.
9. F. Lin, C.-P. Zhang, W.-C. Song, L.-W. Jiao, and S.-S. Chen, "Flux observer of induction motor based on extended state observer," Proceedings of the Chinese Society for Electrical Engineering, vol. 23, no. 4, pp. 145–147, 2003.
10. X. Shi and S. Chang, "Extended state observer-based time-optimal control for fast and precise point-to-point motions driven by a novel electromagnetic linear actuator," Mechatronics, vol. 23, no. 4, pp. 445–451, 2013.
11. L. Chan, F. Naghdy, and D. Stirling, "Extended active observer for force estimation and disturbance rejection of robotic manipulators," Robotics and Autonomous Systems, vol. 61, no. 12, pp. 1277–1287, 2013.
12. Y.-S. Lu, "Sliding-mode disturbance observer with switching-gain adaptation and its application to optical disk drives," IEEE Transactions on Industrial Electronics, vol. 56, no. 9, pp. 3743–3750, 2009.
13. H. K. Khalil and L. Praly, "High-gain observers in nonlinear feedback control," International Journal of Robust and Nonlinear Control, vol. 24, no. 6, pp. 993–1015, 2014.
14. Y.-Q. Zhu, J.-Y. Dong, S.-H. Chen, F.-P. Pan, and D.-H. Li, "Extended state observer based control system design for boiler-turbine unit," Journal of the University of Electronic Science and Technology of China, vol. 43, no. 5, pp. 700–705, 2014. X. Guan and G.-T. Zheng, "Integrated design of space telescope vibration isolation and attitude control,"Journal of Astronautics, vol. 34, no. 2, pp. 214–221, 2013.
15. Y. Xiubin, J. Li, and J. Guang, "Image matching analysis and experiment of spaceborne camera swayed in the end

of orbit life," Chinese Journal of Space Science, vol. 33, no. 3, pp. 337–345, 2013.

16. K. Wang, M.-Z. Luo, K. Li, Y. Cao, and Q.-J. Zhang, "Control algorithm optimization of permanent magnet synchronous motor based on state constraints," Electric Machines and Control, vol. 19, no. 2, pp. 33–38, 2015.
17. J.-F. Mao, A.-H. Wu, G.-Q. Wu, X.-D. Zhang, and G.-X. Wu, "Adaptive sliding mode MPPT control for PMSG-based wind energy conversion systems based on extended state observer," Power System Protection and Control, vol. 42, no. 18, pp. 58–65, 2014.
18. L. Sun, J. Dong, D. Li, and Y. Xue, "Cascaded load-frequency digital electric-hydraulic regulation of turbine via extended state observer," Proceedings of the Chinese Society of Electrical Engineering, vol. 35, no. 7, pp. 1697–1703, 2015.

CHAPTER 8 MODELING AND CONTROL OF THE DISTRIBUTED POWER CONVERTERS IN A STANDALONE DC MICROGRID

Xiaodong Lu and Jiangwen Wan

School of Instrumentation Science and Opto-electronics Engineering, Beihang University, Beijing 100191, China;

ABSTRACT

A standalone DC microgrid integrated with distributed renewable energy sources, energy storage devices and loads is analyzed. To mitigate the interaction among distributed power modules, this paper describes a modeling and control design procedure for the distributed converters. The system configuration and steady-state analysis of the standalone DC microgrid under study are discussed first. The dynamic models of the distributed converters are then developed from two aspects corresponding to their two operating modes, device-regulating mode and bus-regulating mode. Average current mode control and linear compensators are designed accordingly for each operating mode. The stability of the designed system is analyzed at last. The operation and control design of the system are verified by simulation results.

INTRODUCTION

With increasing use of renewable energy sources (RES), the standalone DC microgrid is envisioned as one promising technology to integrate distributed RES, energy storage devices (ESD) and varieties of loads in many power management and distribution applications [1,2,3,4].

In a generic standalone DC microgrid, distributed RES, ESD and loads are interfaced to the DC bus using power electronics converters. There is no specific line regulator; thus, ESD converters or RES converters have to take charge of the bus regulation. Therefore, power converters in the system can be divided into two groups depending on their respective operating modes, bus-regulating mode and device-regulating mode.

In the design of a DC microgrid integrated with distributed power modules, one of the primary concerns is how to avoid the undesirable interaction among different modules. The interaction may appear as two forms: the non-zero output impedance of bus-regulating converters may interfere with the control loops of device-regulating converters, and the non-resistive input impedance of device-regulating converters may affect the control loops of bus-regulating converters. In most applications, the device-regulating converters have to provide strict output voltage regulation to satisfy the voltage requirement of the devices, which, together with the high efficiency of these converters, causes them to perform like constant power sources/loads to the DC bus. Consequently, when the net power of device-regulating converts is towards/

Citation: Xiaodong Lu and Jiangwen Wan, Modeling and Control of the Distributed Power Converters in a Standalone DC Microgrid, doi:10.3390/en9030217.

against the DC bus, the total closed-loop input impedance of the device-regulating converts can be approximated as a positive/negative incremental resistance, with the magnitude varying with the net power. It may result in significant stability problems, since the impedance interaction is the root of the instability in a source-load interconnected system [5].

To solve the stability issues caused by impedance interaction, several passive damping methods [6,7] and active stabilization methods [8,9] have been proposed. Passive damping methods use passive components, such as resistors, capacitors and inductors, to improve the system stability. Active stabilization methods, on the other hand, usually use some advanced control methods to change the output impedance of the line regulator, so as to avoid the impedance interaction.

However, in a standalone DC microgrid, the regulation of the DC bus is implemented by distributed RES or ESD converters. They should not only be able to regulate the rest of the system in a stable manner, but also operate their own devices efficiently and provide the required protection, such as current limitation. As a result, simple linear control methods are preferred in the design of distributed power converters. The purpose of this work is to develop dynamic models of the distributed converters in a standalone DC microgrid, by which average current mode control and linear compensators could be employed in the control of these converters, so as to improve the performance and stability of the system.

This paper is organized as follows. The configuration and operation of the standalone DC microgrid under study are provided in Section 2. In Section 3, dynamic models of the distributed power converters are developed from two aspects, device-regulating mode and bus-regulating mode. Necessary transfer functions for compensator design are derived from these models. The stability of the entire system is analyzed at last. In Section 4, the operation and control design of the system are verified by simulation results. Section 5 concludes this paper.

SYSTEM CONFIGURATION AND OPERATION

Figure 1a shows a simplified architecture of the standalone DC microgrid studied in this work, which is composed of distributed PV arrays, DC loads and battery packs. The target system is based on a 48-V DC bus. These kinds of low voltage DC microgrids have been widely used in residential applications, remote communication stations and data centers [10,11,12]. To reduce the development time and cost, as well as to simplify the operation of the system, as shown in Figure 1b, all distributed power sources or loads are interfaced to the DC bus via identical 500-W non-inverting four-switch buck-boost converters, considering their capabilities of bidirectional power flow and voltage step up/down. Very high efficiency over a wide range of operating voltages has already been demonstrated using this topology [13,14].

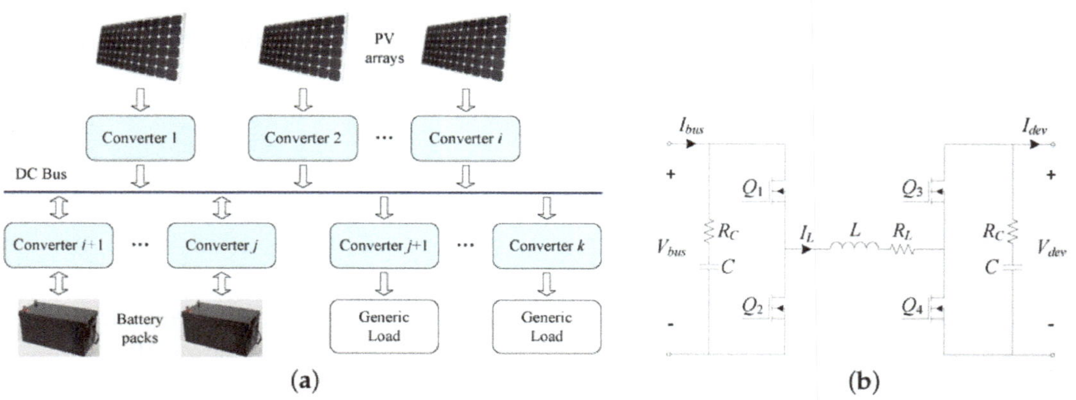

Figure 1. (**a**) Architecture of the standalone DC microgrid. (**b**) Power stage of the adopted bidirectional non-inverting buck-boost converter.

Different operating modes of the standalone DC microgrid are summarized in Table 1. Modes I and II correspond to the case when PV arrays operate in the maximum power point tracking (MPPT) mode, while the DC bus is regulated by the battery packs. Mode III refers to the case when the maximum output power of the PV arrays is greater than the total power demanded by batteries and loads. In this case, to protect the batteries from overcharging, battery charging current is limited; thus, PV converters have to take over the task of bus regulation. In Mode IV, the demanded load power is greater than the total maximum power of PV arrays and battery packs. At this point, load shedding is required in order to maintain the system stability. A simple load shedding scheme based on voltage thresholds can be seamlessly applied [15]; therefore, Mode IV will not be discussed in this paper.

Table 1. Operating modes of the standalone DC microgrid

System Mode	PV Mode	Battery Mode	Load Mode
Mode I	MPPT (device)	Discharging (bus)	Normal (device)
Mode II	MPPT (device)	Charging (bus)	Normal (device)
Mode III	Off-MPPT (bus)	Charging (device)	Normal (device)
Mode IV	MPPT (device)	Discharging (bus)	Shedding (device)

From Table 1, it can be concluded that in each operating mode of the standalone DC microgrid, all distributed power converters can be divided into two groups depending on their respective responsibilities: bus-regulating converters and device-regulating converters. PV converters or battery converters take charge of the bus regulation based on the states of the batteries and the voltage level of the DC bus. The rule of the transitions between bus-regulating mode and device-regulating mode is shown in Figure 2. Battery converters switch from the bus-regulating mode to the device-regulating mode when the battery voltage V_{bat} reaches the fully-charged voltage $V_{bat,ref}$, while the charging current I_{charge} is still positive, and switch back to the bus-regulating mode when the bus voltage V_{bus} is less than its reference value $V_{bus,ref}$. In the case of PV converters, they switch from the device-regulating mode to the bus-regulating mode when $V_{bus} \geq 1.1V_{bus,ref}$ and switch back to the device-regulating mode when $V_{bus} \leq 1.05V_{bus,ref}$. It should be noted that there is an intermediate state during the transition when both battery converters and PV converters are working in the device-regulating mode. During that state, V_{bus} keeps changing until the next threshold is reached.

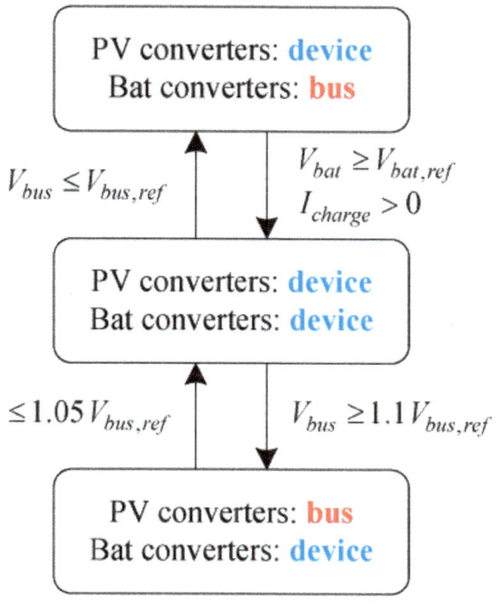

Figure 2. Diagram of the transitions between bus-regulating mode and device-regulating mode.

Based on the above classification, a demonstration circuit diagram of the DC microgrid is shown in Figure 3a, where *m* converters work in the bus-regulating mode and *n* converters work in the device-regulating mode. The well-known droop control method [16] is adopted in the bus regulation to guarantee that several distributed converters can operate on the bus simultaneously. Assuming the control loops are well designed in each converter, the bus port of a bus-regulating converter behaves as a voltage source $V_{bus,i}$ in series with a droop resistance $R_{droop,I}$ $R_{droop,i}$, while the device port of a device-regulating converter behaves as a constant voltage source Vdev,j Vdev,j. Meanwhile, considering the high efficiency of these converters, the device ports of bus-regulating converters and the bus ports of device-regulating converters can be regarded as constant power sinks/sources depending on the direction of the power flow.

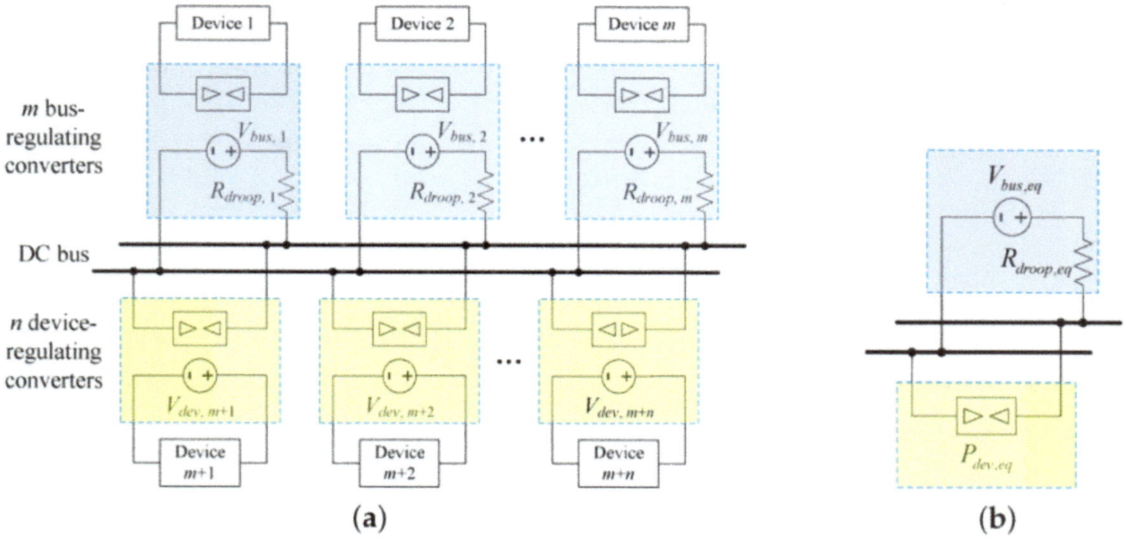

Figure 3. (**a**) Circuit diagram of the DC microgrid where *m*converters work in the bus-regulating mode and *n* converters work in the device-regulating mode. (**b**) Equivalent circuit diagram.

When only the effect of each converter with respect to the bus is taken into account, an equivalent circuit diagram of the standalone DC microgrid is obtained in Figure 3b. All of the bus-regulating power modules can be represented as a Thevenin equivalent source ($V_{bus,eq}$, $R_{droop,eq}$), where $R_{droop,eq} = R_{droop,1} \parallel \cdots \parallel R_{droop,m}$, and $V_{bus,eq} = R_{droop,eq} \sum_{i=1}^{m}(V_{bus,i}/R_{droop,i})$. All of the device-regulating power modules can be combined into a power source/sink $P_{dev,eq} = \sum_{j=1}^{n} P_{dev,j}$.

DYNAMIC MODELING AND CONTROL DESIGN

The modeling and control design of the distributed converters in a standalone DC microgrid will be discussed from two aspects: device-regulating mode and bus-regulating mode.

Closed-Loop Input Impedance of the Device-Regulating Converters

The small-signal averaged circuit model of a device-regulating converter is shown in Figure 4, where Z_{dev} represents the input impedance of the device. The adopted bidirectional buck-boost converter is supposed to work in the buck mode when $V_{bus}>V_{dev}$ and work in the boost mode when $V_{bus}<V_{dev}$. Only the continuous conduction mode (CCM) is considered here for simplification, since the device-regulating converters work in CCM for most cases. In the circuit model, the output impedance of bus-regulating converters is assumed to be zero, while the effect of the non-zero output impedance of bus-regulating converters on the dynamics of a device-regulating converter will be discussed in Section 3.3.

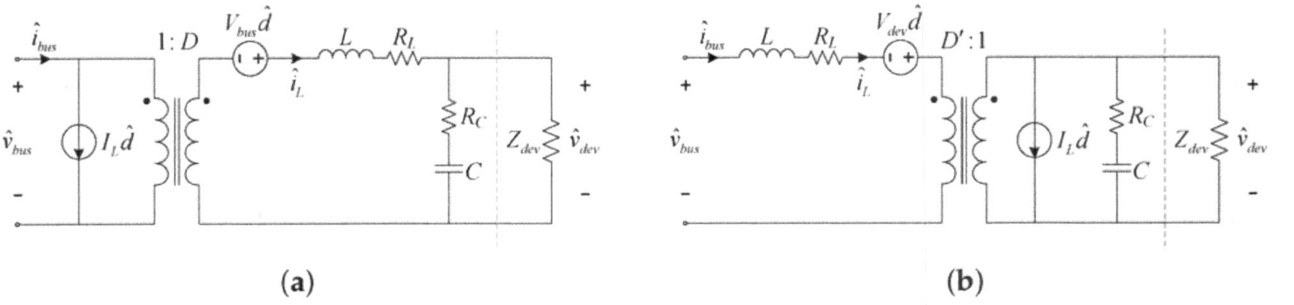

Figure 4. Averaged small-signal circuit of a device-regulating converter when it operates in **(a)** the buck mode or in **(b)** the boost mode.

In the device-regulating converter, the average current mode control method is adopted to simplify the feedback control design, as well as to provide the required current limitation. The small-signal control block diagram of a device-regulating converter is shown in Figure 5. $G_{id}(s)$and $G_{vd}(s)$ are the transfer functions from control variable d^ to inductor current i^L and device voltage v^dev. $G_{ig}(s)$ and $G_{vg}(s)$ are the transfer functions from input voltage v^bus to inductor current i^L and device voltage v^dev. $H_i(s)$ and $H_v(s)$ are the current and the voltage sensing gains. $G_{ci}(s)$ and $G_{co}(s)$ are the compensators of the inner current loop and outer voltage loop, respectively, and G_m is the gain of the pulse width modulator. The derivations of $G_{id}(s)$, $G_{vd}(s)$, $G_{ig}(s)$ and $G_{vg}(s)$ for a device-regulating converter working in the buck mode or boost mode are expressed in Table 2, where $Z_{load} = Z_C \parallel Z_{dev}$, $Z_C = 1/sC + R_C$ and $Z_L = sL + R_L$. In the control design, the target crossover frequency of the inner current loop is $f_{ci} = f_{sw}/10$, while that of the outer voltage loop is $f_{co} = f_{ci}/10$, where f_{sw} is the converter switching frequency. The detailed design of the average current mode control for a device-regulating converter is a classic topic [17,18], so it will not be discussed here.

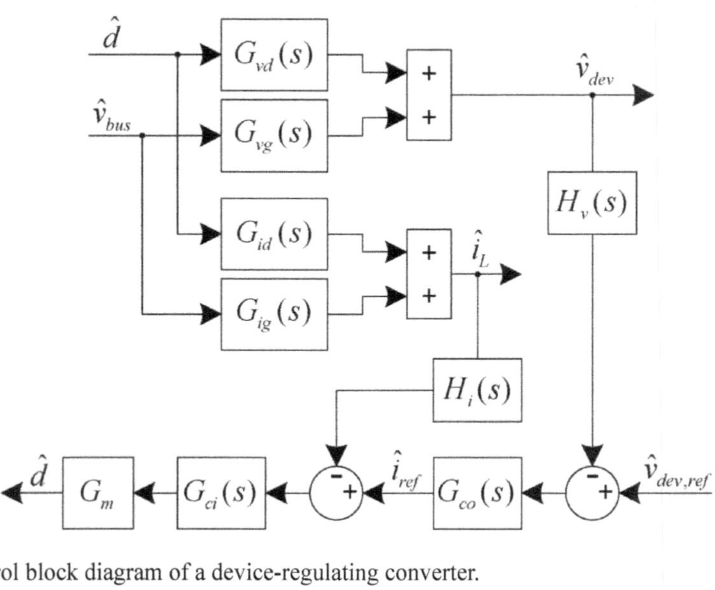

Figure 5. The small-signal control block diagram of a device-regulating converter.

Table 2. The derivations of $G_{id}(s)$, $G_{vd}(s)$, $G_{ig}(s)$ and $G_{vg}(s)$.

Transfer Function	Buck Mode	Boost Mode
$G_{id}(s)$	$\dfrac{V_{bus}}{Z_L+Z_{load}}$	$\dfrac{V_{dev}+D'I_LZ_{load}}{Z_L+D'^2Z_{load}}$
$G_{vd}(s)$	$G_{id}(s)Z_{load}$	$(D'G_{id}(s)-I_L)Z_{load}$
$G_{ig}(s)$	$\dfrac{D}{Z_L+Z_{load}}$	$\dfrac{1}{Z_L+D'^2Z_{load}}$
$G_{vg}(s)$	$G_{ig}(s)Z_{load}$	$G_{ig}(s)D'Z_{load}$

To derive the closed-loop input impedance of a device-regulating converter, the control loops depicted in Figure 5 are expressed as Equation Set 1

$$\begin{cases} \hat{d}G_{vd}(s)+\hat{v}_{bus}G_{vg}(s)=\hat{v}_{dev} \\ \hat{d}G_{id}(s)+\hat{v}_{bus}G_{ig}(s)=\hat{i}_L \\ -\hat{v}_{dev}H_v(s)G_{co}(s)=\hat{i}_{ref} \\ (\hat{i}_{ref}-\hat{i}_LH_i(s))G_{ci}(s)G_m=\hat{d} \end{cases} \quad (1)$$

The converter input current i^bus can be expressed as the linear combination of i^L and d^ as below:

$$\hat{i}_{bus}=M\hat{i}_L+j\hat{d} \quad (2)$$

where M=D,j=IL in the buck mode and M=1,j=0 in the boost mode. Combining Equations 1 and 2, the closed-loop input impedance of a device-regulating converter is derived as:

$$Z_{in}=\frac{\hat{v}_{bus}}{\hat{i}_{bus}}=\frac{1+G_{vd}H_vG_{co}G_{ci}G_m+G_{id}H_iG_{ci}G_m}{MG_{ig}(1+G_{vd}H_vG_{co}G_{ci}G_m)-(j+MG_{id})G_{vg}H_vG_{co}G_{ci}G_m-jG_{ig}H_iG_{ci}G_m} \quad (3)$$

Figure 6 shows a comparison between the derived closed-loop input impedance Z_{in} and the incremental resistance R_{in} of a device-regulating converter when it operates at full power (500 W), where subscripts buck and boost represent that the converter works in the buck or boost mode. It can be seen that the incremental resistance R_{in} is a good approximation to Z_{in} below the crossover frequency (f_{co} = 1 kHz) of the converter's outer loop. Key parameters of the bidirectional buck-boost converter are shown in Table 3. Other parameters used in this example are $V_{dev,buck}$ = 24 V and $V_{dev,boost}$ = 72 V. The incremental resistance R_{in} is derived based on $R_{in}=-V_{bus}2/P_{dev}$, which assumes no losses exist in a device-regulating converter, and it behaves as a constant power load/source. P_{dev} is positive when the power is transferred from the DC bus to the device.

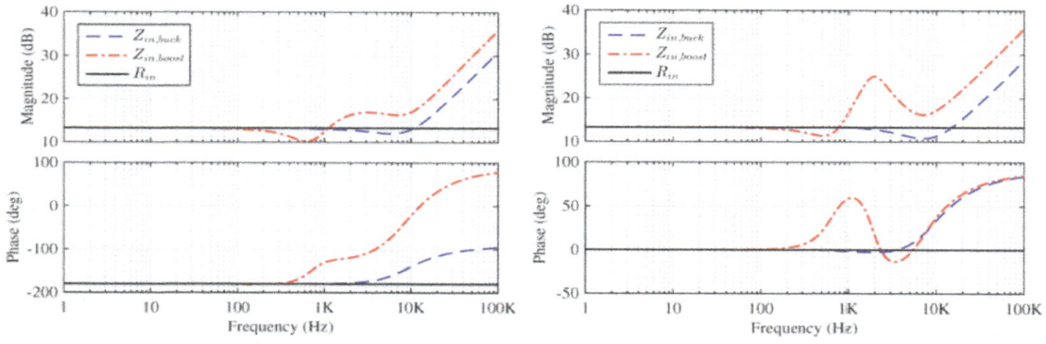

Figure 6. Comparison between the closed-loop input impedance Z_{in} and incremental resistance R_{in} of a device-regulating converter when **(a)** $P_{dev}>0$ and **(b)** $P_{dev}<0$.

Table 3. Key parameters of the bidirectional buck-boost converter

L	R_L	C	R_C	f_{sw}	G_m	$H_{i(s)}$	$H_{v(s)}$
100 μH	10 mΩ	1 mF	150 mΩ	100 kHz	1	$\dfrac{1}{1+\frac{s}{\pi f_{sw}}}$	$\dfrac{1}{1+\frac{s}{\pi f_{sw}/5}}$

In a standalone DC microgrid, n device-regulating converters operate on the bus simultaneously. Thus, the total closed-loop input impedance $Z_{in,dev}$ of these converters can be expressed as the parallel combination of their respective input impedance Z_{in}

$$\frac{1}{Z_{in,dev}} = \sum_{j=1}^{n} \frac{1}{Z_{in,j}} \approx \sum_{j=1}^{n} \left(-\frac{P_{dev,j}}{V_{bus}^2}\right) = -\frac{P_{dev,net}}{V_{bus}^2} \tag{4}$$

From Equation 4, it can be concluded that the input impedance $Z_{in,dev}$ is determined by the net power $P_{dev,net}$ of the device-regulating converters. When $P_{dev,net}<0$, which means the direction of the net power is towards the DC bus, $Z_{in,dev}$ can be approximated as a positive resistance at frequencies below fcofco. On the contrary, typical negative resistance appears when $P_{dev,net}>0$.

Modeling and Control Design of the Bus-Regulating Converters

In a standalone DC microgrid, PV or battery converters work as the bus-regulating converters. Since a relatively low bus voltage of 48 V is chosen in this work, both PV and battery converters can easily be configured working in the buck mode. Figure 7 shows a circuit diagram where m bus-regulating converters and n device-regulating converters are interfaced to the DC bus. $Z_{in,1}$–$Z_{in,n}$ represent the closed-loop input impedance of device-regulating converters. $Z_{bus,1}$–$Z_{bus,m+n}$ represent the line impedance.

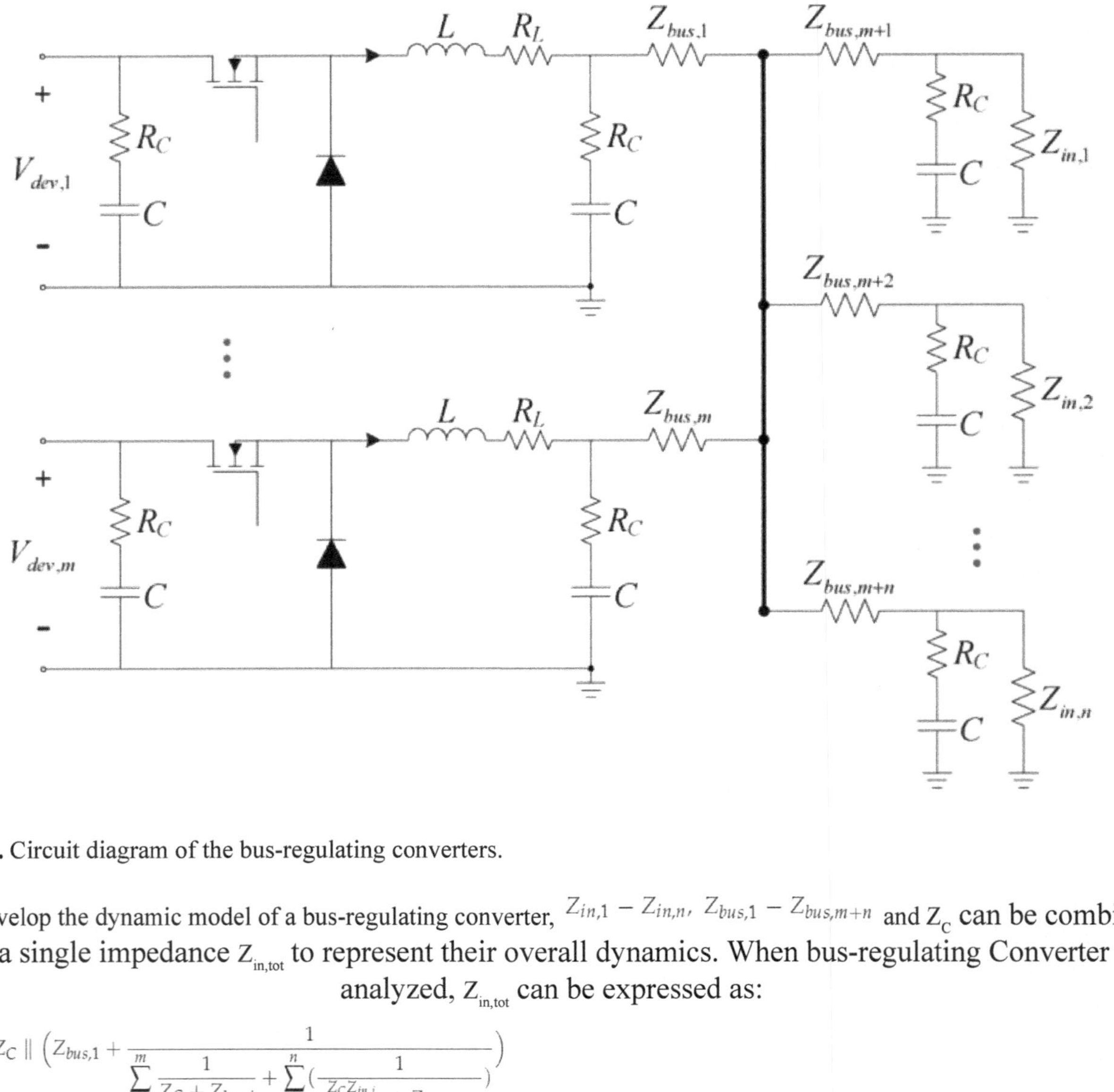

Figure 7. Circuit diagram of the bus-regulating converters.

To develop the dynamic model of a bus-regulating converter, $Z_{in,1} - Z_{in,n}$, $Z_{bus,1} - Z_{bus,m+n}$ and Z_C can be combined into a single impedance $Z_{in,tot}$ to represent their overall dynamics. When bus-regulating Converter 1 is analyzed, $Z_{in,tot}$ can be expressed as:

$$Z_{in,tot} = Z_C \parallel \left(Z_{bus,1} + \frac{1}{\sum_{i=2}^{m} \frac{1}{Z_C + Z_{bus,i}} + \sum_{j=1}^{n} \left(\frac{1}{\frac{Z_C Z_{in,j}}{Z_C + Z_{in,j}} + Z_{bus,m+j}}\right)}\right) \tag{5}$$

Actually, a simplified expression of $Z_{in,tot}$ can be derived as $Z_{in,tot} = Z_{C,tot} \parallel Z_{in,dev}$ if the bus impedance is neglected, where Z_C,tot is the total impedance of all bus capacitors.

The averaged small-signal circuit of a bus-regulating converter is depicted in Figure 8; R_{dev} is the equivalent small-signal resistance of the battery pack (R_{bat}) or PV array (R_{pv}). Both the CCM and the discontinuous conduction mode (DCM) are taken into account in the small-signal model, since the power processed by the bus-regulating converters varies with the net power $P_{dev,net}$ of device-regulating converters.

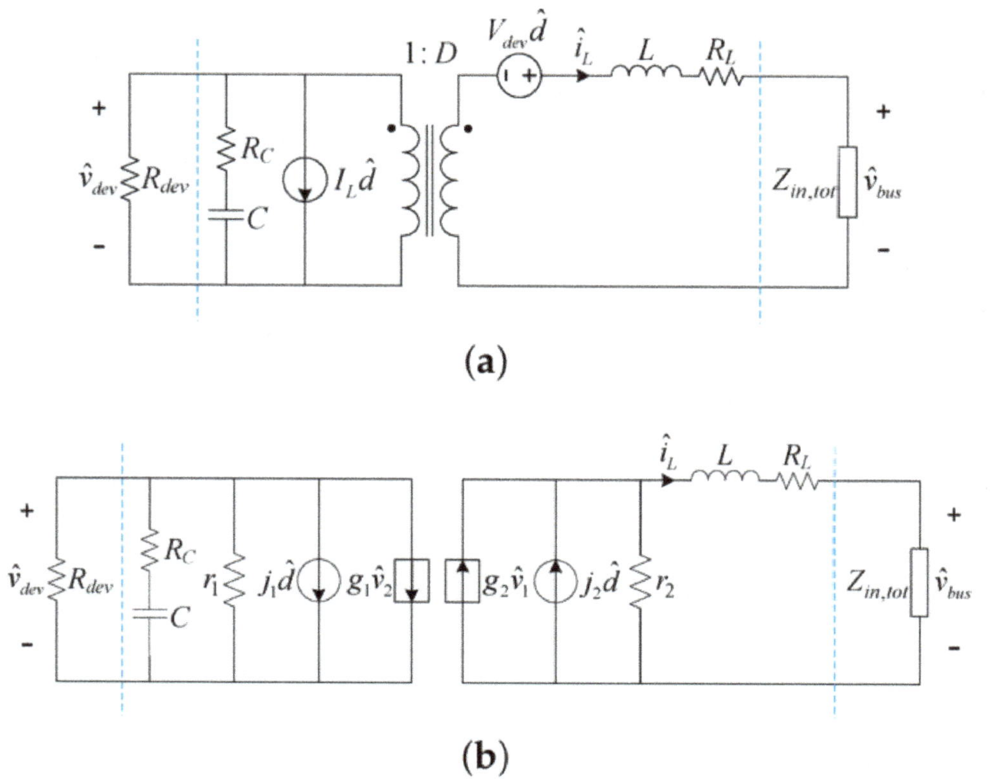

Figure 8. Averaged small-signal circuit of a bus-regulating converter when it works in (**a**) continuous conduction mode (CCM) or in (**b**) discontinuous conduction mode (DCM).

In this work, the small-signal model of the battery pack is represented as a voltage source v^batv^batconnected in series with its internal resistance R_{bat}. The small-signal model of the PV array is represented as a current source $i^\wedge_{pv}$ paralleled with a resistance R_{pv}, in which R_{pv} is obtained by the linearization of the nonlinear current-voltage curve of the PV array at its steady-state operating point. Values of R_{pv} used in this work are shown in Figure 9, which are obtained from a PV array model consisting of two series-connected Conergy P 175M PV modules.

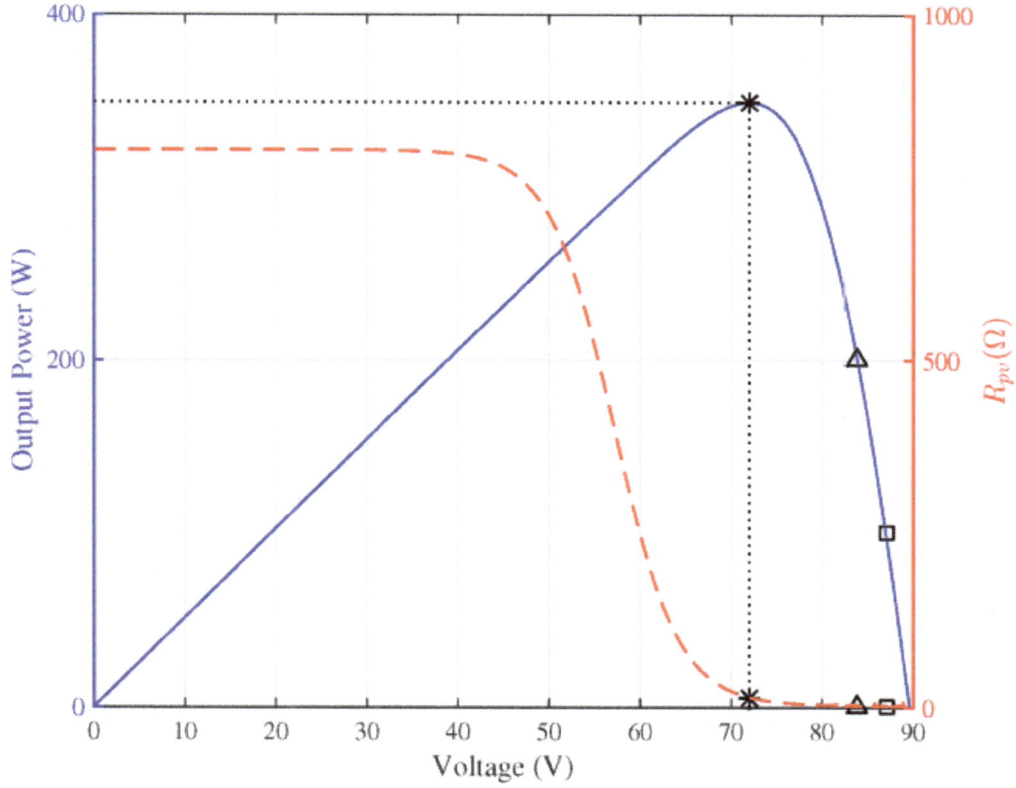

Figure 9. Linearized resistance R_{pv} of a PV array consisting of two series-connected Conergy P 175M PV modules when the solar irradiance is 1000 W/m22 and the temperature is 25 ooC.

When a bus-regulating converter works in CCM, the transfer functions of control to inductor current G_{id},CCM(s) and inductor current to bus voltage G_{vi}, CCM(s) are derived as follows:

$$G_{id,CCM}(s) = \frac{\hat{i}_L}{\hat{d}} = \frac{V_{dev} - DI_L Z_g}{D^2 Z_g + Z_L + Z_{in,tot}} \tag{6}$$

$$G_{vi,CCM}(s) = \frac{\hat{v}_{bus}}{\hat{i}_L} = Z_{in,tot} \tag{7}$$

where $Z_g = R_{dev} \parallel Z_C$. In DCM, the transfer functions of $G_{id,DCM}(s)$ and $G_{vi,DCM}(s)$ are obtained as follows:

$$G_{id,DCM}(s) = \frac{\hat{i}_L}{\hat{d}} = \frac{j_2 Z_{load} - g_2 j_1 Z_g Z_{load}}{(1 + g_1 g_2 Z_g Z_{load})(Z_L + Z_{in,tot})} \tag{8}$$

$$G_{vi,DCM}(s) = \frac{\hat{v}_{bus}}{\hat{i}_L} = Z_L + Z_{in,tot} \tag{9}$$

where $Z_g = R_{dev} \parallel Z_C \parallel r_1$ and $Z_{load} = r_2 \parallel (Z_L + Z_{in,tot})$. The parameters used in DCM are shown in Table 4, where $R_e = [(2/M - 1)^2 - 1]R/4$, , and M is the voltage conversion ratio [19].

Table 4. Parameters of the small-signal DCM model for the buck converter

g_1	j_1	r_1	g_2	j_2	r_2
$\frac{1}{R_e}$	$\frac{2(1-M)V_{dev}}{DR_e}$	R_e	$\frac{2-M}{MR_e}$	$\frac{2(1-M)V_{dev}}{DMR_e}$	$M^2 R_e$

The control block diagram of the bus-regulating converters is shown in Figure 10. To provide the required current limitation, average current mode control is also tried in the bus-regulating converters. The droop control method is adopted in the outer control loop for the purpose of current sharing.

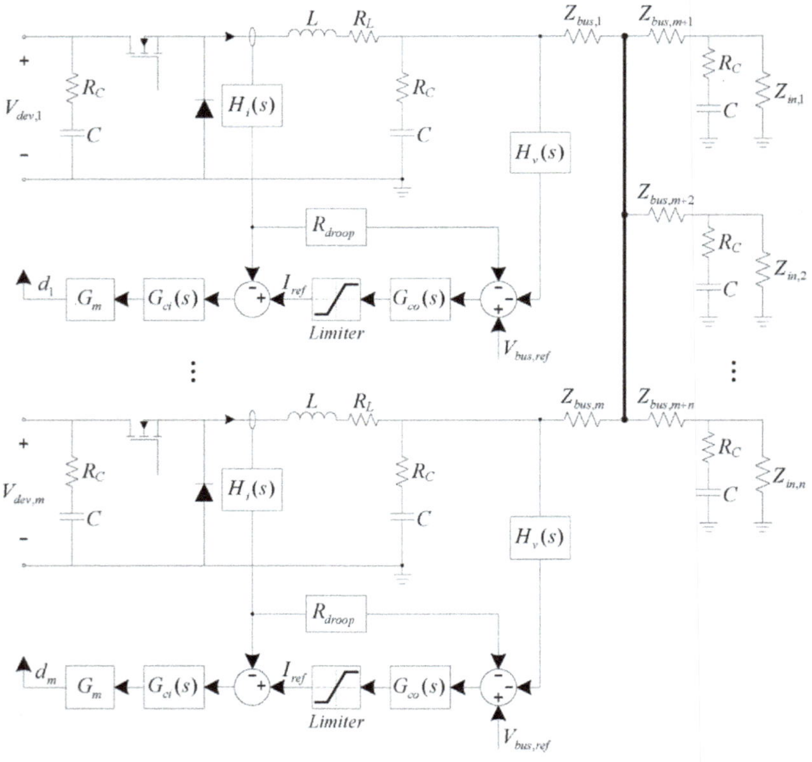

Figure 10. The control block diagram of bus-regulating converters.

The resulting expressions for the inner current-loop gain $T_i(s)$ and outer voltage-loop gain $T_v(s)$ are shown below:

$$T_i(s) = G_{id}(s) H_i(s) G_{ci}(s) G_m \tag{10}$$

$$T_v(s) = \frac{T_i(s)/H_i(s)}{1 + T_i(s)} \left(G_{vi}(s) H_v(s) + H_i(s) R_{droop} \right) G_{co}(s) \tag{11}$$

where $G_{ci}(s)$ and $G_{co}(s)$ are the compensators for the inner current loop and outer voltage loop, respectively. To achieve the desired crossover frequency and phase margin, different inner compensators $G_{ci,CCM}(s)$ and $G_{ci,DCM}(s)$ are designed according to the conduction mode of the inductor current. Whereas, the outer loop $T_v(s)$ is compensated using the same $G_{co}(s)$. The detection of the conduction mode is implemented using the method in [20].

It is a common practice to assume a resistive load in the control design of switching regulators. In this work, control loops $T_i(s)$ and $T_v(s)$ are designed based on $Z_{in,dev}$ when the net power $P_{dev,net}$ of device-regulating converters is negative, since in this case, the total closed-loop input impedance $Z_{in,dev}$ of device-regulating converters behaves as a positive resistance at low frequencies. Bode plots of $T_i(s)$ and $T_v(s)$ of a bus-regulating converter when it operates at −300 W (CCM) and −20 W (DCM) are shown in Figure 11, where $G_{ci,CCM}(s)$, $G_{ci,DCM}(s)$ and $G_{co}(s)$ are realized by PI compensators. Key parameters used in this example are $m = 2$, $n = 6$, $V_{dev} = 60$ V, $R_{dev} = 20$ mΩ, $Z_{bus} = 10$ mΩ and $R_{droop} = 25$ mΩ. It can be seen that no right half-plane (RHP) zeroes or poles appear in the loop gains, so the stability and dynamic behavior of the converter could be analyzed using the Bode plots. As shown in the figure, the inner-loop crossover frequencyfcifci is 10 kHz, and the outer-loop crossover frequency f_{co} is 1 kHz, which all meet the design requirements, $f_{ci}=f_{sw}/10$ and $f_{co}=f_{ci}/10$. The phase margins of $T_{i,CCM}$ and $T_{i,DCM}$ are 65° and 88°, while the phase margins of $T_{v,CCM}$ and $T_{v,DCM}$ are 70° are and 64°, respectively.

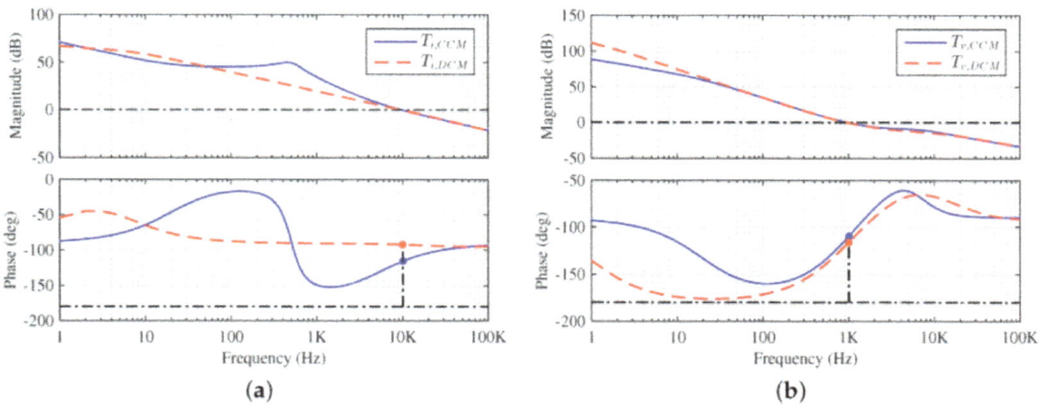

Figure 11. Bode plots of (**a**) $T_i(s)$) and (**b**) $T_v(s)$ when $P_{dev,net}<0$.

However, when $P_{dev,net}$ is positive, $Z_{in,dev}$ behaves as a negative resistance at low frequencies, which results in one RHP zero in Ti(s)Ti(s) and one RHP pole in $T_v(s)$. The RHP zero and pole could be calculated numerically, and the effects of the RHP zeroes (circles) and poles (stars) can be observed in the Bode plots shown in Figure 12. In this figure, $T_i(s)$ and $T_v(s)$ are calculated when the bus-regulating converter operates at 300 W (CCM) and 20 W (DCM), and the PI compensators designed above are still employed in the control loops.

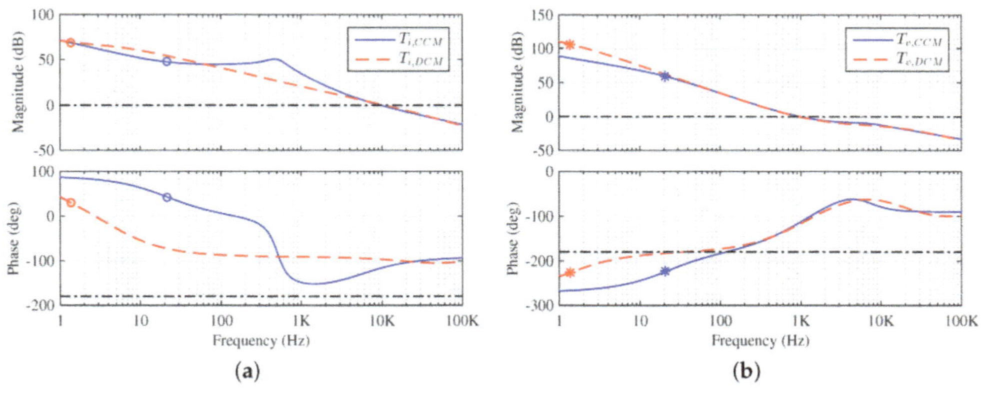

Figure 12. Bode plots of (**a**) $T_i(s)$ and (**b**) $T_v(s)$ when $P_{dev,net}>0$.

So far, it has been demonstrated that loop gains Ti(s)Ti(s) and Tv(s)Tv(s) of a bus-regulating converter when $P_{dev,net} > 0$ differ substantially from those when $P_{dev,net} < 0$. When $P_{dev,net} > 0$, Bode plots are no longer applicable to the stability analysis, since the RHP zero or pole appears. In order to analyze the stability of the bus-regulating converter, the Nyquist criterion has to be adopted. Actually, it is not difficult to find that the inner current loop $T_i(s)$ is no longer stable when $P_{dev,net} > 0$ by using the Nyquist criterion. However, the stability of a multi-loop controlled converter is not determined by each loop alone, but by both loops working together. For the stability analysis of a bus-regulating converter, two loop gains $T_1(s)$ and $T_2(s)$ are measured, as shown in Figure 13. Loop gain $T_1(s)= \hat{e}_{y1}/\hat{e}_{x1}$ is measured at Point "1", while loop gain $T_2(s)= (s) = \hat{e}_{y2}/\hat{e}_{x2}$ is measured at Point "2", where $\hat{e}_{x1}$ and $\hat{e}_{x2}$ and e^x$_2$ e^x$_2$ are the small-signal perturbations injected for the measurement of the loop gain and e^y$_1$e^y$_1$ and e^y$_2$e^y$_2$ are their corresponding responses, respectively. These two loop gains have been verified regarding their importance in the stability analysis of a multi-loop controlled converter in [21].

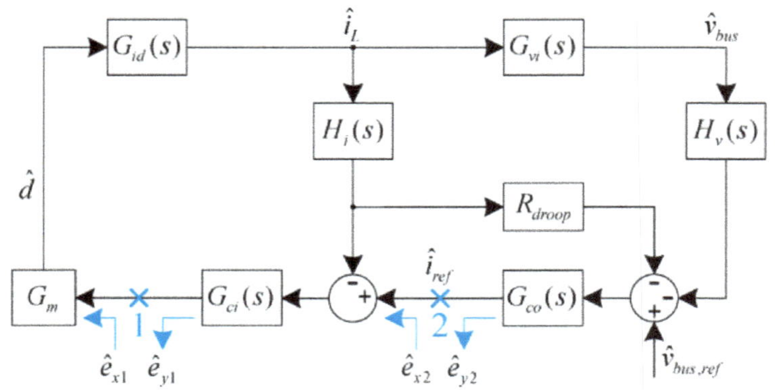

Figure 13. Measurement of the loop gains $T_1(s)$ and $T_2(s)$.

Loop gains T1(s)T1(s) and T2(s)T2(s) are expressed as follows:

$$T_1(s) = T_i(s) + G_m G_{id}(s)(G_{vi}(s)H_v(s) + H_i(s)R_{droop})G_{co}(s)G_{ci}(s) \tag{12}$$

$$T_2(s) = \frac{T_i(s)/H_i(s)}{1+T_i(s)}(G_{vi}(s)H_v(s) + H_i(s)R_{droop})G_{co}(s) \tag{13}$$

It is easy to find that actually $T_2(s)$ equals $T_v(s)$; thus, one RHP pole exists in $T_2(s)$. As for $T_1(s)$, second-order poles appear at the original one due to the cascading of PI compensators. The statement of the Nyquist stability criterion for a continuous system is Z=P−N, where *Z* is the number of unstable closed-loop poles, *P* is the number of unstable open-loop poles and *N* is the number of counter-clockwise encirclements that the Nyquist plot of the loop gain makes around the $(-1, j_0)$ point. Nyquist plots of $T_1(s)$ and $T_2(s)$ are shown in Figure 14. Both $T_1(s)$ and $T_2(s)$ encircle the (−1,j0)(−1,j0) point counter-clockwise exactly once (N=1). As one unstable pole exists in both $T_1(s)$ and $T_2(s)$ (P=1), the closed-loop system is proven to be stable (Z=0) when $P_{dev,net}$ is positive.

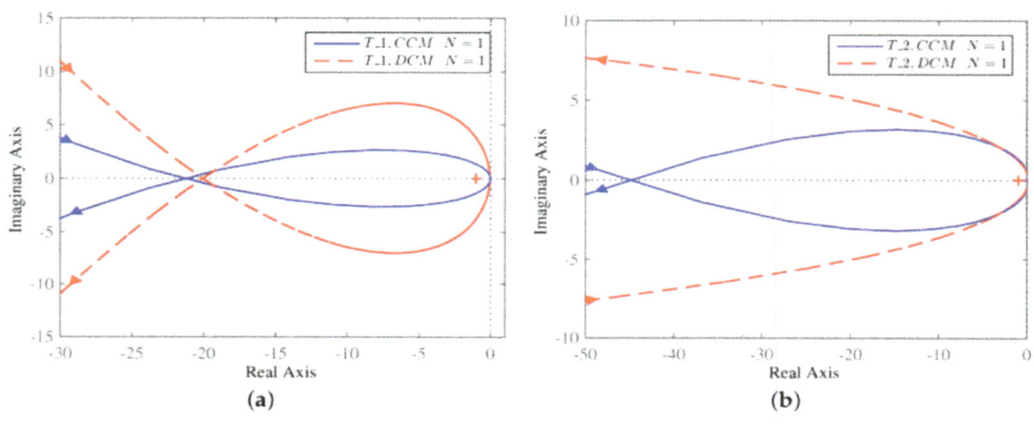

Figure 14. Nyquist plots of **(a)** $T_1(s)$ and **(b)** $T_2(s)$ when $P_{dev,net}>0$.

While the bus-regulating converter is proven to be stable when $P_{dev,net}>0$, knowledge of the system relative stability is still necessary for us. As a result, gain and phase margins of $T_1(s)$ and $T_2(s)$ that are defined by the above Nyquist plots are translated into the Bode plots in Figure 15. It can be seen that the control bandwidth of the inner loop is around 10 kHz, while that of the outer loop is 1 kHz, as desired. Fortunately, good system performance can be ensured by the gain and phase margins.

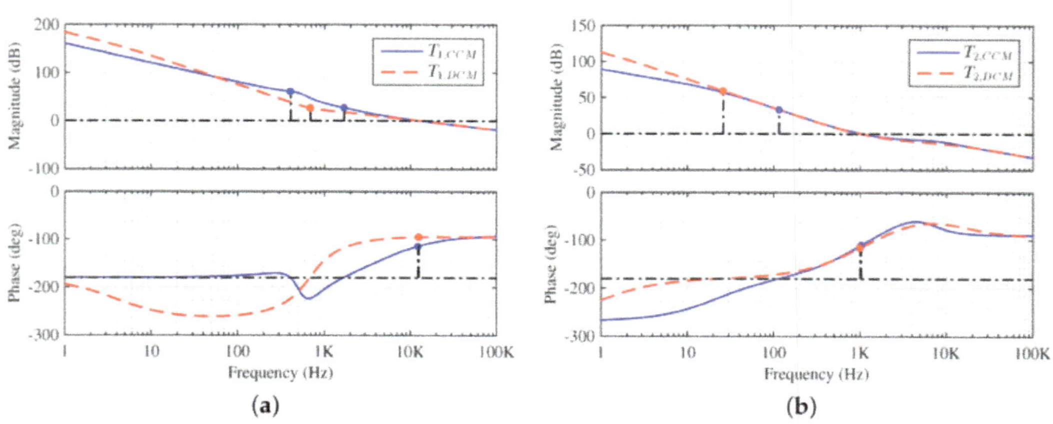

Figure 15. Bode plots of **(a)** $T_1(s)$ and **(b)** $T_2(s)$ with gain and phase margins when $P_{dev,net}>0$.

Furthermore, to regulate the DC bus voltage, battery and PV converters have to work over a wide range of operating conditions. To guarantee the performance of the bus-regulating converters, loop gains $T_1(s)$ and $T_2(s)$ are evaluated under several representative operating conditions. Three power levels (100 W, 200 W and 350 W) are picked out for the PV converters when the solar irradiance is 1000 W/m22 and the temperature is 25 ooC. The PV output voltages V_{pv} and the small-signal linearized resistances R_{pv} corresponding to the three power levels are marked in squares (100 W), triangles (200 W) and stars (350 W) in Figure 9. Three power levels (100 W, 300 W and 500 W) are also picked out for the battery converters. The battery voltage R_{bat} and its internal series resistance R_{bat} are assumed to be constant (V_{bat} = 60 V, R_{bat} = 20 mΩ) when different power is processed, since their variations are much less compared to the PV array.

Bode plots of $T_1(s)$ and $T_2(s)$ corresponding to the representative operating conditions are shown in Figure 16. It should be appreciated that neither the gain nor the phase changes significantly over the power levels, device voltages or small-signal device resistances, especially at the crossover frequency. Similar results could also be obtained under other operating condition, which allows one to conclude that the simple PI compensators designed for the bus-regulating control loops can be implemented under different operating conditions within a reasonable range.

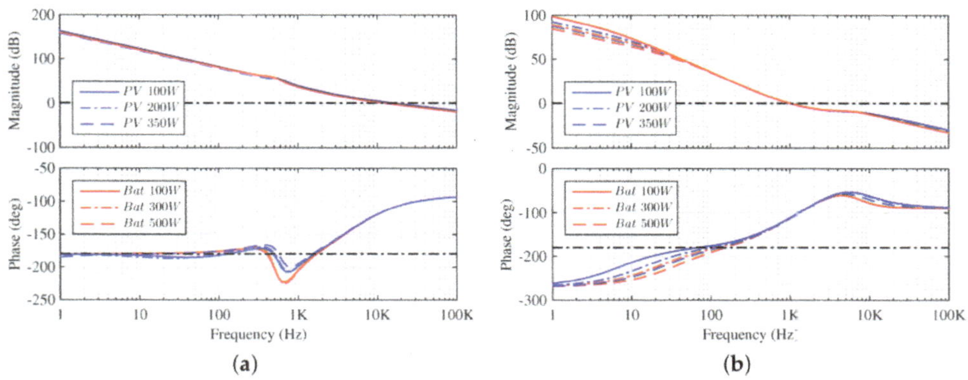

Figure 16. Bode plots of **(a)** $T_1(s)$ and **(b)** $T_2(s)$ when the bus-regulating converter works at different steady-state operating points.

Stability Analysis of the DC Microgrid

Based on the control loops designed in the above subsection (Figure 13), the closed-loop output impedance Z_{out} of a bus-regulating converter can be derived as follows:

$$\begin{cases} -(\hat{i}_L H_i(s) R_{droop} + \hat{v}_{bus} H_v(s)) G_{co}(s) = \hat{i}_{ref} \\ (\hat{i}_{ref} - \hat{i}_L H_i(s)) G_{ci}(s) G_m = \hat{d} \\ \hat{d} G_{id}(s) G_{vi}(s) = \hat{v}_{bus} \end{cases} \tag{14}$$

$$Z_{out} = -\frac{\hat{v}_{bus}}{\hat{i}_L} = \frac{H_i(R_{droop}G_{co}+1)G_{ci}G_mG_{id}G_{vi}}{1+H_vG_{co}G_{ci}G_mG_{id}G_{vi}} \tag{15}$$

Actually, the magnitude of Z_{out} is verified to be quite close to R_{droop} due to the adoption of the droop control method and average current mode control.

A small-signal equivalent circuit of the standalone DC microgrid can then be obtained in Figure 17, where *m* converters work in the bus-regulating mode and *n* converters work in the device-regulating mode. Small-signal Thevenin sources ($\hat{v}_{bus,i}$, $Z_{out,i}$) are used to model the closed-loop dynamics of the bus-regulating converters.

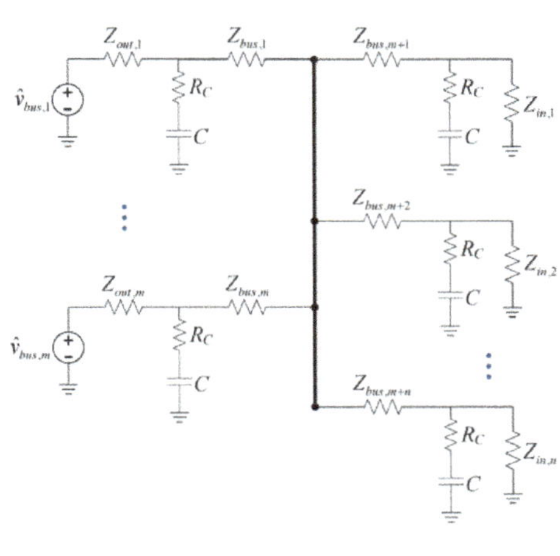

Figure 17. Small-signal equivalent circuit of the standalone DC microgrid.

However, when the closed-loop input impedance $Z_{in,1}$–$Z_{in,n}$ is derived in Subsection *3.1*, it is assumed that the output impedance $Z_{out,1} - Z_{out,m}$ of bus-regulating converters is zero. Therefore, it needs to be determined that whether the transfer functions ($G_{id}(s)$, $G_{vd}(s)$ that were used to derive $Z_{in,1}$–$Z_{in,n}$ in Subsection *3.1* are modified by the none-zero output impedance $Z_{out,1} - Z_{out,m}$. Middlebrook's extra element theorem (EET) [5,19] is employed here to determine how the addition of $Z_{out,1} - Z_{out,m}$ alters these transfer functions. Take $G_{vd}(s)$ as an example. The modified transfer function can be expressed as follows:

$$G_{vd}(s) = \left(G_{vd}(s)\big|_{Z_{out,tot}=0}\right)\frac{1+\frac{Z_{out,tot}}{Z_N}}{1+\frac{Z_{out,tot}}{Z_D}} \tag{16}$$

where $G_{vd}(s)|_{Z_{out,tot}=0}$ is the transfer function $G_{vd}(s)$ derived in Section 3.1 and $Z_{out,tot}$ is the combination of $Z_{out,1} - Z_{out,m}$, $Z_{bus,1} - Z_{bus,m+n}$ and Z_C. When device-regulating Converter 1 is analyzed, Zout,totZout,tot can be expressed as below:

$$Z_{out,tot} = Z_C \parallel \left(Z_{bus,m+1} + \frac{1}{\sum_{i=1}^{m}\frac{1}{\frac{Z_C Z_{out,i}}{Z_C+Z_{out,i}}+Z_{bus,i}} + \sum_{j=2}^{n}\left(\frac{1}{Z_C+Z_{bus,m+j}}\right)}\right) \tag{17}$$

Additionally, a simplified expression of $Z_{out,tot}$ is derived as below if the bus impedance is neglected:

$$\frac{1}{Z_{out,tot}} = \sum_{i=1}^{m}\frac{1}{Z_{out,i}} + \frac{1}{Z_{C,tot}} \tag{18}$$

The definition and derivation of the impedance Z_N and Z_D for a device-regulating converter are described in Table 5.

Table 5. Definition and derivation of Z_N and Z_D for a device-regulating converter

	Z_N	Z_D
Definition	$\frac{\hat{v}_{bus}}{\hat{i}_{bus}}\Big\|_{\hat{v}_{dev}\to 0}$	$\frac{\hat{v}_{bus}}{\hat{i}_{bus}}\Big\|_{\hat{d}=0}$
Buck mode	$-\frac{V_{bus}}{DI_L}$	$\frac{1}{D^2}\left(Z_L + \frac{Z_{dev}Z_C}{Z_{dev}+Z_C}\right)$
Boost mode	$Z_L - \frac{V_{dev}D'}{I_L}$	$Z_L + D'^2\frac{Z_C Z_{dev}}{Z_C + Z_{dev}}$

From Equation 16, it can be easily obtained that when the inequalities of $\|Z_{out,tot}\| \ll \|ZN\| \ll \|Z_N\|$ and $\|Z_{out,tot}\| \ll \|Z_D\|$ are satisfied, $G_{vd}(s) \approx G_{vd}(s)|Z_{out,tot}=0$, which means the control loops of the device-regulating converter are not interfered with obviously. Therefore, the comparison between $\|Z_{out,tot}\|$ and $\|Z_N\|$, $\|Z_D\|$ needs to be implemented. A standalone DC microgrid, which consists of eight distributed power modules, is taken as an example, where two battery converters work in the bus-regulating mode; three PV converters and three load converters work in the device-regulating mode. The power processed by each converter is listed as follows: P_{bat} = 300 W, P_{pv} = 300 W and P_{load} = 500 W. The comparison result is shown in Figure 18, where both the load converter (buck mode) and the PV converter (boost mode) are analyzed. It can be concluded that the control loops of the device-regulating converter are not affected significantly by the non-zero output impedance of the bus-regulating converters, since a wide separation can be observed between $\|Z_{out,tot}\|$ and $\|Z_N\|$, $\|Z_D\|$. Therefore, the closed-loop input impedance $Z_{in,1}$–$Z_{in,n}$,nderived in Subsection *3.1* is proven to be a good approximation to the value when $\|Z_{out,tot}\|$ is taken into account.

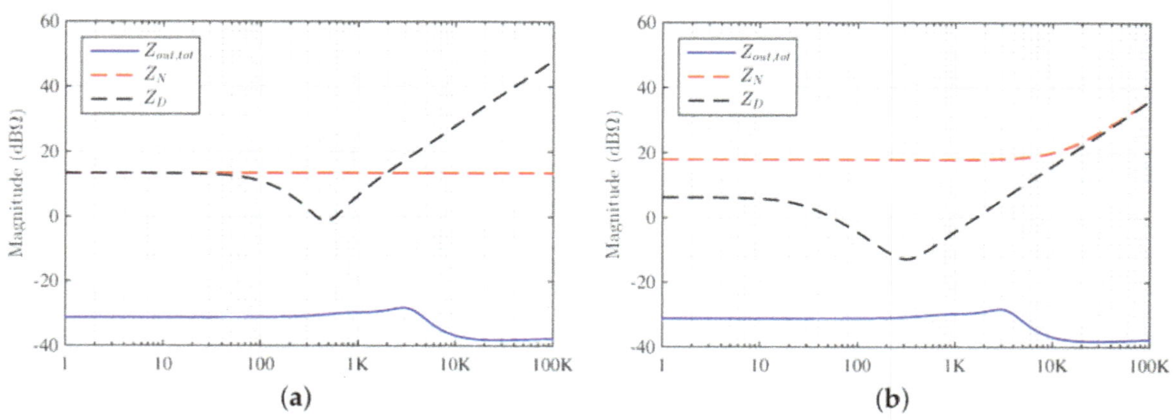

Figure 18. Comparison between $\|Z_{out,tot}\|$ and $\|Z_N\|$, $\|Z_D\|$ when (**a**) the load converter or (**b**) the PV converter is analyzed.

Since it has already been verified in Section 3.2 that the stability and performance of the designed bus-regulating control loops are not affected considerably by the input impedance of device-regulating converters, a low degree of interaction among the distributed converters can be predicted in the designed standalone DC microgrid, and the system is stable.

SIMULATION RESULTS

In order to verify the control design of distributed power converters in a standalone DC microgrid integrated with distributed PV arrays and battery packs, system simulations have been carried out using SimPowerSystems/Simulink. A schematic diagram of the simulated standalone DC microgrid is shown in Figure 19. The simulated system consists of three distributed PV arrays, two battery packs and three resistive loads. All of these distributed devices are connected to a 48-V DC bus via identical 500-W bidirectional buck-boost converters.

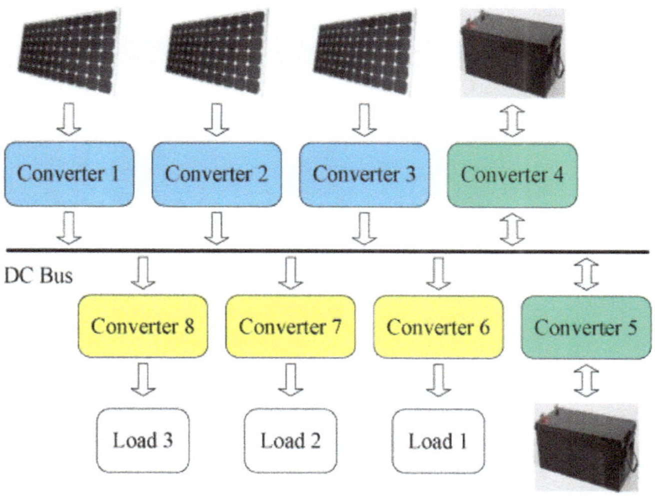

Figure 19. Schematic diagram of the simulated standalone DC microgrid.

The configuration of the simulated standalone DC microgrid is shown in Table 6. For the consideration of a stable operation of the system, the maximum discharging power of the battery packs, as well as the maximum output power of the PV arrays, is greater than the total power demand of the loads. In this simulation, a PV array model is chosen from the library of SimPowerSystems, which is composed of two Conergy P 175M modules connected in series. To emulate the fully-charged status of the battery in a short time, a simple model of the battery pack is adopted, which consists of an ideal voltage source in series with a constant internal resistance.

Table 6. Configuration of the standalone DC microgrid in simulation

	Type	Conergy P 175M
PV array	Configuration	$N_p = 1, N_s = 2$
	Open-circuit voltage V_{oc}	89.6 V
	Short-circuit current I_{sc}	5.17 A
	MPP voltage V_{mpp}	72 V
	MPP current I_{mpp}	4.86 A
	Maximum power P_{mpp}	350 W
Battery Pack	Nominal voltage	60 V
	Fully charged voltage $V_{bat,ref}$	65 V
	Internal Resistance R_{bat}	100 mΩ
Load	Rating of Load 1	12 V, 200 W
	Rating of Load 2	72 V, 300 W
	Rating of Load 3	24 V, 400 W

In the simulation, operation events of the distributed devices over a time span of 1 s are scheduled in Table 7. During this time, a sequence of events takes place, emulating different operating modes (Mode I to III) of the standalone DC microgrid, as mentioned in Table 1, and possible transitions between these modes. Since this work is focused on the dynamic modeling and control, to facilitate the observation of the dynamic responses of the system, all of the operation events are scheduled in a short time span of 1 s, which is made possible since the control bandwidth of every distributed converter is designed to be 1 kHz.

Table 7. Operation events scheduled in the simulation

Time (s)	Solar Irradiance
0 to 0.3	gradually increases from 100 W/m^2 to 900 W/m^2
0.3	Steps from 900 W/m^2 to 1000 W/m^2
0.3 to 0.5	keeps constant at 1000 W/m^2
0.5 to 0.7	gradually decreases from 1000 W/m^2 to 0
0.7 to 1	keeps constant at 0
	Battery Open-circuit Voltage
0 to 0.2	keeps constant at 60 V
0.2 to 0.4	gradually increases from 60 V to 64.8 V
0.4 to 1	keeps constant at 64.8 V
	Loads
0	Loads 1, 2 and 3 are all on
0.1	Load 2 is switched off
0.5	Load 3 is switched off
0.8	Loads 2 and 3 are switched on

Simulation results are shown in Figure 20, where several variables of interest are sampled at the frequency of 100 kHz.

- From $t = 0$ to $t = 0.395$ s: DC bus voltage V_{bus} is regulated by the battery converters, while PV converters work in the MPPT mode. During this time, solar irradiance increases from 100 W/m22 to 1000 W/m22.

Correspondingly, the simulated DC microgrid works in Mode I from 0 to 0.178 s, but switch to Mode II at 0.178 s. A smooth transition can be observed in Figure 20c thanks to the separate control designs for the CCM and DCM. Besides, smooth responses can be seen in Figure 20b and Figure 20c when step disturbances of the load off and solar irradiance variation occur at the time instants of 0.1 s and 0.3 s, respectively.

- From $t = 0.395$ s to $t = 0.408$ s: To observe the transition from Mode II to Mode III, the battery open-circuit voltage is increased gradually from 60 V to 64.8 V during 0.2 s to 0.4 s. At $t = 0.395$ s, battery converters detect that V_{bat} reaches the fully-charged voltage $V_{bat,ref}$ and, thus, switch to the device-regulating mode to protect the batteries from overcharging. During this interval, PV converters still work in the MPPT mode. As a consequence, V_{bus} increases steadily.
- From $t = 0.408$ s to $t = 0.659$ s: PV converters switch to the bus-regulating mode when V_{bus} reaches 1.1Vbus,ref1.1Vbus,ref at $t = 0.408$ s. Smooth responses can be seen in Figure 20e and Figure 20f when the step disturbance of the load off occurs at 0.5 s. Solar irradiance starts to decrease from $t = 0.5$ s. From $t = 0.653$ s, the total maximum power generated by the PV arrays starts to be less than the power demand of the device-regulating converters. Consequently, V_{bus} starts to drop. At $t = 0.657$ s, V_{bus} drops down to $1.05V_{bus,ref}$; PV converter switch back to the MPPT mode accordingly. At $t = 0.659$ s, V_{bus} drops down to $V_{bus,ref}$; the regulation of V_{bus} is taken over by the battery converters again.
- From $t = 0.659$ s to $t = 1$ s: Loads 2 and 3 are switched on simultaneously at 0.8 s. Smooth step responses can be observed in Figure 20b and Figure 20c.

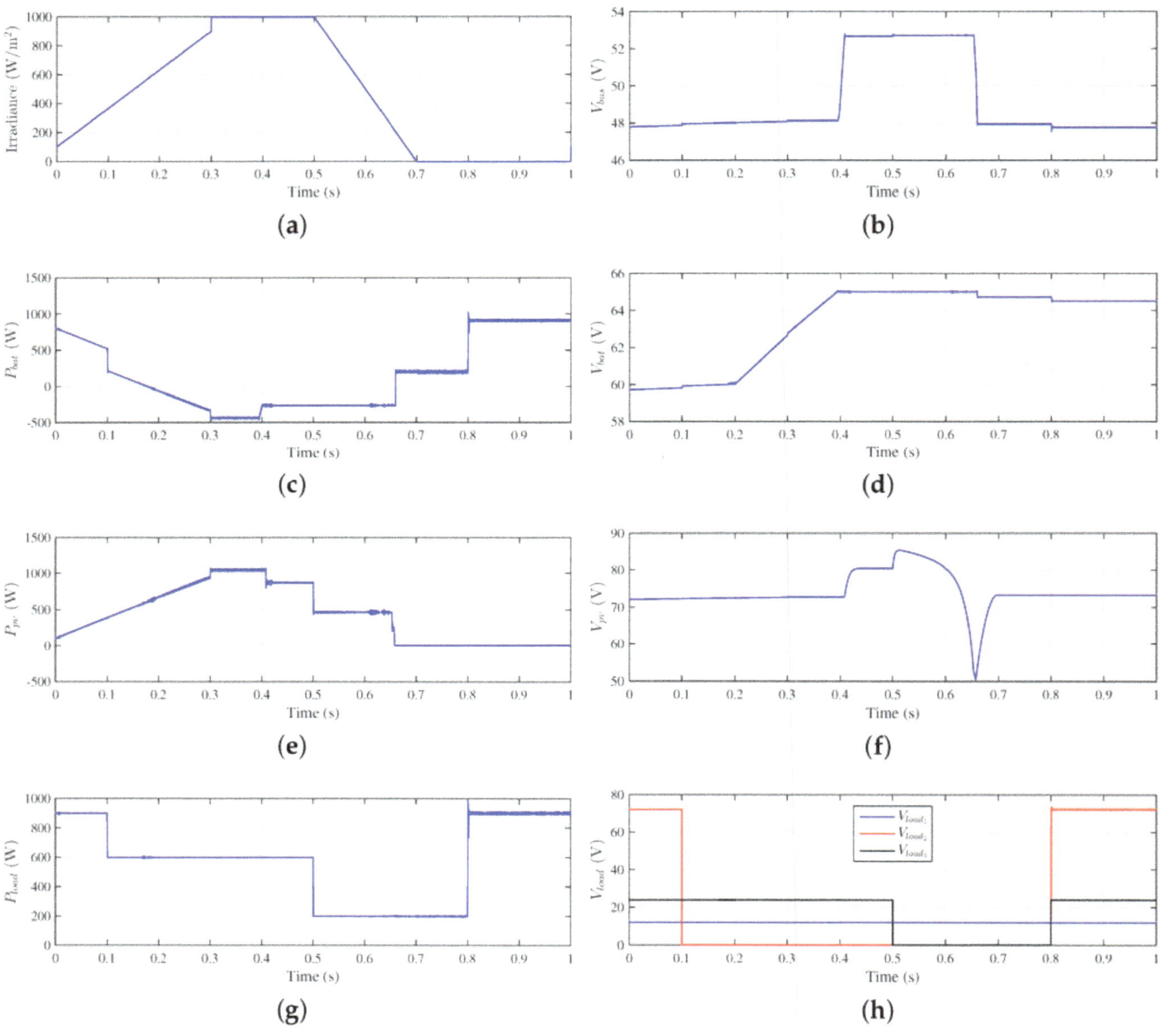

Figure 20. Simulation results. (**a**) Solar irradiance variation. (**b**) DC bus voltage V_{bus}. (**c**) Total charging/discharging power P_{bat} the two battery packs. (**d**) Battery voltage V_{bat}. (**e**) Total power generation P_{pv} of the three PV arrays. (**f**) PV array voltage VpvVpv. (**g**) Total load power P_{load}. (**h**) Voltages Vload1,2,3 at three load points.

CONCLUSIONS

To mitigate the interaction among distributed power modules in a standalone DC microgrid, a modeling and control design procedure for the distributed converters was proposed in this work. First, the system configuration and steady-state analysis of the standalone DC microgrid under study were discussed. Next, the dynamic model of the distributed converters was developed from two aspects, which correspond to their two operating modes, device-regulating mode and bus-regulating mode. Average current mode control and linear compensators were then designed for each corresponding operating mode. The stability of the designed system was analyzed at last. The operation and control design of the system were demonstrated by simulation results using SimPowerSystems/Simulink.

AUTHOR CONTRIBUTIONS

Xiaodong Lu carried out the main research work and wrote the full manuscript. Jiangwen Wan provided constructive comments in the analysis and modeling. All authors have read and approved this manuscript.

REFERENCES

1. Schonberger, J.; Duke, R.; Round, S.D. DC-Bus Signaling: A Distributed Control Strategy for a Hybrid Renewable Nanogrid. *IEEE Trans. Ind. Electron.* **2006**, *53*, 1453–1460.
2. Diaz, N.L.; Dragicevic, T.; Vasquez, J.C.; Guerrero, J.M. Intelligent Distributed Generation and Storage Units for DC Microgrids - A New Concept on Cooperative Control Without Communications Beyond Droop Control. *IEEE Trans. Smart Grid* **2014**, *5*, 2476–2485.
3. Longe, O.M.; Ouahada, K.; Ferreira, H.C.; Chinnappen, S. Renewable Energy Sources microgrid design for rural area in South Africa. In Proceedings of IEEE PES Innovative Smart Grid Technologies Conference (ISGT), Washington, DC, USA, 19–22 February 2014; pp. 1–5.
4. Sechilariu, M.; Locment, F.; Wang, B. Photovoltaic Electricity for Sustainable Building. Efficiency and Energy Cost Reduction for Isolated DC Microgrid. *Energies* **2015**, *8*, 7945–7967.
5. Middlebrook, R.D. Input filter consideration in design and application of switching regulators. In Proceedings of IEEE Industry Applications Society (IAS) Annual Meeting, Chicago, IL, USA, 11–14 October 1976; pp. 366–382.
6. Cespedes, M.; Xing, L.; Sun, J. Constant-Power Load System Stabilization by Passive Damping. *IEEE Trans. Power Electron.* **2011**, *26*, 1832–1836.
7. Erickson, R.W. Optimal single resistor damping of input filters. In Proceedings of IEEE Applied Power Electronics Conference and Exposition (APEC), Dallas, TX, USA, 14–18 March 1999; pp. 1073–1079.
8. Wang, J.; Howe, D. A power shaping stabilizing control strategy for DC power systems with constant power loads. *IEEE Trans. Power Electron.* **2008**, *23*, 2982–2989.
9. Rahimi, A.; Emadi, A. Active damping in DC/DC power electronics converters: A novel method to overcome the problems of constant power loads. *IEEE Trans. Ind. Electron.* **2009**,*56*, 1428–1439.
10. Li, W.; Mou, X.; Zhou, Y.; Marnay, C. On voltage standards for DC home microgrids energized by distributed sources. In Proceedings of Power Electronics and Motion Control Conference (IPEMC), Harbin, China, 2–5 June 2012; pp. 2282–2286.
11. McMenamin, D. Case studies supporting -48 VDC as the power input of choice for computer equipment deployed in the telecom network. In Proceedings of Twentieth International Telecommunications Energy Conference (INTELEC), San Francisco, CA, USA, 4–8 October 1998; pp. 261–265.
12. White, R. Computers in the central office-a primer on powering equipment from –48 V. In Proceedings of IEEE Applied Power Electronics Conference and Exposition (APEC), Anaheim, CA, USA, 15–19 February 1998; pp. 902–908.
13. Linares, L.; Erickson, R.W.; MacAlpine, S.; Brandemuehl, M. Improved Energy Capture in Series String Photovoltaics via Smart Distributed Power Electronics. In Proceedings of Applied Power Electronics Conference and Exposition (APEC), Washington, DC, USA, 15–19 February 2009; pp. 904–910.
14. Stahl, G.; Rodriguez, M.; Maksimovic, D. A high-efficiency bidirectional buck-boost DC-DC converter. In Proceedings of IEEE Applied Power Electronics Conference and Exposition (APEC), Orlando, FL, US, 5–9 February 2012; pp 1362–1367.
15. Kwasinski, A.; Onwuchekwa, C. Dynamic behavior and stabilization of DC microgrids with instantaneous constant-power loads. *IEEE Trans. Power Electron.* **2011**, *26*, 822–834.
16. Luo, S.; Ye, Z.; Lin, R.L.; Lee, F. A classification and evaluation of paralleling methods for power supply modules. In Proceedings of IEEE Power Electronics Specialists Conference (PESC), Charleston, SC, USA, 27 June–1 July 1999; pp. 901–908.
17. Dixon, L.H. Average current-mode control of switching power supplies. In *Unitrode Power Supply Design Seminar Handbook*; Unitrode Corporation: Merrimack, NH, USA, 1990.
18. Tang, W.; Lee, F.C.; Ridley, R.B. Small-signal modeling of average current-mode control.*IEEE Trans. Power Electron.* **1993**, *8*, 112–119.
19. Erickson, R.W.; Maksimovic, D. *Fundamentals of Power Electronics*, 2nd ed.; Kluwer: New York, NY, USA, 2001; pp. 362–368, 381–385, 420–427.
20. Moon, S.; Corradini, L.; Maksimovic, D. Accurate mode boundary detection in digitally controlled boost power factor correction rectifiers. In Proceedings of IEEE Energy Conversion Congress and Exposition (ECCE), Atlanta, GA, USA, 12–16 September 2010; pp. 1212–1217.
21. Ridley, R.B.; Cho, B.H.; Lee, F.C. Analysis and interpretation of loop gains of multiloop-controlled switching regulators. *IEEE Trans. Power Electron.* **1988**, *3*, 489–498.

CHAPTER 9

PRINCIPAL COMPONENTS IN MULTIVARIATE CONTROL CHARTS APPLIED TO DATA INSTRUMENTATION OF DAMS

Emerson Lazzarotto[1], Liliana Madalena Gramani[2], Anselmo Chaves Neto[2], Luiz Albino Teixeira Junior[3]

[1]Universidade Estadual do Oeste do Paraná - UNIOESTE, Brazil

[2]Universidade Federal do Paraná, Brazil

[3]Universidade Federal da Integração Latino Americana - UNILA, Brazil

ABSTRACT

A high number of instruments that assess various quality characteristics of interest that have an inherent variability monitors hydroelectric plants. The readings of these instruments generate time series of data on many occasions have correlation. Each project of a dam plant has characteristics that make it unique. Faced with the need to establish statistical control limits for the instrumentation data, this article makes an approach to multivariate statistical analysis and proposes a model that uses principal components control charts and statistical T^2 and Q to explain variability and establish a method of monitoring to control future observations. An application for section E of the Itaipu hydroelectric plant is performed to validate the model. The results show that the method used is appropriate and can help identify the type of outliers, reducing false alarms and reveal instruments that have higher contribution to the variability.

INTRODUCTION

The control chart is one of statistical quality control techniques most used and can be very useful in controlling the instrumentation of a dam. The facility of online supplying of data with high frequency provides a huge mass of data that generates a series of control charts. However, it is necessary to interpret these data to produce additional knowledge beyond simple time series data.

Citation: Emerson Lazzarotto, Liliana Madalena Gramani, Anselmo Chaves Neto, Luiz Albino Teixeira Junior. Principal components in multivariate control charts applied to data instrumentation of dams. Independent journal of management & production (IJM&P), v. 7, n. 1, January - March 2016. DOI: 10.14807/ijmp.v7i1.369

Hotelling (1947) introduced multivariate control charts at the time of the Second World War and, from the development of computers that enabled its implementation, had great development. Its use has spread because of the need for quality control of several variables not is adequately treated with univariate tools such as Shewhart control charts, especially when there is correlation between variables. Once the variables do not behave independently of one another, they should be considered together and not separately (MASON; YOUNG, 2002).

A typical control chart plots the averages of measurements of a quality characteristic as a function of time. The chart has a center line that represents the target of the quality characteristic value if there were no variability and the other two lines, the upper and lower control limits that are determined statistically. Such control charts are called Shewhart control charts (MONTGOMERY, 2013).

There are several techniques to address the multivariate control type Shewhart T^2 charts, among them the charts, cumulative sums (CUSUM), exponentially weighted moving average (EWMA), principal component analysis

(PCA), partial least squares (PLS), and non-parametric or distribution free techniques (BERSIMIS; PSARAKIS; PANARETOS, 2007). The PCA can be extremely useful in quality control applications, transforming a set of correlated variables into a new set of uncorrelated variables that can be easier to be monitored via control charts (JACKSON, 1991).

The monitoring of structural behavior of hydroelectric dams because the economic, social and environmental importance of plants requires that dams are monitored by means of tools for assessing the safety and ensure the operation of power generation. Usually a large dam has hundreds or even thousands of instruments that measure various quality characteristics of the dam. Evaluate the results of measurements instrumentation of a large number of instruments individually may be impractical or even lead to high false alarms and discredit the monitoring system.

It should be given continuous attention to operation and maintenance of the safety of a dam power plant. The tool for this is the instrumentation of the dam. Due to the particularities of each hydropower plant, there are no universal procedures applicable to all dams in respect to the evaluation of the instrumentation. The evaluation and judgment of information available for an experienced engineering team is the best way to contribute to a decision and choose the best action to be implemented (USACE, United States Army Corps of Engineers, 1995).

The main sources of variability of the readings of instruments for monitoring dams are attributed to temperature, reservoir level and aging (ROSSO et al., 1995), (ITAIPU BINACIONAL, 1999; CHENG; ZHENG, 2013; NEDUSHAN, 2002).Two models were created to relate the effects of instrument readings and environmental variables in a dam of China. The methods showed ability to reduce the rate of false alarms and detect defective instruments (CHENG; ZHENG, 2013).

Diagnosed singular values in monitoring dam safety, with a case study on the hydroelectric plant in China, via multivariate analysis of principal components and graphic control Hotelling T^2(GU et al., 2011). In a hydroelectric power in China it was applied a model that extracts principal components of data instrumentation and establish a seasonal hydrostatic in time model between the variables reservoir level, temperature and time effects and principal components (YU et al., 2010).

Techniques for detection of structural damage are proposed using nonlinear principal component analysis and auto-associative neural network method to data monitoring in an arch dam caused by varying environmental conditions through the statistical analysis threshold level for early warning on dam static deformation can be determined (LOH; CHEN; HSU, 2011). The PCA, Q and T^2 statistics are used to detect and distinguish damages in steel plates and turbine blades plane where vibrations caused are measured by sensors attached to the surface (MUJICA et al., 2011).

In the dam instrumentation, control values are those closer to the physical reality, considering the thermal environmental influences that act on structures and more realistic rheological models of concrete and foundation. Control values are used to monitor the performance of the structures by means of readings taken and are warning signs for abnormal situations of structural behavior. Values that were determined earlier in the design phase and the filling of the reservoir are not, in many cases, more applicable during the operation phase (ITAIPU BINACIONAL, 1999). It is necessary to establish operational control values for the instrumentation of hydroelectric plants.

This article aims to propose a multivariate statistical model for monitoring instruments for monitoring dams via control charts and principal components analysis and seeks to separate the effect of environmental variables on the reading of instruments from other sources of variability by use of statistics T^2 and Q establish control values for monitoring future observations. The method is evaluated in a case study applied to real data from monitoring a dam hydroelectric plant.

The article presents in section 2 the theoretical basis of knowledge of control charts and statistics to be used in this work and the place where the case study was applied and its significance. The section 3 describes the data and methods used in the developed model. Section 4 presents and discusses the results of a case study. In the section 5 some conclusions and considerations are presented.

MULTIVARIATE CONTROL CHARTS

T^2 Control Charts

The field of multivariate analysis consists of statistical techniques that consider two or more random variables related to a single entity in an attempt to produce an overall result that takes into account the relationship between the variables (JACKSON, 1991).The multivariate process control is a methodology based on control charts used to monitor the stability of a multivariate process. Stability is achieved when one or more parameters of interest remain stable on samples (MASON; YOUNG, 2002).

One of the first studies to examine correlated variables from the perspective of statistical control using multivariate procedures for military purposes was given by Hotelling (HOTELLING, 1947). This control procedure was based on a statistic that generalizes the Student t statistic which later received the name of Hotelling T^2.

The application of univariate control charts can lead to erroneous and misleading interpretations and that multivariate methods are a good alternative (MONTGOMERY, 2013; JOHNSON; WICHERN, 2007). If the variables are correlated increases the probability of emission of false alarms and not to receive an alert when the multivariate process is out of control (RYAN, 2011).

Let m be samples of p variables, extracted for evaluation, of a process under control, of size n=1, represented by $\underline{X}_1, \underline{X}_2, \dots, \underline{X}_m$, where $\underline{X}_i' = (X_{i1}, X_{i2}, \dots, X_{ip})$, with i=1,2…m, taken from a -variate normal distribution, with mean and covariance matrix Σ (MASON; YOUNG, 2002; MONTGOMERY, 2013).

At almost always Σ and $\underline{u}$ are unknown and in the practice are estimated, respectively, by unbiased estimators $\underline{\bar{X}}$, the sample mean and the positive definite matrix $s = (s_{tf})$, where

$$\underline{X} = \frac{1}{m}\sum_{t=1}^{m} X_t \quad (1)$$

$$S = \frac{1}{m-1}\sum_{t=1}^{m}(\underline{X}_t - \underline{\bar{X}})(\underline{X}_t - \underline{\bar{X}})' \quad (2)$$

and s_{ij} is the covariance between variables I and j. Multivariate statistical generalization of t^2 is the statistic called the Hotteling T^2 or, only, T^2, measuring the distance from an observation vector $\underline{X}$ to the mean vector $\underline{\bar{X}}$ weighted by the covariance matrix s and is given by

$$T^2 = (\underline{X} - \underline{\bar{X}})'[S]^{-1}(\underline{X} - \underline{\bar{X}}). \quad (3)$$

It is pointed to the existence of two phases in statistical quality control process. In the phase I (retrospective), the control limits are established and tested to the data available. It should be under levels considered statically under control. In the second phase (perspective) control limits are established from the same preliminary data and are used to monitor future data (MONTGOMERY, 2013; RYAN, 2011). In the case of reading instrumentation monitoring of dams, which will be of interest in this work, the value of T^2 the statistic evaluated in $\underline{X}$ in the phase I by (Montgomery, 2013), (JOHNSON; WICHERN, 2007; RYAN, 2011), where a vector observation $\underline{X}$ is not independent of the estimators $\underline{\bar{X}}$ and s, is given by equation (3). The upper control limit (UCL)in the case of n=1, is recommended to be calculated based on a beta distribution (MASON; YOUNG, 2002), in this case

$$UCL^{I}_{T^2} = \frac{(m-1)^2}{m}\beta_{\frac{p}{2},\frac{m-p-1}{2},1-\alpha} \quad (4)$$

Where $\beta_{\frac{p}{2},\frac{m-p-1}{2},1-\alpha}$ represents the 1=α quantile of the distribution β with $\frac{p}{2}$ and $\frac{m-p-1}{2}$ degrees of freedom. The upper control limit (UCL) of phase II, when the parameters are estimated in a previous sample and a vector observation $\underline{X}$ is independent of the estimators $\underline{\bar{X}}$ and , is given by

$$UCL^{II}_{T^2} = \left[\frac{p(m \mid 1)(m \quad 1)}{m(m-p)}\right]F_{p,m-p,1-\alpha} \quad (5)$$

where $F_{p,m-p,1-\alpha}$ represents 1-α the quantile of the F distribution with p and m-p degrees of freedom. The p dimensional ellipsoid 100(1- α)% prediction of a future observation is given by all vectors $\underline{X}$ satisfying

$$(\underline{X} - \underline{\bar{X}})'[S]^{-1}(\underline{X} - \underline{\bar{X}}) \leq UCL^{II}_{T^2} \quad (6)$$

where m is the number of samples (time).

An important issue in the treatment process with individual observations is the way of estimating the covariance matrix. The usual estimator is given by equation (2), however, there are various ways to estimate the covariance matrix, for example, the covariance matrix estimated by successive differences, is given by

$$S_D = \frac{1}{2(m-1)}\sum_{t=2}^{m}(\underline{X}_t - \underline{X}_{t-1})(\underline{X}_t - \underline{X}_{t-1})'. \quad (7)$$

This matrix was proposed by (HOLMES; MERGEN, 1993) and (CHOU; MASON; YOUNG, 1999) made a comparison between five types of covariance matrix estimates and showed that the common estimator is preferred for outliers detection.

Principal Components Analysis (PCA)

When the number of variables to be analyzed increases, the parameter that evaluates the average number of samples required to detect changes in the process is deteriorated, in this way, if suspected that the process variability is not equally distributed among all variables, it is useful to use other methods (MONTGOMERY, 2013). Techniques 'reduction' data are based on the principle of creating sets of latent variables that capture the significant variation 'hidden' in the data. The change that the sets of latent variables extract of the process variables is of fundamental importance for the evaluation of product quality, process security and, more generally, if the process is in statistical control (KRUGER; XIE, 2012). A method which can extract features in the data can be useful in dam safety study. As the instrumentation readings are a result of the combination of several factors, methods of multivariate data analysis can provide the following advantages: 1) more profitable by reducing the number of individual analysis, 2) greater ability to explain and separate the variability due to one because of random variability attributable since the random variabilities are, by definition, uncorrelated from one instrument to another and 3) to identify patterns of behavior (NEDUSHAN, 2002). The PCA is a multivariate data analytical technique in which a number of related variables are transformed into a set of uncorrelated variables that are linear combinations of the original variables, where it is expected to explain the variability between variables with a smaller number of variables (JACKSON, 1991). Its industrial application has contributed to the multivariate statistical process control, since only a few of multivariate control charts can serve as an index of process performance. PCA improves the early detection of failures in relation to the univariate graphs (KOURTI, 2005). The eigenvectors of the covariance matrix form the columns of the orthogonal matrix U the spectral decomposition of S, so that

$$U'SU = L \quad (8)$$

where L is a diagonal matrix of eigenvalues representing the variance of each principal component (JACKSON, 1991). Therefore, one can transform p correlated variables $x=(x_1,x_2,\dots x_p)$ in p new uncorrelated variables $z=(z_1,z_2,\dots z_p)$ through the transformation

$$z = U'x. \quad (9)$$

Is also true that

$$x = Uz \quad (10)$$

However, when using a number k<p principal component, then takes the sub-matrix of order p×k of U and the sub-vector of order k×1 in product Uz, yielding an approximation to x will be represented by $\hat{x}$. There is no consensus in the literature regarding the amount and the criteria for determining the number of components to be retained. A series of criteria for choice is presented in Jackson (1991), in this work the choice was based on the percentage of variance explained and the ability to detect out of the limit values as compared with the T^2 control chart. Here, this choice is relativized because the not retained components will also be evaluated on the Q statistic. The fact that the PCA produce independent variables have the advantage of making it possible to compare the false alarm rate of statistical control procedures of multivariate quality with univariate procedures such as Shewhart charts, because, according Montgomery (2013), the true probability type I error, if the variables are independent to the set control procedure is $\alpha' = 1 - (1 - \alpha)^p$, where p is the number of variables and there is no closed formula otherwise.

Regression Analysis for Missing Data

When working with large databases is relatively common not to have all the desired data. There are several reasons for this fact. In the case of automatic data acquisition, electronic problems could cause the loss or unreliability in receiving the information. In the case of non-automated acquisition, there may be several forms of human errors that cause no part of the data to be evaluated. In multivariate case, it is suggested that missing data for a variable are obtained by using a regression procedure, e.g. linear, in which the variable with missing data is regressed on the other variables. A model of least squares can be used for the parameters of the variables in the linear prediction model (MASON; YOUNG, 2002). Another way to treat this problem is given by the regression and time series forecasting established by well-known methodology of Box & Jenkins. The ARIMA (auto regressive integrated moving average) models can be used to model time series data that has, making future predictions and cover missing data (BOX; JENKINS; REINSEL, 2008). In general ARMA models (p,q) has the form

$$z_t = \phi_1 z_{t-1} + \phi_2 z_{t-2} + \cdots + \phi_p z_{t-p} + a_t - \theta_1 a_{t-1} - \theta_2 a_{t-2} - \cdots - \theta_q a_{t-q}. \quad (11)$$

A measure of the suitability of a model to a time series is given by measuring the mean square error (MSE), given by

$$MSE = \frac{\sum_{t=1}^{n}(y_t - \hat{y}_t)^2}{n - r - 1} \quad (12)$$

where y_i is the observed value $\hat{y}_i$, is the predicted value, n is the number of observations and r-p+q+d is the number of parameters of the ARIMA model or the number of independent variables used in the linear regression model.

The Q statistic

When is formulated a model of the principal components in which the projection data has been standardized, an observation, consisting of a vector x of p variables, can be write as $x = Uz$, if not taken all principal components, has only an approximation $\hat{x} = \hat{U}\hat{z}$, so that

$$x = \hat{U}\hat{z} + (x - \hat{x}). \quad (13)$$

The first term on the right side of the equation (13) represents the contribution of k(k<p) principal component and the second term the amount that is not explained by the principal component model, the residual. There are two types of outliers associated with each of these terms, the first term of an outlier that would be detected even if the principal component model was not applied and the second term indicates that an outlier any observation vector that cannot be adequately characterized the subset of principal components chosen (JACKSON, 1991). The statistic is defined by

$$Q = (x - \hat{x})'(x - \hat{x}). \quad (14)$$

Q is also sometimes known by SPE (squared prediction error). A preview of the vector $(x - \hat{x})$ in the space of principal components in relation to the space of components retained is seen in the Figure 1.This statistic represents the changes that are not explained by the principal component model. It is a measure of the difference between the sample mean and its projection on the PC model (MUJICA et al., 2011).

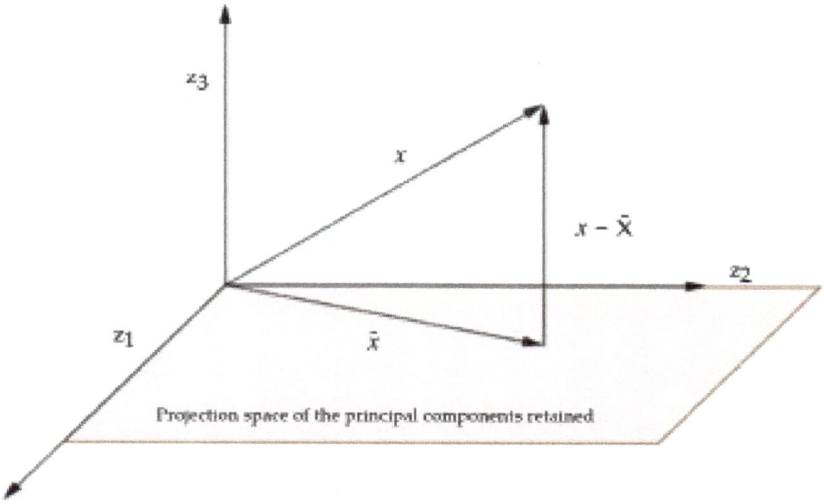

Figure 1: Geometric interpretation of the residual vector in the space of principal components.

The upper control limit of Q, denoted by Q_α, according Mudholkar and Jackson (1979) is

$$Q_\alpha = \theta_1\left[\frac{z_\alpha\sqrt{2\theta_2 h_0^2}}{\theta_1} + \frac{\theta_2 h_0(h_0 - 1)}{\theta_1^2} + 1\right]^{1/h_0} \tag{15}$$

where z_α is the value that corresponds to 100(1-α)percentile of the standard normal distribution, α is the probability of type I error (will be fixed at 0.0027 here) and

$$\theta_t = \sum_{j=k+1}^{p} l_j^t \tag{16}$$

for t=1,2,3,l_i is the ith eigenvalue of the covariance matrix and

$$h_0 = 1 - \frac{2\theta_1\theta_3}{3\theta_2^2}. \tag{17}$$

The Itaipu dam and the importance of instrumentation Recently, in May 2014, the Binational Itaipu Dam, located on the Paraná River between Brazil and Paraguay, completed 30 years since the beginning of the generation and the world leader in production of clean and renewable energy. Itaipu is responsible for producing 17% of electricity in the Brazilian market and 75% of the Paraguayan market (JORNAL INTERNO DE ITAIPU, 2014). In 30 years of operation, Itaipu has generated 2.16 billion MWh, enough to supply the entire world for 38days and Brazil for almost five years energy.

Itaipu is currently the second largest dam in the world in installed capacity, 14,000MW 20 in generating units, overcome by Three Gorges Dam in China which has installed capacity of 22,400 MW. Nevertheless, Itaipu still exceeded Three Gorges in annual energy production until 2013 (JORNAL INTERNO DE ITAIPU, 2014; GAZETA DO POVO, 2014; GLOBO, 2014). To maintain these important economic numbers from the point view of environmental issues and the impact on the entire community that is close not only to the dam, but also to the entire area of the reservoir, it is necessary to ensure that the behavior of the structures is under control. In countries where hydropower potential has been exploited extensively, this development started over 50 years ago and many dams are aging and it is necessary tc maintain the security conditions which may prevent costly repairs and that are often conflicting with the production of energy (XXIII INTERNATIONAL COMITEE OF LARGE DAMS, 2009). The improvement of dams not only ensures the safe operation and performance, but also improves its efficiency (XXIII INTERNATIONAL COMITEE OF LARGE DAMS, 2009). The dams should have adequate instrumentation for monitoring their performance. The goal is to control the operation of the dam safely under any condition. The instrumentation shall allow the measurement of the structural behavior and the physical condition of the dam (USACE, UNITED STATES ARMY CORPS OF ENGINEERS, 1995).

DATA AND METHODS

This work was developed with real data from the instrumentation section E the Itaipu hydroelectric plant, shown in Figure 2. Faced with various instruments present, the piezometers were chosen because of the amount and the importance of their role in measuring uplift pressures in the dam. The E section has seven piezometers. As the frequency of readings from the period of reservoir filling in 1984 varied, was chosen to analyze the period from 2001 until 2013 when the readings were approximately biweekly for all piezometers. This period generated 319 readings for each instrument from now on called piezometers $P_1, P_2, P_3, \ldots P_7$.

The reading of each piezometer is an independent and identically distributed random variable. For phase I, the adjustment test (retrospective) of the model, were selected 300 readings and for Phase II validation (perspective) of the model, we selected the remaining 19 readings. It is important to mention that in this phase I data are considered under the control of a statistical point of view. The instrument P_2 had some missing data during the phase I. In this way, the linear and by time series regressions were applied and the MSE equation (12) was used to choose the best model for filling the missing data.

Figure 2: Dam of Itaipu and section E featured.

Variables that are in very different scales and domain variation should be standardized (JOHNSON; WICHERN, 2007). The reason for this procedure is that the original variables can have scales and domain quite distinct giving false interpretation of its real magnitude variability, which is avoided by standardizing (MONTGOMERY, 2013).

The steps of the method consist of:

- Gathering and data standardization;
- Prediction of missing values;
- Test the multivariate normality;
- Choosing the Type I error probability and the estimate of the covariance matrix;
- Construct the graphic T^2 for the full set of variables;
- Extraction and selection of the number of principal components and normality test;
- Construction of the graph T^2 and the ellipse control of the principal components in phases I and II;
- Calculation of residual and Q statistics;
- Construction of Q-chart;
- Interpretation of results.

RESULTS

Among the forecast models for missing data to P_2 was chosen a model ARIMA (6,0,6) with MSE-0.065, the lowest compared to other ARIMA regression models and linear regression on the other variables to adjust 24 missing data between observations 234-257 adjusted under 233 the preceding. To evaluate the hypothesis of multivariate normality was used a test of adjustment X_p^2 described in Mingoti (2005) and Johnson and Wichern (2007). For each vector $X' = (X_1, X_2, \ldots, X_7)$ containing the standard readings of piezometers was calculated

$$(X-\bar{X})'S^{-1}(X-\bar{X}) \leq \chi^2_{7;95\%} \quad (18)$$

Where $\chi^2_{7;95\%}$ corresponds to the 95% percentile of the distribution X_7^2. Once exactly 95% of the first phase of the sample satisfied the condition, we can accept the hypothesis that the data come from a multivariate normal distribution at 95% confidence level.

For comparison, univariate charts Shewhart control of the sample mean were built for phase I, $UCL - \bar{x} + 3\hat{\sigma}$, $LCL - \bar{x} - 3\hat{\sigma}$ e $\hat{\sigma}$ is, according Montgomery (2013) an unbiased estimator for σ, given by $\hat{\sigma} = \bar{R}/d_2$, where $\bar{R}$ is the average amplitude and ,d_2=1.128 so that the false alarm rate is α=0.27%. The Table 1 shows the number of observations outside the control limits (OCL) to univariate Shewhart charts respective of P_1 to P_7 instruments to 300 data of phase I. Note the impossibility of univariate treatment with the high number of observations OCL. To select the estimator covariance matrix to be used in T^2 chart for the data on the seven standard piezometers were tested usual covariance matrix (2) and the matrix successive differences (7). The Table 2 shows the number of observations outside the control limit (OCL) for T^2 chart for each type of tested covariance matrix for a type I error probability fixed at α=0.0027 for every variable that corresponds the limits 30 of Shewhart charts, then, according Montgomery (2013), the real probability of type I error if the variables are independent, for all control procedure is $\alpha' = 1 - (1 - \alpha)^7 = 0.0187$. It should be noted, of course, the original variables are not independent.

Nevertheless the multivariate treatment becomes feasible, since the probability that 8 or more observations in the universe of 300 are at random above $LSC^{I}_{T^2}$ if the data originates a multivariate normal distribution is 0.2026 for $\alpha' = 0.0187$ it cannot reject the hypothesis that the process is in statistical control at 95% confidence. Because the matrix (7) is more sensitive to small deviations from the mean, for the purpose of this study, the matrix (2) was selected.

Table 1: Observations outside the control limits (OCL) for univariate charts

Inst.	*OCL*	Inst.	*OCL*	Inst.	*OCL*	Inst.	*OCL*	Inst.	*OCL*	Inst.	*OCL*	Inst.	*OCL*
P_1	239	P_2	178	P_3	203	P_4	82	P_5	149	P_6	190	P_7	166

Table 2: Number of observations (OCL) to each covariance matrix for the T^2 chart

Matrix	*OCL*	**Observations**	*UCL*
Usual	8	25,68,249,250,264,268,269,270	16,53
Successive Differences	300	**All**	

Extracting the principal components of the data set of phase I, a simulation was performed by selecting k=2,3,4,5,6 principal components that explain a percentage of variability as shown in Table 3.

Table 3: Variability explained by principal components

Component	Eigenvalue	Variability (%)	
		Explain	Accumulated
1	3.55585	50.80	50.80
2	1.45125	20.73	71.53
3	1.02258	14.61	86.14
4	0.50179	7.17	93.31
5	0.26410	3.77	97.08
6	0.13599	1.94	99.02
7	0.06845	0.98	100.00

The eigenvectors of the matrix U of the spectral decomposition of S are shown in Table 4. Data from scores of k principal components were retained and multivariate normality was tested by the test X^2_k at a confidence level of 95% that obtained the acceptance of the hypothesis normality, according second column of Table 5.

Table 4: Eigenvectors of the sample covariance matrix S

0.1029	0.6931	0.3506	0.1859	0.0651	0.3719	0.0839
0.3485	−0.4289	0.1239	−0.7007	−0.3266	0.2820	0.0454
0.4481	0.3431	0.1700	0.1886	0.1189	0.5824	−0.5135
0.1458	−0.4160	0.7757	0.2213	0.3095	−0.2426	−0.0161
0.4588	0.1106	0.0659	0.4222	−0.6996	−0.2263	0.2324
0.4156	0.1771	0.4145	0.0887	0.1807	0.5500	0.4006
0.4867	0.0274	−0.2342	0.0477	0.5036	0.1391	0.6574

Table 5: T^2 chart results of k PC's obtained from the retention k PC's

k	%	$\alpha'(\%)$	Phase I				Phase II			
			$UCL^I_{T^2}$	$OCL^I_{T^2}$	OBS.	Prob.	$UCL^{II}_{T^2}$	$OCL^{II}_{T^2}$	OBS	Prob.
2	96.67	0.539	10.30	0	–	100%	10.44	0	–	100%
3	96.00	0.808	11.54	5	249,250,263, 268,269	9.81%	11.81	7	249,250,268,269, 317,318,319	1.6%
4	96.00	1.07	12.91	5	249,250,263, 268,269	22.58%	13.12	7	249,250,263,268, 269,318,319	6.01%
5	95.67	1.34	14.15	7	68,249,250, 263,258,269,270	11.13%	14.37	8	58,249,252,268, 269,270,318,319	6.78%
6	94.67	1.61	15.35	7	68,249,250, 264,258,269,270	21.21%	15.60	8	58,249,252,264, 268,269,270,318	14.63%

For example, the upper control limit for statistical T_2^2 (scores of PC's) in phase I was calculated from equation (4) and resulted in UCL_T^2=10.30 and all observations of this phase did not exceed the control limit. For phase II, the upper control limit of statistical T_2^2 calculated from equation (5) resulted UCL_T^2=10.44. For the remaining numbers of components upper control limits are in the fourth and octave column of the Table 5. The fifth and ninth columns of Table 5 contain the number of observations outside the control limits (OCL) for the phase I ($OCL^I_{T^2}$)and for the phase II ($OCL^{II}_{T^2}$) and these observations (OBS) are in the sixth and tenth column for T_k^2 charts in the phases I and II. Finally, seventh, and eleventh columns of Table 5 contain the probability that the amount of OCL obtained is in control statistical at 95%, that is, values less than 5% should reject the hypothesis that the data are under control.

If we use k=2,…,6 of the 7 principal components, then taking the sub-matrix of order 7×k of U and the subvector of order k×1 in the product Uz, yielding an approximation for x, denoted by $\hat{x}$, using the equations (8),(9) and (10). The value obtained for the upper control limit of statistical Q using equations (16), (17) and (15) was Q_α-9.95 when k-2, the others values of Q_α for k-3,…,6 are in the second column of Table 6. The remaining columns of Table 6 show the amount of QCL_E, QCL_U and what are these observations.

Table 6: Results for Q chart of the p-k principal components.

k	Q chart of the $p-k$ PC's				
	Q_α	OCL_I	OBS.	OCL_{II}	OBS.
2	9.95	4	249,250,263,269	4	249,250,263,269
3	4.41	2	25,264	2	25,264
4	2.09	3	25,140,249	3	25,140,249
5	0.96	1	25,138	2	25,138,310
6	0.39	4	25,51,135,170	4	25,51,135,170

The control chart of statistical T_4^2 including the data of the two phases is shown in the Figure 3 and the Figure 4. displays the control of the ellipse 95% confidence data T_4^2 for the first two principal components, the point in red represents the single observation out of control, which is outside the ellipse control we take the components 2 and 3. These plots were constructed for α=0.0027 and the usual covariance matrix, equation (2), at 95% confidence level.

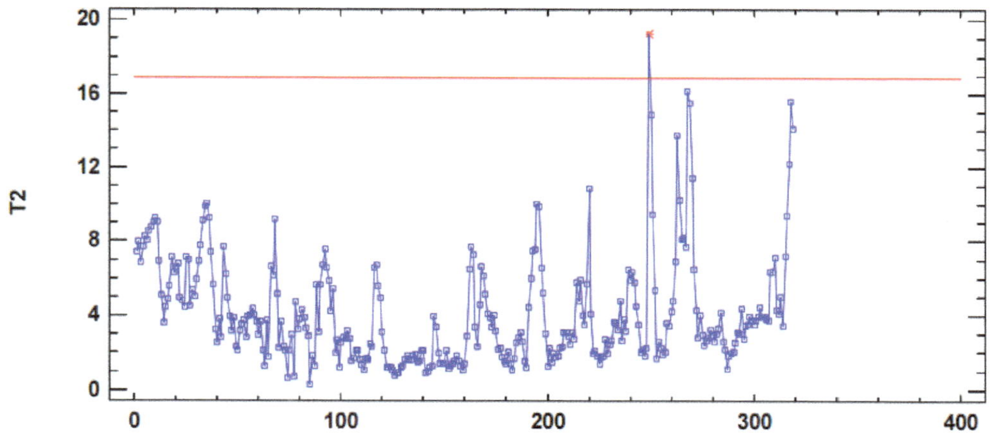

Figure 3: T^2 chart for 4 principal components retained in phases I and II.

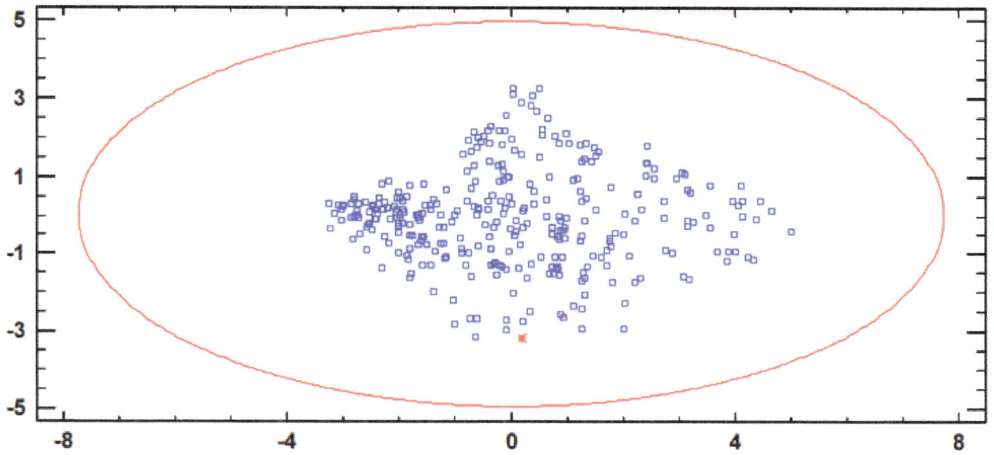

Figure 4: Ellipse control scores the first 2 components of 4 PC's retained.

The Figure 5 shows the behavior of the statistic Q in the period of analysis, for k=4 calculated by equation (14).

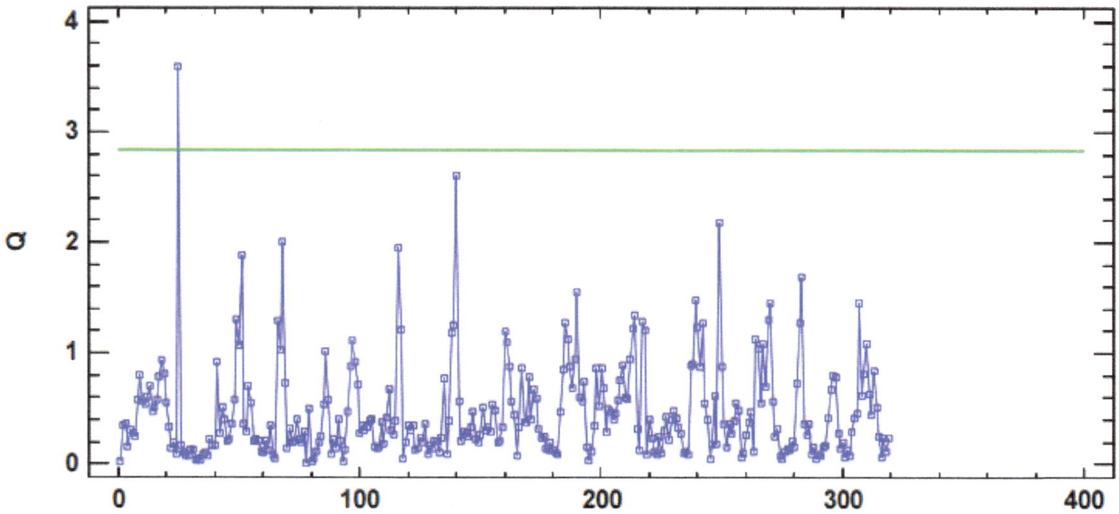

Figure 5: Values of the Q statistic for k=4 principal components in the phases I and II.

CONCLUSIONS

This paper attempted to establish a method for dealing with control charts for dam monitoring instruments. In practice, given the large number of instruments in a large dam and the correlation between them, the individual monitoring of each instrument can be unfeasible, either by excessive graphics to analyze or the large number of false alarms that can discredit the system. The proposed method involves multivariate analyzes and summarizes the analysis of a set of instruments in the statistics T^2 and Q combined with PCA for explaining, respectively, the inherent variability (assignable causes) and random sources in the system. The objective was to reduce the work using multivariate analysis, reducing false alarms to statistically under control levels and identify differences in observations outside the control limits for T^2 and Q statistics.

Is worth mentioning that if it he had chosen to analyze seven Shewhart univariate charts for the mean and the ±38 control limits would be obtained between 82 and 239 values out of control (see Table 1) and the problem of establishing the limit values for instruments cannot be treated by this technique. The results show that the principal components model combined with the Q statistic best fit data of phase I when are taken, at least, four principal components, because in this case the observations listed as out of control to the T^2 chart appearing as out of control to the T^2 chart of k principal components or Q chart (compare Table 2, Table 5 and Table 6) and showed that, in the case study applied the multivariate monitoring of piezometers, located in section E of the Itaipu hydroelectric plant, system is in statistical control at 95% confidence, independently of the contracted model, i.e., T^2 chart or principal components combined with Q statistic.

Another benefit of the combined use of these statistics with 4 principal components was that the observations identified as out of control, one can make a distinction between them. Among the observations given as out of control for the Q statistic for 4 PC's (see Table 6), observation 25 is related to the maximum global value of the instrument p_7, observation 140 is associated with the local maximum value in the instrument p_6 (both outliers are apparent) and observation 249 is associated with the global maximum p_4 of which is the instrument with smaller variance.

The adopted model enables an interpretation of the PC's as a consequent variability of environmental factors (inherent) to the model. It is essential to understand that the variability of the principal components is originated from these factors and also control the random variability that may be linked to outliers (Q statistic). The use of principal components has another advantage which is to overcome the problem of singularities. In this case study, for example, the determinant of the covariance of all the original variables matrix was 0.0065. The singularities are associated with the existence of eigenvalues near zero that can generate computational problems in the inversion of the covariance matrix and the consequent calculation of T^2 statistical.

The first four principal components explain more than 90% of the variability. It was observed that the first principal component has an interpretation in terms of an average contribution of each instrument for the overall variability and that this depends on the elevation layer and in which they are. According to the first column of Table 4, it can be seen that the p_7 instrument is has the greatest contribution to this principal component and its location is before a concrete injection curtain and at a lower elevation, i.e., the local theoretically more susceptible to uplift pressures. The second instrument with greater contribution in this principal component p_5 is located in a joint with lower elevation, as shown in Figure 6. Since the instrument that has almost no effect on this component is p_1 and its location is after injection curtain at the top elevation of instruments study. This confirms the efficiency of concrete injecting curtain at the dam.

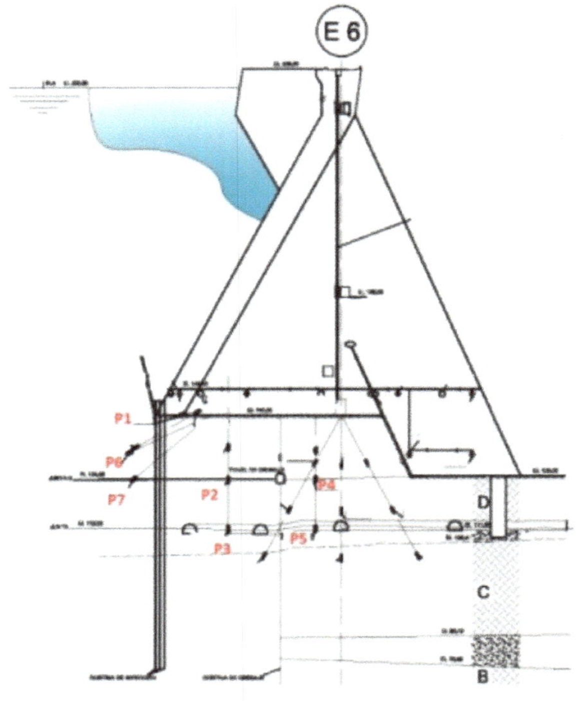

Figure 6: Piezometers in section E of Itaipu Dam

Source: adapted from Itaipu Binational.

Suggestions for future work involving this type of approach may be the use of non-parametric statistics, the variation of the rate of false alarms α and the analysis of other instruments can also enable the discovery of new knowledge, as well as seek interpretation for other principal components.

ACKNOWLEDGEMENTS

The team CEASB - Advanced Center for Studies on Dam Safety, in the accord between Itaipu Binational and the Itaipu Technological Park, for the collaboration and reliance.

REFERENCES

1. BERSIMIS, S.; PSARAKIS, S.; PANARETOS, J. (2007) 'Multivariate statistical process control charts: an overview', Quality and Reliability engineering international, v. 23, p. 517-543, doi:10.1002/qre.829.
2. forecasting and control, 4ª ed., Hoboken, New Jersey: John Wiley & Sons, Inc. CHENG, L.; ZHENG, D. (2013) 'Two online dam safety monitoring models based on the process of extracting environmental effect', Advances in Engineering Software, v. 57,pp. 48-56, doi:10.1016/j.advengsoft.2012.11.015.

3. CHOU, Y. M.; MASON, R. L.; YOUNG, J. C. (1999) 'Power comparisons for a Hotelling's T2 statistic', Commun. Statist. Simulation Comput., v. 28, p. 1031-1050, doi:10.1080/03610919908813591.
4. GAZETA DO POVO (2014) Gazeta do Povo, 08 Maio, [Online], Available: HYPERLINK www.gazetadopovo.com.br/economia/especial-itaipu-30- anos/conteudo.phtml?tl=1&id=1466225&tit=Gigante-pela-propria-natureza [08 Maio 2014].
5. GLOBO (2014) G1 Globo, [Online], Available: HYPERLINK www.g1.globo.com/pr/oeste-sudoeste/noticia/2014/05/seria-inimaginavel-o-paraguaie-o-brasil-sem-itaipu-diz-diretor-geral.html [08 maio 2014].
6. GU, C. S.; ZHAO, E. F.; JIN, Y.; SU, H. Z. (2011) Singular value diagnosis in dam safety monitoring effect values, Sciencie China Technological Sciences, v. 54, n. 5, May, p. 1169-1176, doi 10.1007/s11431-011-4339-7.
7. HOLMES, D. S.; MERGEN, A. E. (1993) Improving the performance of the T2 control chart, Quality Engineering, v. 5, n. 4, March, p. 619-625, DOI:10.1080/08982119308919004.
8. HOTELLING, H. (1947) Multivariate quality control, Techniques of statistical analysis, p. 111-184.
9. ITAIPU BINACIONAL (1999) Estudos para a reavaliação dos valores de controle para a instrumentação civil das estruturas e fundações (2660.50.15402.P), Foz do Iguaçu.
10. ITAIPU BINACIONAL (1999) Estudos para a reavaliação dos valores limites para a instrumentação civil das estruturas e fundações (2662.50.15401-P), Foz do Iguaçu.
11. JACKSON, J. E. (1991) A user's guide to principal components, New York: John Wiley & Sons, Inc.
12. JOHNSON, R. A.; WICHERN, D. W. (2007) Applied Multivariate Statistical Analysis, 6th. ed., Upper Saddle River - NJ: Pearson Prentice Hall.
13. JORNAL INTERNO DE ITAIPU (2014) JIE, 08 maio, [Online], Available: HYPERLINK jie.itaipu.gov.br/index.php?secao=noticias_itaipu&q=node/6&pagina=4 [08 maio 2014].
14. KOURTI, T. (2005) Application of latent variable methods to process control and multivariate statistical process control in industry, International Journal of adaptive control and signal processing, v. 19, p. 213-246, doi: 10.1002/acs.859.
15. KRUGER, U.; XIE, L. (2012) Statistical monitoring of complex multivariate process with applications in industrial process control, Chichester: John Wiley & Sons.
16. LOH, C. H.; CHEN, C. H.; HSU, T. Y. (2011) Application of advanced statistical methods for extracting long-term trends in static monitoring data from an arch dam, Structural Health Monitoring, v. 10, n. 6, November, p. 587-601, doi: 10.1177/1475921710395807.
17. MASON, R. L.; YOUNG, J. C. (2002) Multivariate statistical process control with industrial applications, Philadelphia - Pennsylvania: ASA SIAM - American statistical Association - Society for industrial and applied mathematics.
18. MINGOTI, S. A. (2005) Análise de dados através de métodos de estatística multivariada: uma abordagem aplicada, Belo Horizonte: UFMG.
19. MONTGOMERY, D. C. (2013) Introdução ao Controle Estatístico de Qualidade, 4th edition, Rio de Janeiro: LTC.
20. MUDHOLKAR, G. S.; JACKSON, J. E. (1979) Control procedures for residual associated with principal components analysis., Technometrics, v. 21, n.3, August, p. 341-349.
21. MUJICA, L. E.; RODELLAR, J.; FERNÁNDEZ, A.; GUEMES, A. (2011) Q-statistic and T2 statistic PCA-based measures for damage assessment in structures, Structural Health Monitoring, v. 10, n. 5, September, p. 539-553, doi: 10.1177/1475921710388972.
22. NEDUSHAN, B. A. (2002) Multivariate statistical analysis of monitoring data for concrete dams, Montreal: PhD Thesis, McGill University - Department of Civil Engineering and Apllied Mechanics.
23. ROSSO, J. A.; FIORINI, A. S.; PORTO, E. C.; SILVEIRA, J. F. (1995) Barragem de Itaipu - Lições aprendidas com o comportamento térmico das estruturas tipo gravidade aliviada e contraforte, XXII Seminário Nacional de Grandes Barragens, São Paulo.
24. RYAN, T. P. (2011) Statistical Methods for Quality Improvement, John Wiley & Sons.
25. USACE, UNITED STATES ARMY CORPS OF ENGINEERS (1995) Enginnering and Design - Instrumentation of Embankment Dams and Levees, Washington, DC. XXIII INTERNATIONAL COMITEE OF LARGE DAMS (2009) Hydropower current developments - Question 88, Dams and Hydropower, Brasília, 6-8.
26. XXIII INTERNATIONAL COMITEE OF LARGE DAMS (2009) Question 90, Upgrading of existing dam, Brasília, 7.
27. YU, H.; WU, Z. R.; BAO, T. F.;ZHANG, L. (2010) Multivariate analysis in dam monitoring data with PCA, Sciencie China Technological Sciences, v. 53, n. 4, April, p. 1088-1097, doi: 10.1007/s11431-010-0060-1.

CHAPTER 10

STRUCTURE AND FABRICATION DETAILS OF AN INTEGRATED MODULARIZED MICROFLUIDIC SYSTEM

Qingchang Tian [a,b] , Ying Mu [a], Yanan Xu[a] , Qi Song[a] , Bingwen Yu[a] , Congcong Ma[a] , Wei Jin[a] , Qinhan Jin[a]

[a]Research Center for Analytical Instrumentation, Institute of Cyber Systems and Control, State Key Laboratory of Industrial Control Technology, Zhejiang University, Hangzhou 310058, Zhejiang, PR China

[b]College of Life Sciences, Zhejiang University, Hangzhou 310058, Zhejiang, PR China

ABSTRACT

This article contains schemes, original experimental data and figures for an integrated modularized microfluidic system described in "An integrated microfluidic system for bovine DNA purification and digital PCR detection [1]". In this data article, we described the structure and fabrication of the integrated modularized microfluidic system. This microfluidic system was applied to isolate DNA from ovine tissue lysate and detect the bovine DNA with digital PCR (dPCR). The DNA extraction efficiency of the micro device was compared with the efficiency of benchtop protocol.

Specifications Table

Subject area	Biology
More specific subject area	microfluidic
Type of data	Table,figure
How data was acquired	Amplification plots and standard curve plots of qPCR. Digital PCR fluorescent imagines.
Data format	Raw, analyzed.
Experimental factors	Use of bovine tissue lysate and ovine tissue lysate to make up different weight ratio (WR) (w/w).

Citation: Qingchang Tian, Ying Mu, Yanan Xu, Qi Song, Bingwen Yu, Congcong Ma, Wei Jin, Qinhan Jin, Structure and fabrication details of an integrated modularized microfluidic system, Data in Brief, Volume 5, December 2015, Pages 461-467, ISSN 2352-3409, http://dx.doi.org/10.1016/j.dib.2015.09.036.

Experimental features	Tissue lysate was introduced into microdevice for DNA isolation and different WR tissue lysates were detected by digital PCR.
Data source location	Hangzhou, Zhejiang, China.
Data accessibility	Data in public repository.

Value of the Data

- The structure and fabrication details of an integrated modularized microfluidic system are given.
- Amplification plots and DNA yields resulting from on-chip isolation can be compared with the benchtop procedure.
- Data from qPCR were compared with data from dPCR.

DATA, EXPERIMENTAL DESIGN, MATERIALS AND METHODS

Microdevice Design and Fabrication

Integration is increasingly recognized as an important technical challenge in lab-on-a-chip device. More efficient and lower cost integrated microfluidic systems decrease or eliminate reliance on traditional lab equipment. Chin et al. [2] integrated new procedures for manufacturing, fluid handling and signal detection in microdevice into a single 'mChip' assay to replicate all steps of ELISA. Easley et al. [3] developed an integrated microfluidic genetic analysis system that could extract and purify DNA from crude whole blood sample, carry out PCR-based amplification, following by separation and detection in a manner that allows for microliter samples to be screened for infectious pathogens with sample-in–answer-out results.

We have integrated a NA extraction module and a dPCR reaction module on this chip. As shown in Fig. 1, three washing buffer and sample lysate were preloaded in the Teflon tube one by one with air spacers (Fig. 1c). The NA capture zone was a teardrop-shaped three-way tube (inlet, outlet 1 and outlet 2) (Fig. 1d) with a sizeable magnet. The dPCR module (Fig. 1e) had a dPCR region with 650 reaction chambers (200 μm diameter and 230 μm high) in a square area (15.0 mm×15.0 mm) (Fig. 1f) and a μfilter with annular duct (200 μm in width, 40 μm in depth).

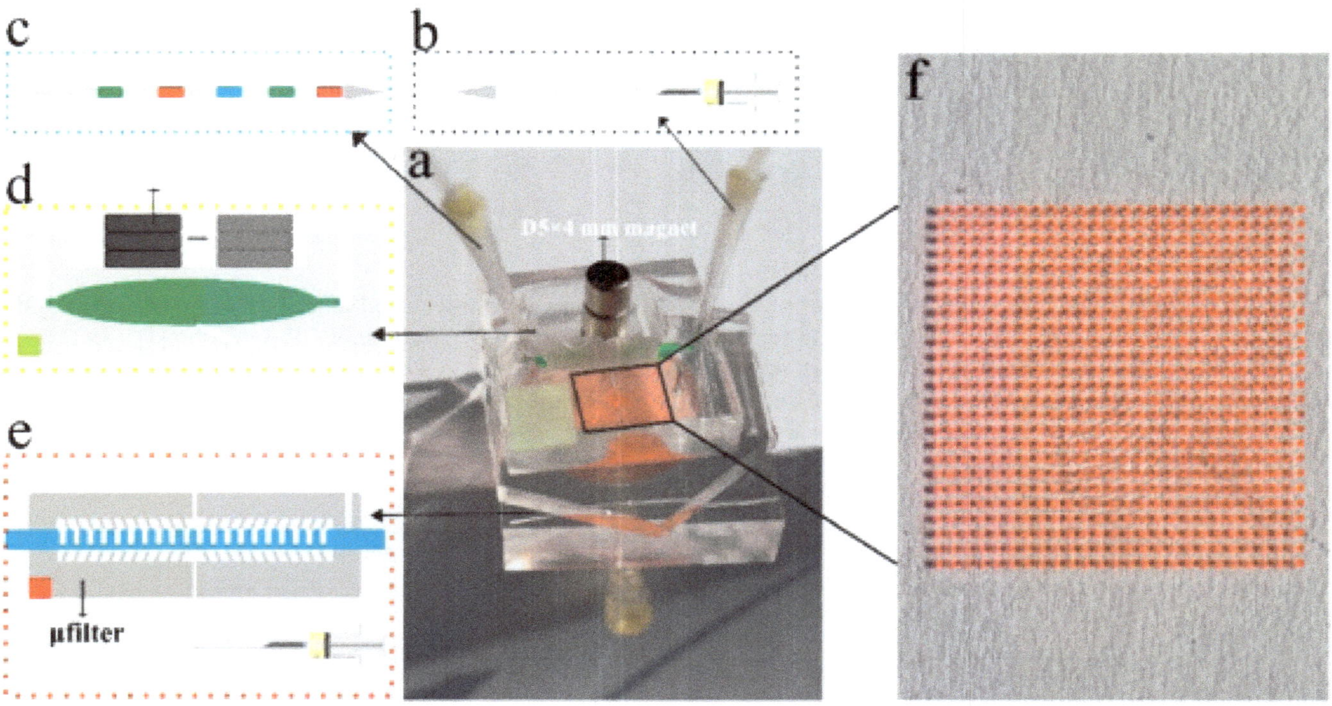

Figure 1: The integrated NA analysis system. (a) The integrated NA analysis system has two distinct functional modules, NA extraction module (green) and dPCR module (red). (b) A Teflon tube connected a syringe to provide negative pressure. (c) A Teflon tube with different reagents (washing buffer and sample lysate) for NA extraction. (d) The NA extraction module with a magnet sliding in a groove. (e) The dPCR module contained a dPCR region and a μfilter with annular duct (tilted «posts» on the μfilter demonstrated the annular duct). (f) The dPCR region with 650 reaction chambers in a square area 15.0 mm×15.0 mm.

For the DNA isolation form tissue, we introduced 70% (v/v) ethanol, 70% (v/v) ethanol, buffer MP3, and lysate orderly by adjusting the knob of pipettor. As shown in Fig. 2, the volume of these reagents and the space of two reagent segments could be controlled by the range that we adjusted the knob of pipettor.

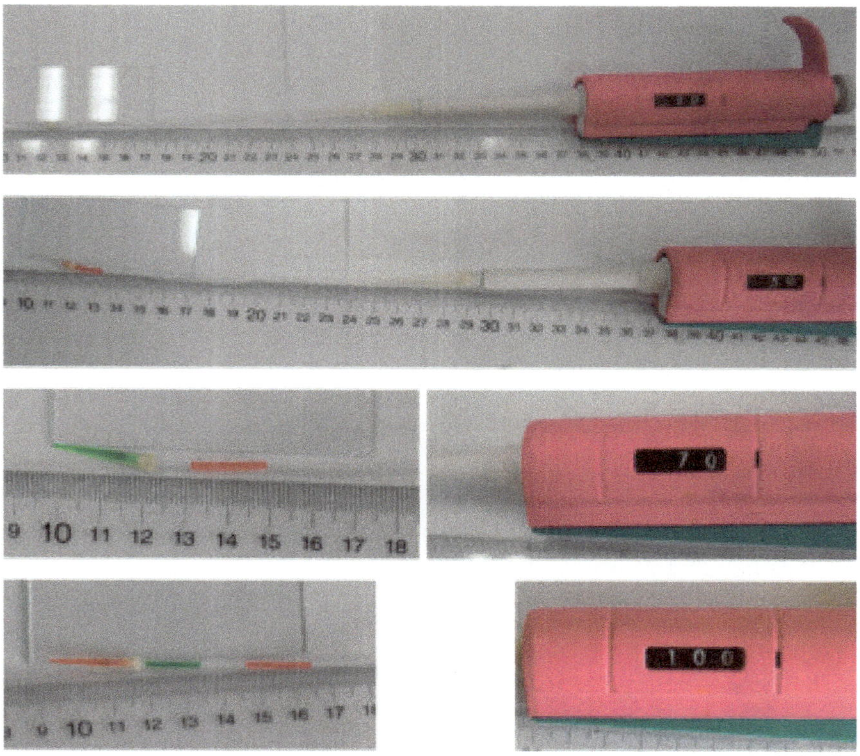

Figure 2: Reagents were loaded into Teflon tube with the help of pipettor.

Fluidic isolation of the NA extraction region and dPCR region was essential for the purposes of avoiding the incompatibility of the NA extraction with the amplification process. Because of the pull force from the negative pressure from the other side of the capture zone, the reagents could not seep into the channel that connected the dPCR module and NA extraction module, in case the reagents disturbed the subsequent dPCR reaction. As shown in Fig. 3, reagents solution were introduced into the NA isolation area from the Teflon tube by the syringe. The Mag-Bind particles were collected in the NA isolation area by the magnet, and the waste liquid was disposed into the syringe but the reagents did not seep into outlet 2. The μfilter provided negative pressure to sample loading and a syringe was used to pull the reagents for NA isolation. Then PCR mix was introduced into the NA isolation area from the Teflon tube, and took the DNA on the Mag-Bind particles into the dPCR part.

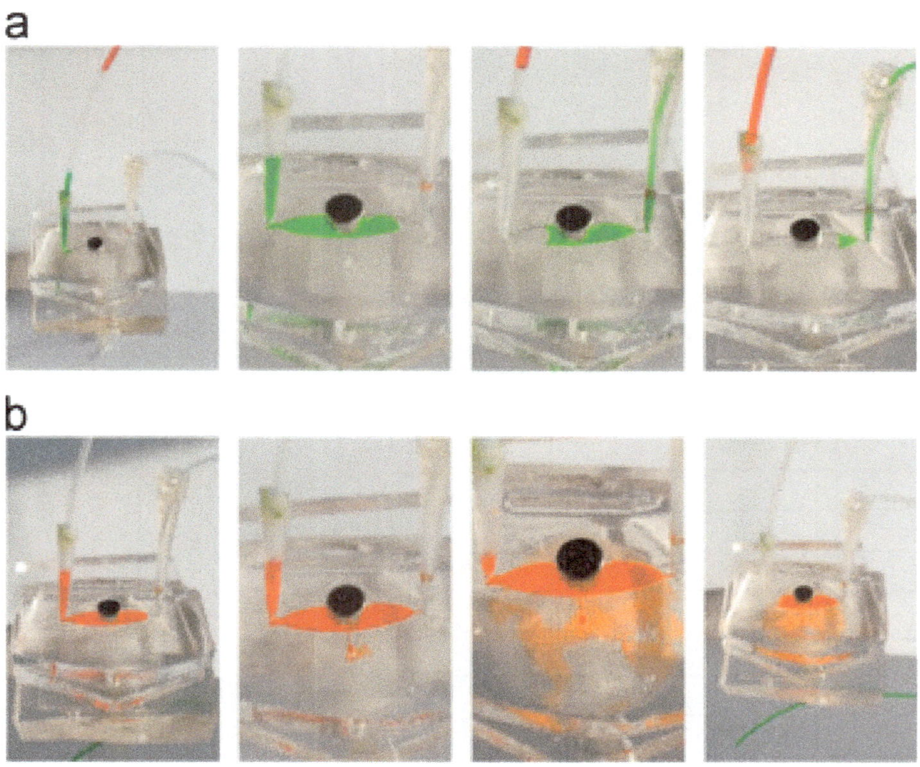

Figure 3: Schematic diagram of the process for flow control. (a) Sample waste and washing buffer (green) were sucked in the syringe avoiding contaminating the dPCR part by outlet 2. (b) PCR mix (red) eluted the DNA on magnetic particles and took them into the chambers under the negative pressure of μfilter.

DNA Extraction

The purification of NA from specimens was laborious and time-consuming until BOOM et al. [4] developed a rapid and simple method. Compared to conventional methods, the microdevices have shown advantages in low reagent and sample consumption, enhanced sensitivity, increased speed, etc. [5], [6] and [7].

When 20 μL lysed sample with the Mag-Bind particles was pulled into the chip and passed through the capture zone by the negative pressure from other side, the Mag-Bind particles with NAs were collected by a magnet. Then the three sections of reagent in Teflon tube were pulled into the capture zone and washed away proteins and/or other contaminants on the Mag-Bind particles. Finally, DNA was eluted by PCR mix and introduced into the dPCR region for subsequent dPCR analysis.

But when the Mag-Bind particles were collected together and washed by 70% (v/v) ethanol, the Mag-Bind particles were clumped together and resisted washing and elution. We introduced a groove for magnet sliding above the NA capture zone to drive the magnetic particles movement around and improve the efficiency of washing and elution. The elution efficiency of PCR mix in a 0.2 mL tube and microdevice was determined by gel electrophoresis (Fig. 4). The genomic DNA and PCR results emerged in the same lane. The flow rates for wash reagents was 10 μL/min. Purified NAs obtained from benchtop and microdevice were quantified using real-time PCR with the Real Time PCR Bovine and Ovine DNA Detection Kit (TAKARA, RR913). Differences on quantified NA between benchtop and microdevice results were evaluated using the Mann–Witney *U* statistical test (SPSS 19.0, P=0.602>0.05) (Table 1). The data showed that there was non-significant differences between on-chip isolation compared to the benchtop procedure in DNA yields (Fig. 5). The bovine meat in different mixture with the ovine meat were detected by qPCR (Fig. 6).

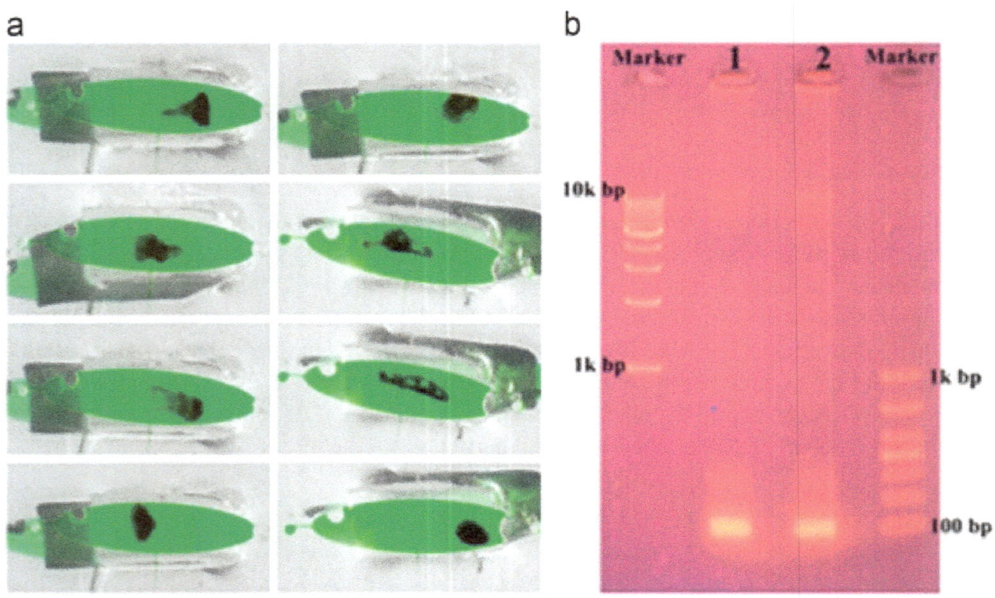

Figure 4: The elution efficiency of PCR mix. (a) The magnetic particles (black) in the capture zone were driven by the magnetic for paddling the reagent. (b) Genomic DNA on magnetic particles was eluted by PCR mix. Lane 1 was the PCR result of the Genomic DNA from 0.2 mL tube, and lane 2 was the PCR result of the Genomic DNA from microdevice (the PCR result was 89 bp).

Table 1: The statistical analysis result of the real-time PCR

Gradients	Samples	DNA concentration (ng/μL)	Average	Standard deviation	Results of Mann–Witney *U*statistical test
WR=10−3	1	0.7351	0.72	0.0423	0.602>0.05
	1	0.6745			
	1	0.7491			
	1	0.7621			
	1	0.6726			
	2	0.8374	0.71	0.127	
	2	0.8599			
	2	0.6016			
	2	0.6384			
	2	0.617			

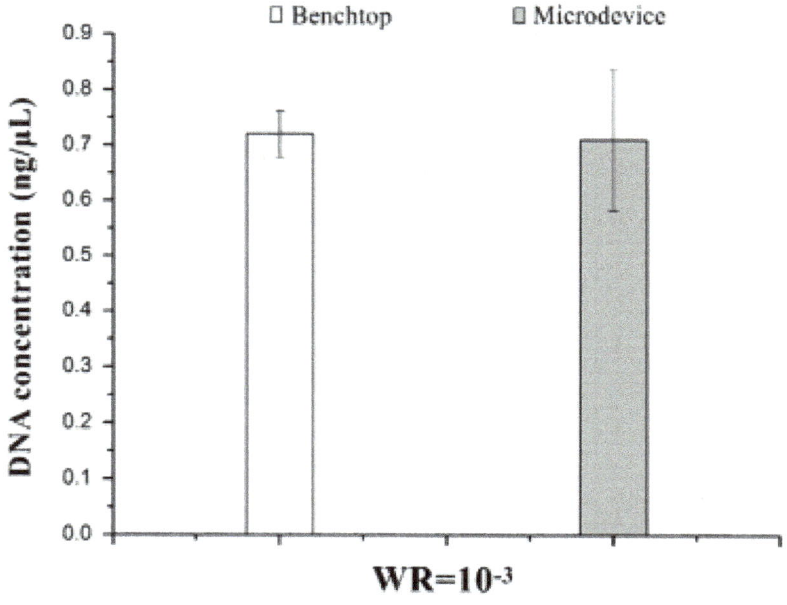

Figure 5: Average concentrations obtained from two gradients with the microdevice compared to the benchtop method. Microdevice and benchtop experiments were performed in triplicate. Error bars depict 95% CI.

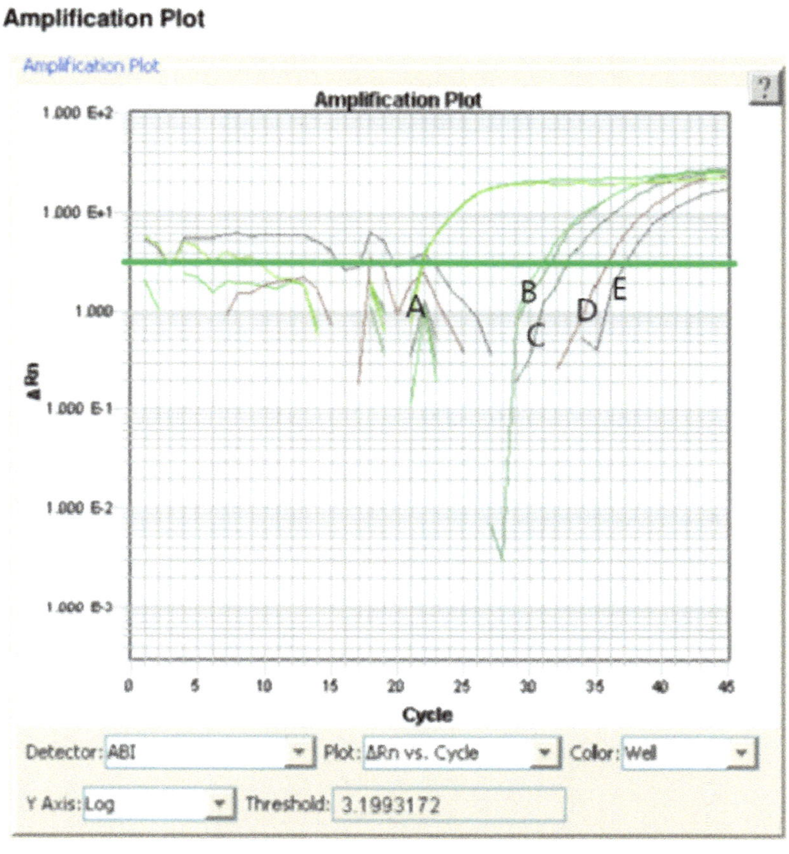

Figure 6: Amplification plot of different WR. A, WR=10^{-3}; B, WR=10^{-7}; C, WR=10^{-8}; D, WR=10^{-9}; E, WR=10^{-10}.

Conflict of Interest Statement

The authors declare that they have no competing financial interests.

ACKNOWLEDGMENTS

This work was supported by the National Natural Science Foundation of China (No.31270907), National Key Foundation for Exploring Scientific Instrument (No.2013YQ470781) and the Autonomous Research Project of the State Key Laboratory of Industrial Control Technology, China (No. 1501).

REFERENCES

1. Q. Tian, Y. Mu, Y. Xu, Q. Song, B. Yu, C. Ma, W. Jin, Q. Jin, An integrated microfluidic system for bovine DNA purification and digital PCR detection, Anal. Biochem. (2015), 10.1016/j.ab.2015.08.030.
2. C.D. Chin, T. Laksanasopin, Y.K. Cheung, D. Steinmiller, V. Linder, H. Parsa, J. Wang, H. Moore, R. Rouse, G. Umviligihozo, E. Karita, L. Mwambarangwe, S.L. Braunstein, J. van de Wijgert, R. Sahabo, J.E. Justman, W. El-Sadr, S.K. Sia, Microfluidicsbased diagnostics of

infectious diseases in the developing world, Nat. Med. 17 (2011) 1015–1019.

3. C.J. Easley, J.M. Karlinsey, J.M. Bienvenue, L.A. Legendre, M.G. Roper, S.H. Feldman, M.A. Hughes, E.L. Hewlett, T.J. Merkel, J.P. Ferrance, J.P. Landers, A fully integrated microfluidic genetic analysis system with sample-in-answer-out capability, Proc. Natl. Acad. Sci. U.S. Am. 103 (2006) 19272–19277.
4. R. Boom, C.J. Sol, M.M. Salimans, C.L. Jansen, P.M. Wertheim-van Dillen, J. van der Noordaa, Rapid and simple method for purification of nucleic acids, J. Clin. Microbiol. 28 (1990) 495–503.
5. D.R. Reyes, D. Iossifidis, P.A. Auroux, A. Manz, Micro total analysis systems. 1. Introduction, theory, and technology, Anal. Chem. 74 (2002) 2623–2636.
6. P.A. Auroux, D. Iossifidis, D.R. Reyes, A. Manz, Micro total analysis systems. 2. Analytical standard operations and applications, Anal. Chem. 74 (2002) 2637–2652.
7. P.S. Dittrich, K. Tachikawa, A. Manz, Micro total analysis systems. Latest advancements and trends, Anal. Chem. 78 (2006) 3887–3908.

CHAPTER 11

A COMPREHENSIVE DATA ACQUISITION AND MANAGEMENT SYSTEM FOR AN ECOSYSTEM-SCALE PEATLAND WARMING AND ELEVATED CO_2 EXPERIMENT

M. B. Krassovski[1] , J. S. Riggs[3] , L. A. Hook[1] , W. R. Nettles[2] , P. J. Hanson[2] , and T. A. Boden[1]

[1]Carbon Dioxide Information Analysis Center, Oak Ridge National Laboratory, Oak Ridge, TN 37831-6290, USA

[2]Climate Change Science Institute, Oak Ridge National Laboratory, Oak Ridge, TN 37831-6290, USA

[3]Integrated Operations Support Division, Oak Ridge National Laboratory, Oak Ridge, TN 37831-6290, USA

ABSTRACT

Ecosystem-scale manipulation experiments represent large science investments that require well-designed data acquisition and management systems to provide reliable, accurate information to project participants and third party users. The SPRUCE project (Spruce and Peatland Responses Under Climatic and Environmental Change, http: //mnspruce.ornl.gov) is such an experiment funded by the Department of Energy's (DOE), Office of Science, Terrestrial Ecosystem Science (TES) Program. The SPRUCE experimental mission is to assess ecosystem-level biological responses of vulnerable, high carbon terrestrial ecosystems to a range of climate warming manipulations and an elevated CO_2 atmosphere. SPRUCE provides a platform for testing mechanisms controlling the vulnerability of organisms, biogeochemical processes, and ecosystems to climatic change (e.g., thresholds for organism decline or mortality, limitations to regeneration, biogeochemical limitations to productivity, and the cycling and release of CO_2 and CH_4 to the atmosphere). The SPRUCE experiment will generate a wide range of continuous and discrete measurements. To successfully manage SPRUCE data collection, achieve SPRUCE science objectives, and support broader climate change research, the research staff has designed a flexible data system using proven network technologies and software components. The primary SPRUCE data system components are the following:

- **data acquisition and control system – set of hardware and software to retrieve biological and engineering data from sensors, collect sensor status information, and distribute feedback to control components;**
- **data collection system – set of hardware and software to deliver data to a central depository for storage and further processing;**

Citation: M. B. Krassovski, J. S. Riggs, L. A. Hook, W. R. Nettles, P. J. Hanson, and T. A. Boden. A comprehensive data acquisition and management system for an ecosystem-scale peatland warming and elevated CO2 experiment. Geosci. Instrum. Method. Data Syst. Discuss., 5, 177–205, 2015. DOI: 10.5194/gid-5-177-2015

- **data management plan – set of plans, policies, and practices to control consistency, protect data integrity, and deliver data.**

This publication presents our approach to meeting the challenges of designing and constructing an efficient data system for managing high volume sources of in situ observations in a remote, harsh environmental location. The approach covers data flow starting from the sensors and ending at the archival/distribution points, discusses types of hardware and software used, examines design considerations that were used to choose them, and describes the data management practices chosen to control and enhance the value of the data.

INTRODUCTION AND BACKGROUND

Proactive management of data collection within observational networks (e.g., Ameri Flux, http://ameriflux.lbl.gov) or large, multiyear manipulations (FACE – Norby et al., 2002; TDE – Hanson et al., 2003) has often been cited as a desirable precursor to post-measurement analyses and model inter comparisons (Hanson et al., 2004, 2008; Walker et al., 2014). This paper describes the details of and rationale for the data acquisition and management methods for a largescale, decade-long manipulation study of peatland ecosys tems in northern Minnesota using whole-ecosystem warming techniques (Hanson et al., 2011).

Overview of the SPRUCE experimental Site

The SPRUCE (Spruce and Peatland Responses Under Climatic and Environmental Change, http://mnspruce.ornl.gov) experimental field site is a Sphagnum–Picea mariana (black spruce) bog within the US Forest Service (USFS), Northern Research Station's, Marcell Experimental Forest (MEF; 47∘30.1710 N, 93∘28.9700 W). The MEF is located approximately 40 km north of Grand Rapids, Minnesota, USA. Within the 1141 ha MEF research area, the SPRUCE manipulation is located in the S1 watershed, at 47∘30.4760 N, 93∘27.1620 W and is 418 m a.m.s.l. (above mean sea level). While remote, the SPRUCE site is accessible year-round via Itasca County and USFS-maintained roads. The S1 experimental site is an 8.1 ha ombrotrophic bog (a raised dome peat bog in which water and nutrient inputs originate primarily from atmospheric sources). Peat depths generally range from 2 to 3 m over the experimental area but are as deep as 9 m in small pockets (Parsekian et al., 2012). The entire bog is underlain by mineral soil. SPRUCE is an ecosystem-level climate change manipulation experiment designed to evaluate the response of existing biological communities to a range of warming treatments from ambient to +9 °C, in large open-top enclosures. Ecosystem warming is achieved by a combination of deep peat heating and air warming. Soil warming is achieved through electrical resistance heating in an array of vertically installed below-ground heaters. Air warming is achieved through forced air moving across a propane-fired heat exchanger. Details on the heating methods can be found in Barbier et al. (2012). To facilitate air warming the experimental plots are enclosed in a 12 m diameter by 8 m high open-top enclosure. The open-top enclosure allows for natural inputs of precipitation (rain and snow) while limiting total air turnover making it possible to generate an envelope of warm air around the enclosed plot of vegetation. The ambient to +9 °C warming treatments will be duplicated at elevated CO_2 (in the range of 800–900 ppm). Both direct and indirect effects of these experimental perturbations will be analyzed to develop and refine Earth system analysis models. Access to experimental plots is provided by three, 8 ft. wide, aluminum-framed, composite-decked boardwalks above the bog surface. Boardwalks are supported by helical piles anchored into mineral soil. The boardwalks provide support for electrical power, propane and CO_2 distribution lines, and data transmission infrastructure. In total, 17 plots were initially established as open to the environment (Fig. 1); each of them containing a 10 m high instrumentation tower with environmental sensors at 0.5, 1, 2 and 4 m above the ground (nominal heights of typical shrub and tree foliage and branches). Ten plots are thermally isolated from the environment by 8 m high sidewalls, and these plots can be heated at ground level and throughout the vertical space of the enclosure. The sidewalls of isolated plots are glazed with greenhouse panels allowing 80 % of the photosynthetically active radiation to reach the vegetation. (Fig. 2). Of the 10 enclosed plots, 2 are operated without heating (control plots), and the remaining 8 plots are equipped with propane-fired heating and ventilation units to deliver duplicate elevated temperature treatments at 2.25, 4.5, 6.75, and 9.0 °C above recorded readings in control plots. Five of the 10 plots, including each of the temperature treatments, are equipped with CO_2 injection equipment to deliver an elevated CO_2 (eCO_2) treatment (500 ppm above ambient). The seven remaining non-enclosure plots are also instrumented and monitored to provide information on undisturbed, background ambient conditions. A below-ground "corral" constructed of linked sheet piling has been added around each of the 10 enclosed plots. The corral will extend from 0.5 m above the surface to the mineral soil layer (~ 3 m) to isolate and monitoring plot hydrology. Electrical power is supplied to each of the transects of plots by a 208 V, 350 A three-phase service. Each plot is then supplied from a main distribution panel and is limited by a 70 A breaker.

Sensors and Instruments

SPRUCE scientists defined the measurements needed to assess the response of the biological communities to warming and elevated CO_2 treatments. To implement the experimental design a wide range of in situ continuous and discrete measurements were identified as necessary for the control of the experimental conditions and the quantification of ecosystem responses to the treatments.

The instruments and infrastructure needed for these measurements include meteorological tower-based CO_2/H_2O sampling and analysis systems, air temperature and relative humidity probes, precipitation gauges, wind speed and direction instruments, solar radiation sensors, subsurface soil moisture and temperature sensors, water level and conductivity sensors, continuous vegetation sap flow and dendrometer sensors, and companion dataloggers to record and report the data. Present plans call for deployment of approximately 1100 sensors and associated instruments. All equipment assets are registered and tracked in the ORNL (Oak Ridge National Laboratory) Facility Service Center (FSC) system.

The FSC tracks project-determined schedules for preventive maintenance and calibration services, notifies the instrument custodian, and captures details of the accomplished tasks. The FSC system tracks and reports equipment details like maintenance histories, replacement histories, calibration records, and preventive maintenance schedules as well. All reports and records from the FSC system can be exported for integration and coupling with the SPRUCE measurement data.

Figure 1: Aerial view of the S1-bog within the Marcell Experimental Forest and the SPRUCE experimental footprint in April 2015.

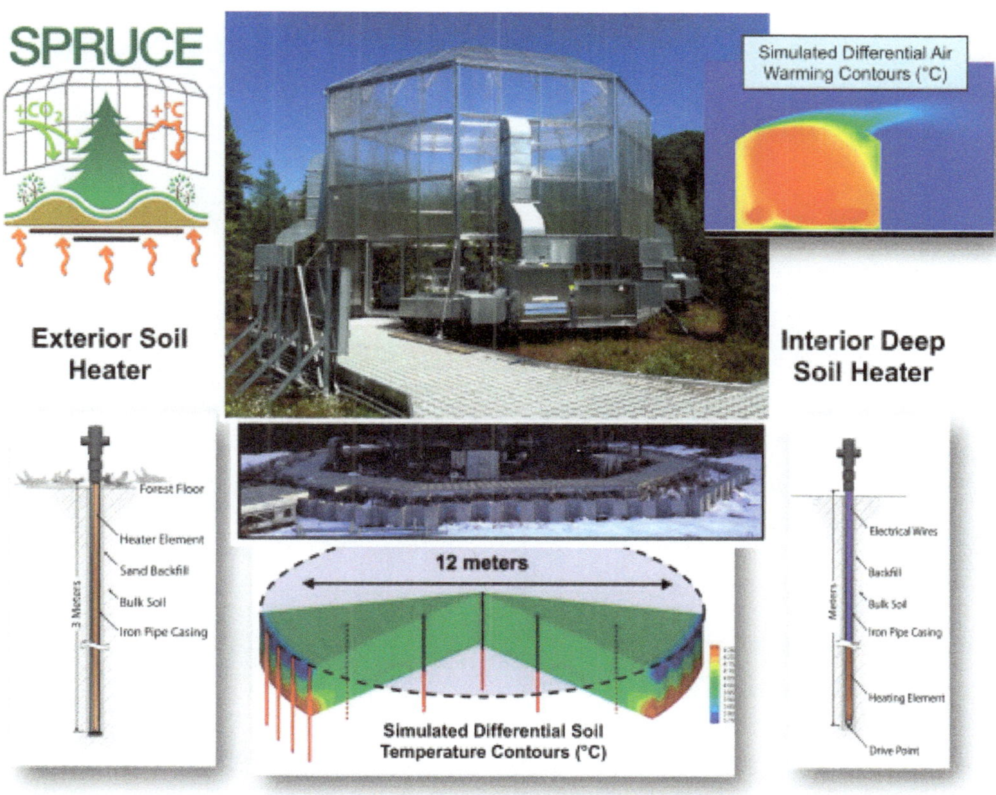

Figure 2: Open-top enclosure on the S1-Bog for air warming of the peatland ecosystem.

The SPRUCE goal of quantifying environmental parameters necessitates that most of the sensors and instruments be periodically calibrated to maintain accuracy within the SPRUCE-defined limits. A calibration schedule for all SPRUCE instrumentation has been established delineating frequency and methods and specifying acceptable operational ranges and error for each type of measurement (http: // mnspruce.ornl.gov/content/spruce-calibration). Recalibration of existing sensors versus their replacement with new sensors does consider cost. Where the cost of the calibration effort (personnel time, materials, shipping) exceeds the cost of the acquisition of new sensors (with calibration), new sensors are acquired. A database is used to track sensors, their deployment, recalibration and replacement when necessary. This calibration program falls under the ORNL's accreditation to ISO (International Organization for Standardization) 9001 standards.

DATA ACQUISITION AND CONTROL SYSTEM

General Description

The SPRUCE data acquisition and control system employs two dataloggers for each of the 16 instrumented plots: one logger controls and monitors an instrument panel dedicated to CO_2 and communications related functions while the second logger monitors an instrument panel designed to collect and distribute meteorological and biological data. Dataloggers in experimental manipulation plots have a dual role:

- measurement data acquisition and reporting, and
- regulation of temperature and CO_2 treatment levels.

One plot has been designated as the experimental temperature reference and all elevated temperature treatments are set relative to the control enclosure values. The control plot dataloggers share data with the elevated temperature plots' dataloggers which have been programmed to administer the CO_2 and warming treatment additions. Heat additions for below-ground warming at −2 m are delivered through electrical resistance heaters, while heat additions for air warming are made via propane-fired heat exchangers within air handling units (four per enclosure). The treatment plots' programs regulate air and soil temperatures at designated differentials: +2.25, 4.5, 6.75, and 9 °C above the control enclosure reference temperature.

Monitoring and control programs are prepared using Campbell Scientific's CRBasic software loaded into the dataloggers. For CO_2 regulation, each plot's CO_2 datalogger collects CO_2 concentration data from its multiport sampling system within the plot, combines them with meteorological data, and feeds these data to a PID (proportional, integral, derivative) regulated mass flow controller charged with injecting the proper amount of CO_2 into the distribution system to achieve the target level. The multiport system measures a vertical profile of CO_2 concentration and water vapor at four heights on the centrally located, 10 m tower and characterizes CO_2 distributions throughout the enclosure. CO_2 control algorithms utilize the 2 m CO_2 concentration observations coupled with external wind speed and air temperature data. CO_2 additions are stopped under very high winds (5 m s^{-1}) and during the coldest winter months (December–March). More specifically, each of the panels uses a Campbell Scientific CR1000 datalogger at its core. All dataloggers connect to the data collection system using network switches and Campbell Scientific model NL115 Ethernet interfaces. Each datalogger has a unique IP address and a unique station name.

Figure 3: Data acquisition panel (left) and CO2 panel (right).

Dataloggers in the control plot panels communicate with all dataloggers in treatment plot panels and share data. Both panels in a single plot share data that are unique to each datalogger panel. All dataloggers run the same CRBasic program. The program is written to take advantage of each datalogger having a unique station name. The program uses the logger's station name and some other distinctive variables to determine which parts of the program are appropriate for execution. The CRBasic program also utilizes an "INCLUDE" instruction that is similar to a subroutine. The "INCLUDE" instruction calls a CRBasic file that resides in the datalogger CPU. The file contains sensor calibration data and treatment settings that are exclusive to each plot. Both panels use the same power supply. The single power supply provides 12 V DC and 24 V AC for the two panels. The 12 V DC supply is backed by an uninterruptable power supply (UPS). The UPS powers the datalogger panels for a minimum of 2 h if there is a loss of the main AC power.

CO_2 Datalogger Panel

The main components of the CO_2 datalogger panel are a CR1000 datalogger, an NL115 Ethernet interface, a network switch, a Campbell Scientific SDM-CD16 AC relay controller, a Li-Cor Li-840 CO_2/H_2O analyzer, solenoid valves, a gas manifold, an Aalborg GFC-17 mass flow controller, a Campbell A574 conductivity probe interface, an air pump, switches, and connectors (Fig. 3). CO_2/H_2O sampling is accomplished with a GAST model DOA-P707-AA diaphragm pump that continuously pulls air from each of the four monitoring ports located on the meteorological tower at 0.5, 1.0, 2.0 and 4.0 m above the surface of the ground. The sample stream flows into a multiport gas manifold where each height is sequentially subsampled every 6 min and pushed into an Aalborg model GFC-17 mass flow controller (MFC) before entering the Li-Cor model Li840A CO2 H2O gas analyzer. The subsample flow rate is set by the MFC to 1 L min^{-1}. The subsample flow rate is manually adjustable by a control potentiometer located on the panel. The CR1000 datalogger is interfaced to a Campbell Scientific SDM-CD16 AC relay controller that opens each of the sampling ports at the appropriate time. After the four sampling locations are measured the data file is updated with values for CO_2, H_2O, sample flow rate, and wind speed along with the measurement time for each of the four sample ports. Measurements of plot surface outflow water are also recorded on the CO_2 datalogger. Water temperature and conductivity values from Campbell Scientific CS547A probes located in the plot overflow reservoir are recorded. A 16- port relay controller is used to activate an automated water sampler. The same relay controller controls a reservoir sump pump. The pump operation is based on water-level data shared from the meteorological datalogger.

Meteorological Datalogger Panel

The main components of the meteorological datalogger panel are a CR1000 datalogger, an NL115 Ethernet interface, two Campbell Scientific AM 16/32 multiplexers, two Campbell Scientific SDM-CVO4 current/voltage output modules, and a power supply unit for sap flow instrumentation and associated connectors (Fig. 3). The meteorological datalogger receives input from a diverse set of sensors located below the surface of the plot and on the central tower. These sensors monitor vegetation biophysical responses and heating system diagnostics. Selected inputs are also used to control the heating systems and CO_2 injections. Measurements recorded by the meteorological datalogger:

- Soil temperature at nine depths – 0, 5, 10, 20, 30, 40, 50, 100, and 200 cm – below the surface are collected with a custom designed integrated temperature probe (W. H. Cooke & Co., Inc.). Soil surface temperature and three hummock temperatures are also measured. These measurements are taken at three locations in the plot. The 200 cm depth temperature values are used in the soil warming control algorithm. Two sets of hummock soil temperatures are also recorded within the plot. Hummock temperature sensors (thermistors) are placed at a height equivalent to the 0 cm subsurface thermistor and at two heights (nominally +10 and +20 cm) within the hummock.
- Air temperature and relative humidity from Vaisala HMP-155 sensors located on the 10 m meteorological tower at heights of 0.5, 1.0 m, 2.0 m east, 2.0 m west and 4.0 m are measured by the panel. The 2.0 m air temperature values are used in the air warming algorithm.
- Soil moisture or the volumetric water content of the hummock peat is measured using Decagon HS-10 probes at two heights within the hummock, 10 cm above the surface of the plot and 10 cm below the top of the hummock. The probes measure the dielectric constant of the media using capacitance/frequency domain technology. These probes were specifically calibrated in representative peat material from the S1 Bog.
- Water table and water temperature are measured with TruTrack model WT-VO water level sensors and capacitive thermistor sensors, respectively. Depth to the water table, relative to the surface of hollow (0 cm), is calculated. Negative values indicate a water table level below the surface and positive values indicate a water table height above the hollow surface. Water temperature is measured near the bottom of the TruTrack sensor.
- Wind speed and direction data from a Gill 2-dimensional sonic anemometer (WindSonic 4) located on top of the meteorological tower is also collected by the panel. Wind speed measurements are used in the CO_2 injection algorithm.
- Total rainfall for 30 min intervals is measured at 6 m a.g.l. (above ground level) with a tipping bucket rain gauge.
- Photosynthetically active radiation (PAR, 400–700 nm, the active range of radiation for vegetation) is measured at 2.5 m a.g.l. with a Li-Cor quantum sensor, silicon photovoltaic detector.
- Vegetation biophysical measurements including sap flow (Granier's TDP Sap Flow System, PlantSensors) will be measured with heat dissipation probes on selected trees and shrubs in each plot, and tree circumference will be measured with automatic band dendrometers

(DR26 Band Dendrometer, ICT International) on selected trees in each plot.

The data acquisition panel logger also performs heating system and CO2 injection control. For each plot, the panel controls the power applied to the below-ground heaters through five outputs of the SDM-CV04 modules that are wired to silicon-controlled rectifiers (SCRs) and monitors the current used by the below-ground heater circuits. Current usage data assist with troubleshooting heater problems and monitoring heater output. A 4–20 mA output of one of the SDM-CVO4 modules is wired to a controller in the master makeup air unit (MAU). This single controller sets the heat output of the furnaces for all four of the MAUs in the plot. Airflow switches located in the four MAUs are monitored for alarm purposes to alert the operator of a blower failure. Another milliamp output of one of the SDM-CVO4 modules is wired to a mass flow controller and sets the injection rate of CO2 gas into the plot based on shared CO_2 data from the CO_2 datalogger panel.

DATA COLLECTION

Network Topology, Wiring and Equipment

The SPRUCE data acquisition system was influenced by a number of factors including facility remoteness and size, the anticipated volumes of data, and the harsh environmental conditions. Having the experimental site in such a remote area, with heavy snowfalls and possible access road blockages, forced us to build a data collection network that requires minimum human intervention and which can be self-supportive for a number of days. Such a system can be built only with a local area network (LAN) as the backbone that incorporates industrial-type components capable of withstanding the harsh environment and extreme temperatures. Another argument in favor of a LAN is the anticipated data volume. Although it is hard to make an exact calculation before the experiment goes full scale, preliminary estimates indicated that all instruments, sensors, and cameras should generate about 100 Gb of data per month. As mentioned in the experiment overview, the experimental site spans more than 8 ha (19 acres) and thus the distances between network components can be rather long. After careful consideration, the choice was made to implement a wired network. Although modern wireless technologies are capable of spanning long distances, the installation would have required additional repeaters to carry the signal from, and between, the experimental plots to the control building (CB). This distance might have become problematic in the long term because areas between the plots, and the plots and the area around the CB, have numerous trees that will continue to grow, creating lineof-sight interference. It would be possible to find the right location for the wireless equipment at present, but it is very likely that transmissions would be weakened or even completely blocked by vegetation in a couple of years. Another reason supporting the use of a wired LAN is the possibility for future high-frequency flux measurements. These types of measurements generate 10 Hz data and require fast and reliable storage. Therefore, it was decided that using a wired LAN to transfer data to a central location in the CB is the best setup.

The LAN component choices were dictated by the distance between the CB and the experimental site, which exceeded the maximum cable length (100 m) for Ethernet over twisted pair, and necessitated the use of fiber optics as the main data transfer media. Multimode fiber optic cable was run from the CB to the midpoint of each of the three docks. The fiber optic cable terminates at the CB with Fiber/Ethernet media converters that pass the data to central storage and monitoring equipment.

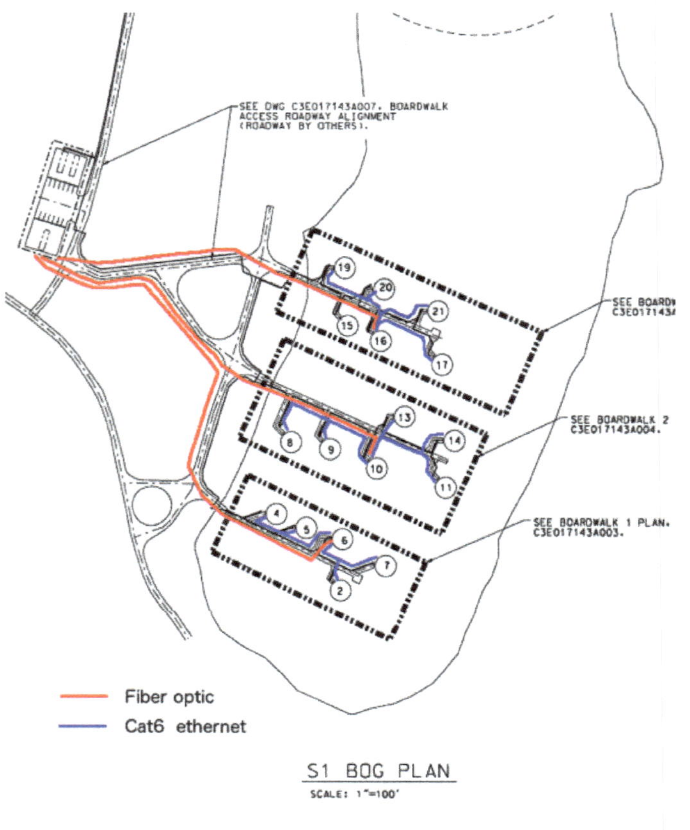

Figure 4: Network layout.

The fiber optic cable is terminated with industrial type Fiber/Ethernet switches on each dock. Category 6a Ethernet cable links each plot to its centrally located Fiber/Ethernet switch. Category 6 is a standard cable for Gigabit Ethernet and is compatible with Category 5/5e and Category 3 cable standards. It is suitable for 10BASE-T, 100BASE-TX (Fast Ethernet) 1000BASE-T/1000BASE-TX (Gigabit Ethernet) and 10GBASE-T (10-Gigabit Ethernet) networks. This allows for flexibility to deal with practically any type of data and avoid very expensive cable upgrades if network speed needs to be improved. Communications inside the plots use Category 5e or 6 plenum cable because it is less bulky, easier to deal with in short lengths, and fully compatible with 6a cable. This layout minimizes the number of connection points and pieces of networking equipment involved making troubleshooting and maintenance easier (Fig. 4). Although all network and data acquisition equipment are located inside enclosures that provide protection from the weather elements, they still may suffer extreme temperature and humidity variations. That is why all networking equipment deployed outside the CB is not of only industrial grade but also has extended operating temperature and humidity ranges. Ethernet switches are rated for an operating environment of −40 to +70 °C and 10–90 % relative humidity. Fiber/Ethernet switches are rated for −20 to +70 °C and 5– 95 % relative humidity.

WAN Connection and Remote Access

Another problem caused by the remote location of the experimental site is the absence of communication lines. The closest phone line is several miles away and the closest wired Internet access is more than 10 mi. away. Estimates showed that the expense to extend wired Internet access to the site was too high and could not be justified at this time. Existing cellular technologies have enough bandwidth to provide the necessary type of data service and could be a good alternative, but a survey of the local coverage showed that signal strength was very inconsistent. Reliable communication is needed to facilitate data transfer from the site and to perform remote system maintenance. Based on the positive experiences of the USFS at the nearby Marcell Research Center with access to the Internet via a satellite link, we explored this possibility.

Over the last several years, satellite Internet service providers have significantly improved connection speed and increased available bandwidth. The upload speed for this type of connection is still significantly lower than download speeds and sensitivity to some weather conditions remains an issue. However, for the SPRUCE project the most critical data connection speed and reliability has to be in the LAN. Once data are transferred to the CB and backed up, there are additional options for moving data off-site for further analyses, quality control, and safekeeping. Examination of services offered by different companies revealed that most of them are oriented towards home users that mostly consume data from the Internet and do not need services geared towards large data uploads or remote access to computer systems.

After extensive consultation with many providers, the one with the highest uplink speed was selected. The established link provides 5 Mb s^{-1} download and 1 Mb s^{-1} upload speeds and a static IP address. Despite expected sensitivity to some weather conditions like heavy rain and snow, this type of connection has proven sufficiently reliable to satisfy all current project needs. To assist the local project team in maintaining the network while minimizing the number of system administrator trips to the site, remote access via TeamViewer software (TeamViewer GmbH) was established. This software allows performing remote maintenance behind a firewall and NAT (network address translation) router and does not require any exceptions in their settings. The TeamViewer software provides file transfer capabilities and utilizes completely secure data channels featuring 2048 bit RSA (Rivest–Shamir– Adleman) key exchange and 256 bit AES (advanced encryption standard) session encoding.

Data Flow, Control and Redundancy

The components described above (sensors, instruments, data acquisition panels, LAN, and networking equipment), combine to create a system that allows the project to efficiently collect data from all plots and to transfer it to the central data storage location (Fig. 5). Data from an instrument or a sensor are recorded by dataloggers located in the data acquisition panels and then copied via the LAN to the data storage server. The data storage server calls each datalogger, every 30 min. If communication is established, the server copies the data acquired since the last call. Each datalogger has removable, compact flash (CF) card storage that serves as an intermediate buffer. The CF storage operates as a re-writable stack, which means that if storage space is not available, newer data will replace older data at the top of the stack. Using CF cards as additional storage in dataloggers provides redundancy and serves as localized backup in case the whole network is down (e.g., power outage). Each CF card can hold several days of data, which is likely sufficient time to restore network communications. For longer outages new CF cards can be manually swapped with the installed CF cards. All dataloggers deployed are Campbell Scientific (CS) CR1000 models commonly used for environmental science experiments and longterm meteorological observation stations. Use of commonly available standard equipment makes it possible to employ a suite of off-the-shelf software produced by CS as well. The main software component is the LoggerNet package. It supports programming, communication, and data movement between dataloggers and PCs. LoggerNet consists of a server application and several client applications integrated into a single product. Use of this package is especially convenient in applications that require telecommunications and scheduled data retrieval within large datalogger networks. Another software package used to control and monitor data collection, sensors and instrument performance is the CS Real-Time Monitoring and Control software (RTMC). RTMC graphically displays data, sets input locations, and toggles ports and flags. A simple yet powerful user interface allows RTMC to use components such as alarms, sliders, graphs, toggles, tables, and gauges to design custom multitab displays that can be published via a web server and used for remote monitoring and control (see example output window; Fig. 6). All collected data are saved in two forms for redundancy and as a backup: as comma separated value (CSV) text files and as a database where each table corresponds to a file. The CS LoggerNet Database software (LNDB) is used to move data from a LoggerNet data cache into a database such as Microsoft SQL Server or MySQL. To reduce costs, the MySQL database was chosen as one of the most widely used opensource relational database management systems. This allows quality control and quality assurance procedures to be performed using flexible SQL queries and facilitates automated generation of necessary data reports. MySQL's built-in replication mechanisms are deployed to create an additional layer of redundancy and backup. A MySQL server at CB acts as a master and a slave server is set up off-site in the SPRUCE experiment office in Grand Rapids, MN. The Fig. 5 diagram shows all components together and gives a full overview of the SPRUCE data system.

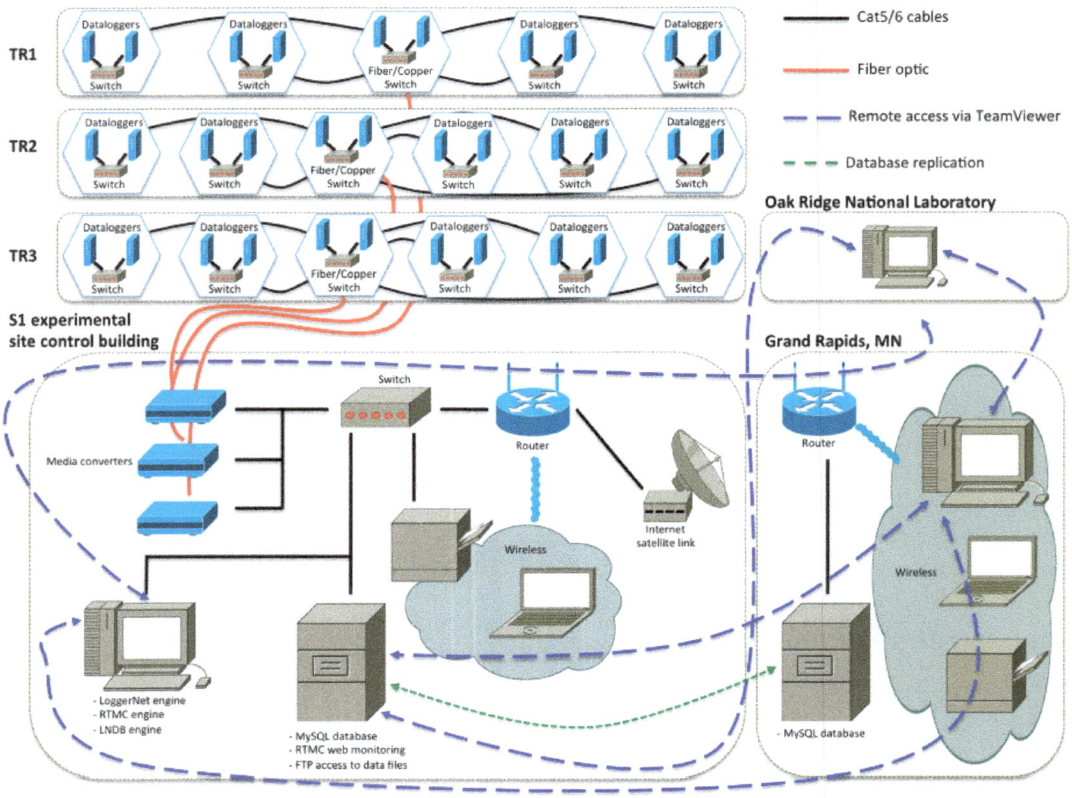

Figure 5: SPRUCE data collection and monitoring system.

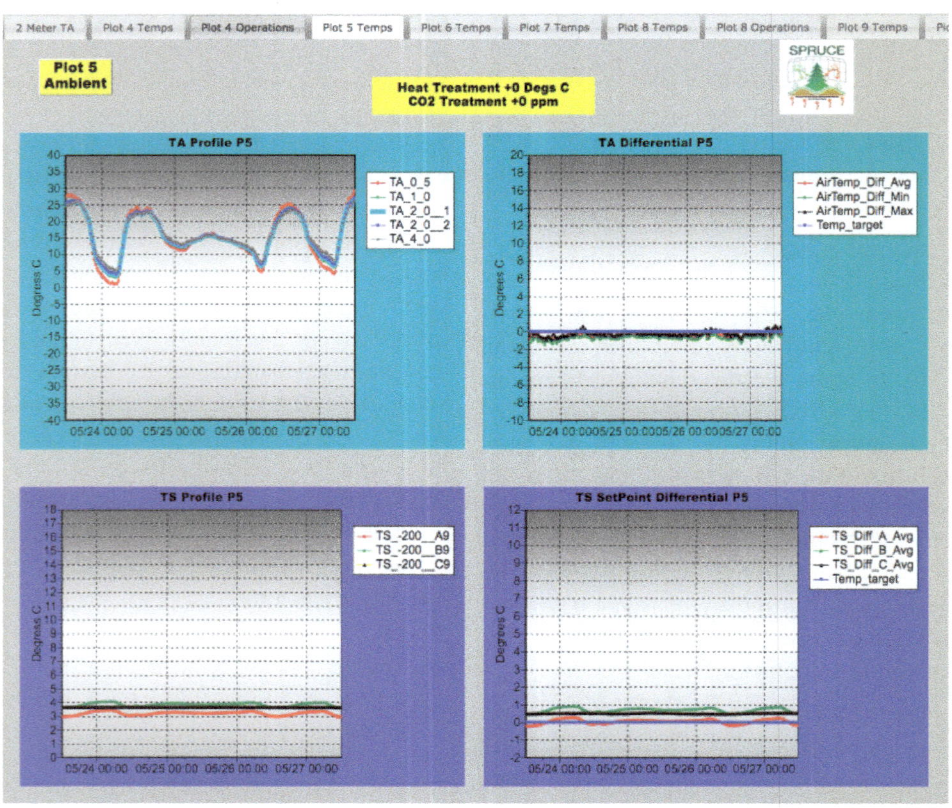

Figure 6: RTMC monitoring graphs.

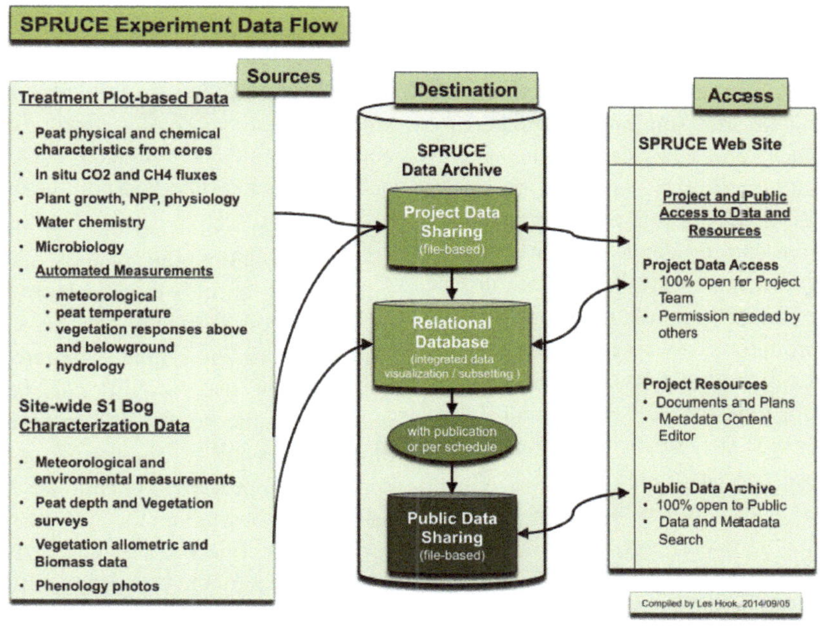

Figure 7: The data flow diagram.

DATA MANAGEMENT

As outlined above, data management is a set of policies, plans, and practices to control consistency, protect integrity, and deliver data to project users and the public. It sets operational definitions before the collection of data begins, ensures that everyone in the system has the same understanding, collects data in a consistent manner, and outlines how to handle the data both during the experiment and after completion. Data management was planned to provide a structured framework for implementing project-defined requirements for maintaining data quality and consistency and controlling data processing from collection through archival and public access. SPRUCE data plans are consistent with the recently issued DOE (Department of Energy), Office of Science, Statement on Digital Data Management (http://science.energy.gov/ber/funding-opportunities/ digital-data-management) and with the Memorandum of Understanding (http://mnspruce.ornl.gov/system/files/sites/ default/files/MOU_USFS%26UT-Battelle.pdf) between the USFS and UT-Battelle.

Data Flow

Figure 7 shows the SPRUCE data flow diagram starting from the sensors and ending at the archiving/distribution points as a representation of the data policy to inform investigators and potential data users of the general flow of data and information before, during, and after the experiment. Data and metadata reporting standards help to facilitate efficient data reporting, processing and analysis, and promote integration of diverse data across measurement domains.

Consistency and Standardization

Data consistency and standardization plans assure that information is uniformly created and maintained as it moves across the network to facilitate future processing of data in comparable forms. SPRUCE adopted several reporting conventions, best represented by three examples. Two of the most important elements are date and time. Time synchronization is achieved by synchronizing LAN time with one of the Internet time servers and then propagated to all dataloggers. The reporting convention for dates and times begins with the time basis of central standard time (CST) – equivalent to local coordinated universal time (UTC) minus 6 h. The start date and time should be reported/archived as the date and time at the beginning of the sampling or measurement or averaging period and always expressed as such in data products for the SPRUCE project and public sharing. The SPRUCE reporting interval is 30 min for continuous time series data. For parameters with sampling/measurement/recording intervals that can be reliably averaged or summed to 30 min, date and time fields should not be missing. For reporting missing measurement data, it was established that there may not be blank data fields. For character fields, "None" is the preferred missing value code. Alternatively, "Not available" or "Not provided" may be more appropriate for the particular context. For integer and decimal format fields, in general, the missing value code should be negative and large enough to not be reasonable as an actual data value: −9999 for all integer and decimal variables. It is acceptable to match the format of the column and use repeated 9s (e.g., decimal: −999.99; integer: −999) or scientific format (e.g., -9.99×10^2). For identifying measured parameters, accepted variable names follow the standards of established research communities (e.g., AmeriFlux) and are likely familiar to data users, for example, with temperature as TA for air temperature, TS for soil temperature, and TW for water temperature. Names also incorporate information about measurement height or depth and duplicates. For instance, the first duplicate air temperature at 2.0 m is abbreviated as TA_2_0_1. Plant species codes are derived from the Natural Resources Conservation Service PLANTS Database (NRCS, 2010).

Data Process Planning

Data processing best practices were identified to support data collection, processing, quality assurance, and archiving activities. For example, guidance is provided for researchers regarding data entry, transfers, and transformations to ensure data integrity is maintained. Guidance is provided for managing and protecting hard copy and electronic media records during and after the project. Scripted programs are encouraged for processing, quality assurance, and analysis of data as they provide documentation and traceability, and can be easily revised and re-executed as needed. Project conventions are specified for naming archive data products and for preferred archive data file formats and file name structures. The purpose of identifying best practices for data processing is to ensure the consistent quality of data products as de- fined in the SPRUCE Data Policy. More specifically, multiple levels of data quality are defined and managed for the SPRUCE experiment. Level 0 data are products of unspecified quality that have been subjected to minimal processing in the field and/or in the laboratory (e.g., raw data, hard copy data sheets, scanned data sheets, and notebooks). This may, for example, be data from an instrument logger, expressed in engineering units or using nominal calibrations, or high resolution data before aggregating to a selected interval. All Level 0 data should be submitted to the Data Archive for long-term storage but they are not shared. Instead, Level 1 data represent an internally consistent data product that has been subjected to standard quality checks and data management procedures. Level 1 data are suitable for sharing and analyses within the project. Level 2 data undergo additional quality checks and the addition of detailed metadata descriptions to enable their use externally as a consistent data product that has undergone interpretative and diagnostic analysis by SPRUCE. Level 2 data are intended to be shared with a the broad research community and the public well beyond the projects active time frame.

Data Documentation, Archiving and Sharing

One of the main goals of the data policy is to expeditiously share experimental data among researchers, the broader scientific community, and the public. Advanced planning furthers efforts to identify, collect, and report consistent data and metadata and to facilitate timely data analysis, sharing, integration, and synthesis by all users. The pre-experiment S1 Bog characterization data are available as file-based products for download including results of peat depth and volume studies, vegetation surveys and allometry measurements (http://mnspruce.ornl.gov/content/ public-data-download). S1 environmental monitoring data collected continuously from 2010 are available as a complete time series for project use but in annual increments of consistent products specifically for modelers and the public. Experimental treatment plot-based data products including discrete pre-treatment characterization results for vegetation growth, peat physical and chemical properties, water chemistry, and microbiological assays are and will be available as files. Integrated plot-based automated collection of treatment response variables will be accessed through relational database (RDB) interface for specific queries and visualizations.

This RDB access is not intended for public use. Public access to these products is planned to be as exported files with consistent content for modeler and public use. Data products are documented for users to understand the "who, what, when, where, and why" of the data product to maximize data reuse. Metadata includes the format and a data dictionary with column names, units, and a brief description of each field. The documentation includes any code definitions, methods and materials, links to related products, references to published analysis papers, and a complete citation for the data product with a registered digital object identifier (DOI). The DOI citation should be included as a reference in subsequent publications to properly credit the original data providers.

All data collected during the SPRUCE experiment, all documentation, all results of analysis or synthesis of information, and all model algorithms and codes developed in support of SPRUCE will be submitted to the SPRUCE Data Archive in a timely manner such that data will be available for use by SPRUCE researchers and, following publication, the public. The archive will maintain two levels of data accessibility. The first is for sharing recently collected, derived, and processed data products among SPRUCE project participants including the ORNL research group, the forest service, and collaborators. The second is for access to mature data products by the broader scientific community and public. Public access will be concurrent with open literature or website publication of SPRUCE results. The discovery (identifying and finding) of SPRUCE data products is facilitated through the compilation of descriptive companion metadata records and their inclusion in searchable metadata databases and clearinghouses. The SPRUCE Data Archive (http://mnspruce.ornl.gov/) is implemented as part of the Carbon Dioxide Information Analysis Center (CDIAC) at ORNL. CDIAC is the primary climate-change data and information analysis center of the US DOE and provides the long-term system stability and archive longevity required for the SPRUCE multiyear project and reliable public data access.

SUMMARY

Ecosystem-scale manipulation experiments are getting more complicated and require innovative approaches that help manage high volumes of in situ observations. New largescale, well-designed, and reliable data acquisition and management systems will become common in the future. The presented approach shows an example of such a system that was built in a remote and harsh environmental location. The provided details about sensors, network layout, hardware, software, and data flow as well as the considerations behind them can be used for the design and development of similar systems for other experiments in the future.

ACKNOWLEDGEMENTS

This research and CDIAC were supported by the US Department of Energy, Office of Science, Biological and Environmental Research (BER), and conducted at Oak Ridge National Laboratory (ORNL), managed by UT-Battelle, LLC, for the US Department of Energy under contract DE-AC05-00OR22725. This manuscript has been authored by UT-Battelle, LLC under contract no. DE-AC05-00OR22725 with the US Department of Energy. The publisher, by accepting the article for publication, acknowledges that the United States Government retains a nonexclusive, paid-up, irrevocable, worldwide license to publish or reproduce the published form of this manuscript, or allow others to do so, for United States Government purposes. The Department of Energy will provide public access to these results of federally sponsored research in accordance with the DOE Public Access Plan (http://energy.gov/downloads/doe-public-access-plan).

REFERENCES

1. Barbier, C., Hanson, P. J., Todd Jr., D. E., Belcher, D., Jekabson, E. W., Thomas, W. K., and Riggs, J. S.: Air Flow and Heat Transfer in a Temperature Controlled Open Top Enclosure, Paper #IMECE2012-86352, ASME International Mechanical Engineering Congress and Exposition, 2012, Houston, TX, 2012.
2. Hanson, P. J. and Wullschleger, S. D. (Eds.): North American Temperate Deciduous Forest Responses to Changing Precipitation Regimes, Springer, New York, p. 421, 2003.
3. Hanson, P. J., Amthor, J. S., Wullschleger, S. D., Wilson, K. B., Grant, R. E., Hartley, A., Hui, D., Hunt Jr., E. R., Johnson, D. W., Kimball, J. S., King, A. W., Luo, Y., McNulty, S. G., Sun, G., Thornton, P. E., Wang, S., Williams, M., Baldocchi, D. D., and Cushman, R. M.: Oak forest carbon and water simulations: model intercomparisons and evaluations against independent data, Ecol. Monogr., 74, 443–489, doi:10.1890/03-4049, 2004.
4. Hanson, P. J., Classen, A., Kueppers, L., Luo, Y., McDowell, N. G., Morris, J., Rogers, A., Thornton, P., Ceulemans, R., Dukes, J., Goulden, M., Jackson, R., Knapp, A., Kirschbaum, M., Lewin, K., MacCracken, M., Melillo, J., Ringler, T., and Workshop Participants: Ecosystem Experiments: Understanding Climate Change Impacts On Ecosystems and Feedbacks to the Physical Climate, in: Workshop Report on Exploring Science Needs for the Next Generation of Climate Change and Elevated CO2 Experiments in Terrestrial Ecosytems, 14–18 April 2008, Arlington, Virginia, http://science.energy.gov/~/media/ber/pdf/Ecosystem_ experiments.pdf (last access: 4 November 2015), 2008.
5. Hanson, P. J., Childs, K. W., Wullschleger, S. D., Riggs, J. S., Thomas, W. K., Todd, D. E., and Warren, J. M.: A method for experimental heating of intact soil profiles for application to climate change experiments, Global Change Biol., 17, 1083–1096, doi:10.1111/j.1365-2486.2010.02221.x, 2011.
6. Norby, R. J., Hanson, P. J., O'Neill, E. G., Tschaplinski, T. J., Weltzin, J. F., Hansen, R. T., Cheng, W., Wullschleger, S. D., Gunderson, C. A., Edwards, N. T., and Johnson, D. W.: Net primary productivity of a CO2 enriched deciduous forest and the implications for carbon storage, Ecol. Appl., 12, 1261–1266, 2002.
7. NRCS – Natural Resources Conservation Service: Plants Database, Minnesota subset, http://plants.usda.gov/java/ nrcsStateDownload?statefips=27 (last access: 6 November 2015), 2010.
8. Parsekian, A. D., Slater, L., Ntarlagiannis, D., Nolan, J., Sebesteyen, S. D., Kolka, R. K., and Hanson, P. J.: Uncertainty in peat volume and soil carbon estimated using ground-penetrating radar and probing, Soil Sci. Soc. Am. J., 76, 1911–1918, doi:10.2136/sssaj2012.0040, 2012.
9. Walker, A. P., Hanson, P. J., De Kauwe, M. G., Medlyn, B. E., Zaehle, S., Asao, S., Dietze, M., Hickler, T., Huntingford, C., Iversen, C. M., Jain, A., Lomas, M., Luo, Y., McCarthy, H., Parton, W., Prentice, I. C., Thornton, P. E., Wang, S., Wang, Y.-P., Warlind, D., Weng, E., Warren, J. M., Woodward, F. I., Oren, R., and Norby, R. J.: Comprehensive ecosystem modeldata synthesis using multiple data sets at two temperate forest free-air CO2 enrichment experiments: Model performance at ambient CO2 concentration, J. Geophys. Res.-Biogeo., 119, 937–964, doi:10.1002/2013JG002553, 2014.

CHAPTER 12

AN INTEGRATED INSTRUMENTATION SYSTEM FOR VELOCITY, CONCENTRATION AND MASS FLOW RATE MEASUREMENT OF SOLID PARTICLES BASED ON ELECTROSTATIC AND CAPACITANCE SENSORS

Jian Li[1], Ming Kong[2], Chuanlong Xu[1], Shimin Wang[1] and Ying Fan[3]

[1]Key Laboratory of Energy Thermal Conversion and Control of Ministry of Education, School of Energy and Environment, Southeast University, Nanjing 210096, China

[2]College of Metrology and Measurement Engineering, China Jiliang University, Hangzhou 310018, China

[3]School of Energy and Power Engineering, Lanzhou University of Technology, Lanzhou 730050, China

ABSTRACT

The online and continuous measurement of velocity, concentration and mass flow rate of pneumatically conveyed solid particles for the high-efficiency utilization of energy and raw materials has become increasingly significant. In this paper, an integrated instrumentation system for the velocity, concentration and mass flow rate measurement of dense phase pneumatically conveyed solid particles based on electrostatic and capacitance sensors is developed. The electrostatic sensors are used for particle mean velocity measurement in combination with the cross-correlation technique, while the capacitance sensor with helical surface-plate electrodes, which has relatively homogeneous sensitivity distribution, is employed for the measurement of particle concentration and its capacitance is measured by an electrostatic-immune AC-based circuit. The solid mass flow rate can be further calculated from the measured velocity and concentration. The developed instrumentation system for velocity and concentration measurement is verified and calibrated on a pulley rig and through static experiments, respectively. Finally the system is evaluated with glass beads on a gravity-fed rig. The experimental results demonstrate that the system is capable of the accurate solid mass flow rate measurement, and the relative error is within −3%–8% for glass bead mass flow rates ranging from 0.13 kg/s to 0.9 kg/s.

INTRODUCTION

Pneumatic conveying is one of the most important ways for the transportation of pulverized fuel materials in many fields such as energy, chemical, metallurgy, pharmaceutical and foods industries. It is of great significance to achieve the accurate, online and continuous measurement of the velocity, concentration and mass flow rate of pneumatically conveyed solid particles for the high-efficiency utilization of energy and raw materials, operational safety and optimized system control.

A variety of advanced measurement techniques have been investigated and developed for the flow measurement of pneumatically conveyed solid particles based on different sensing principles [1], such as optical [2,3], digital imaging [4,5], acoustic [6,7], radiometric [8] and electric methods [9,10,11,12,13,14,15,16]. Among them, optical, digital imaging and acoustic techniques are capable of dilute phase

Citation: Li, J.; Kong, M.; Xu, C.; Wang, S.; Fan, Y. An Integrated Instrumentation System for Velocity, Concentration and Mass Flow Rate Measurement of Solid Particles Based on Electrostatic and Capacitance Sensors. Sensors 2015, 15, 31023-31035. DOI: 10.3390/s151229843

particle flow measurement. However, to date few of the proposed instrumentation systems can work in the harsh industrial environment, and these techniques do not function under dense phase conditions, where high solids loading is preferred to minimize the wear of the conveyor, solids undergo breakdown due to friction and collisions, and the transportation energy consumption is high. The radiometric method is, in principle, an optional choice for the gas-solid flow measurement regardless of the particle concentration. However, it has a risk of radiation leakage and is usually costly due to the complicated sensor structure and radiation shielding required. Electric methods, mainly including electrostatic and capacitance sensing techniques, have been increasingly attractive due to their advantages of non-invasiveness, simple structure, low-cost, high reliability and non-radiation technology. Great advances have been made in particle velocity measurement in gas-solid flow by using electrostatic sensors along with signal processing algorithms such as cross-correlation [10,11,12,13,14] and spatial filtering [15,16,17] techniques, and encouraging results have been obtained in both laboratory and field experiments. On the other hand, the capacitance method has been successfully used for the phase concentration measurement in gas-liquid and gas-solid two-phase flows with optimized sensor structures [18,19] based on the variations of equivalent electrical permittivity with different phase concentrations. Electrostatic and capacitance sensors also can be combined with process tomography, i.e., electrostatic tomography (EST) and electrical capacitance tomography (ECT), to realize particle flow parameters measurement and "flow visualization" [9,20,21].

In this paper, an integrated instrumentation system for the velocity, concentration and mass flow rate measurement of dense phase pneumatically conveyed solid particles based on electrostatic and capacitance sensors is developed. Two identical ring electrostatic sensors are used for the particle mean velocity measurement in combination with the cross-correlation technique, while a capacitance sensor with helical surface-plate electrodes, which has relatively homogeneous sensitivity distribution, is employed for the particle concentration measurement and its capacitance is measured by an electrostatic-immune AC-based circuit. The mass flow rate of solid particles is further inferred from the obtained particle velocity and concentration. Experimental results are presented to evaluate the performance of the developed system.

MEASUREMENT PRINCIPLE

The mass flow rate of solid particles over the cross-section of a pneumatic pipe can be inferred from the particle velocity and concentration:

$$M = \rho \cdot A \cdot v \cdot \beta \qquad (1)$$

where M is the mass flow rate of solid particles, ρ is particle density, A is the cross-sectional area of the conveying pipe, v is the mean particle velocity and β is the mean volume concentration. If the particle velocity and concentration can be measured accurately, the solid mass flow rate could be calculated from Equation (1).

Particle Velocity Measurement Based on Electrostatic Sensors

Pneumatically conveyed solid particles are charged due to friction and collisions between the particles and the pipe wall, which is one of the inherent characteristics of gas-solid flows. Based on particle charging, various electrostatic sensors have been developed for particle flow parameter measurement [9,10,11,12,13,15,16,17], especially for the particle velocity measurement when combined with a cross-correlation signal processing algorithm. Two identical electrostatic sensors are essential for the cross-correlation velocity measurement and are usually arranged on the pipe with a certain axial spacing (L). Two corresponding electrostatic signals can be obtained by suitable electronic circuits connected to the sensors. When the particles flow moves along the axial direction of the pipe from the upstream sensor to the downstream sensor, the output signals from the two sensors are similar with each other, but have a time delay (the transit time). From the cross-correlation function of the two signals, the maximum value can be found and its time coordinate corresponds to the transit time (τ). As a consequence, particle velocity can be calculated by:

$$v=L/\tau \qquad (2)$$

In practical application, the sampling frequency should be selected properly according to the frequency characteristic of the electrostatic sensor and the requirement of the relative error in the transit time measurement [22]. The signal bandwidth of the electrostatic signal is dependent on the particle velocity, the width of the sensor electrode and geometric size of the pipe. The sampling frequency should be more than twice the signal bandwidth according to the Nyquist–Shannon sampling theorem. The transit time is deduced according to the maximum valueof the correlation function and itsreal value maynot be located at the discrete points of the correlation function. The relative error in the transit time measurement is related to particle velocity, axial spacing of the two sensors and the sampling frequency. A lower relative error in the transit time measurement requires a higher sampling frequency, but this is not always necessary due to the inherent instability of particle flows. The tolerance of the relative error of the transit time should be set by taking the real application requirements into account. Another important parameter for the cross-correlation calculation is the integration time. It should be more than 10 times the transit time [23]. Although a longer integration time is useful for the accurate measurement of the transit time, velocity measurement is time-consuming and the response of the system becomes slow.

Particle Concentration Measurement with Capacitance Sensor

The equivalent electrical permittivity of a two-phase mixture varies with different constituent contents. When a gas-solid two-phase flow passes through the sensing space of a capacitance sensor, the capacitance between the metal electrodes of the sensor will change with the solid volume concentration. As a consequence, the particle concentration can be derived from thecapacitance change of the sensor.

For the capacitance sensor applied to gas-solid flow measurement, its capacitance (C_e) is determined by the geometric size and the equivalent relative permittivity (ε_e) of the media within its sensing space [24]:

$$C_e = C_{e0} f(\varepsilon_e) \tag{3}$$

where C_{e0} is the capacitance of the sensor when the pipe is full of gas, and $f(\varepsilon_e)$ is a function. Usually, the relative permittivity of gas can be approximately regarded as 1, and thus C_{e0} is completely determined by the geometric size of the sensor. However, the expression of $f(\varepsilon_e)$ is related to the sensor structure, and is not a proportional function except for the parallel-capacitance model [24]. On the other hand, it is also challenging to calculate the equivalent relative permittivity of the gas-solid mixture, which depends on the relative permittivity of gas and solid, and the particle concentration distribution ($\beta_V(x,y,z)$):

$$\varepsilon_e = g(\varepsilon_g, \varepsilon_s, \beta_V(x,y,z)) \tag{4}$$

where ε_g and ε_s are the relative permittivities of gas and solid particles, respectively, and $g(\varepsilon_g, \varepsilon_s, \beta_V(x,y,z))$ is a function. So C_e can be written as:

$$C_e = C_{e0} h(\varepsilon_g, \varepsilon_s, \beta_V(x,y,z)) \tag{5}$$

where $h(\varepsilon_g, \varepsilon_s, \beta_V(x,y,z))$ is a function. At present, it is difficult to achieve the analytical solution of Equation (5). In capacitance sensor applications, its structure and the type of the media are known, so the equivalent capacitance is mainly dependent on the particle concentration and its distribution (flow pattern) over the cross-section of the pipe. This is because the capacitance sensor has aninhomogeneous sensitivity distribution in its sensing space, which may lead to measurement errorsin particle concentration measurement despitethe similar concentration.

To reduce the measurement error, the sensitivity of a capacitance sensor must beperfectly homogeneous. Many trials have been carried out to optimize and develop a suitable sensor structure to ensure ahomogeneous sensitivity distribution [18,25,26,27]. The results prove that the capacitance sensor with helical surface-plate electrodes has a more homogeneous sensitivity, which makes it immune to flow patterns. After calibration, the capacitance sensor with helical surface-plate electrodes can be used for the accurate concentration measurement of solid particles.

DESIGN OF THE INTEGRATED INSTRUMENTATION SYSTEM

Figure 1 shows a schematic diagram of the proposed integrated instrumentation system. The system is comprised of two ring-shaped electrostatic sensors for particle velocity measurement, one capacitance sensor with helical surface-plate electrodes for particle concentration measurement, one measurement circuit based on a digital signal processor (DSP) and a personal computer (PC).

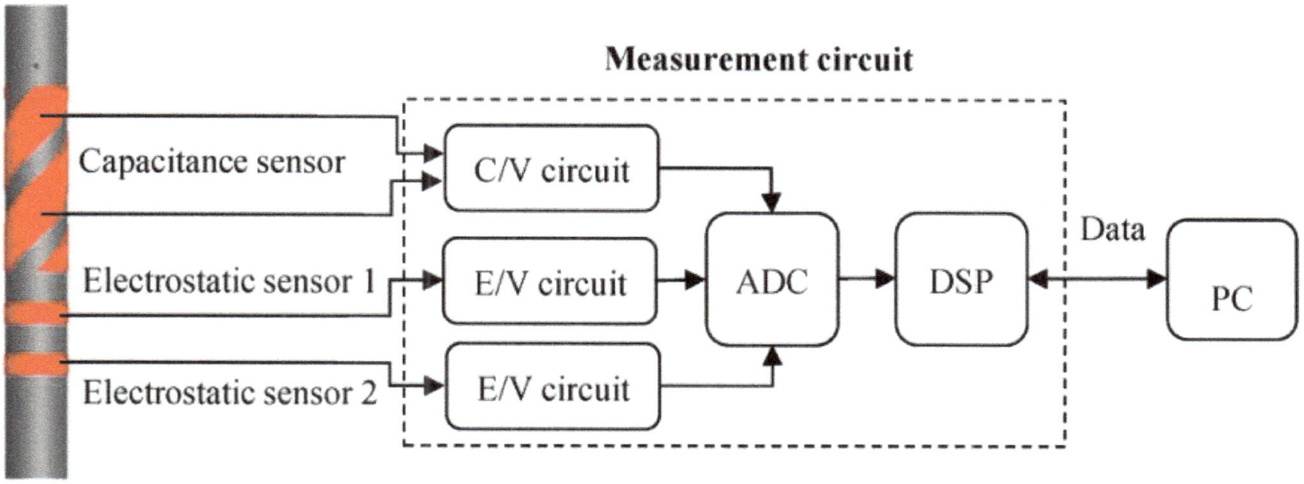

Figure 1: Integrated instrumentation system based on electrostatic and capacitance sensors.

The electrodes of the electrostatic and capacitance sensors are made of copper with a thickness of 0.2 mm, and they are mounted on the outer surface of a plexiglass pipe. All the electrodes are covered in a metal shield to isolate them from any external electromagnetic interference. The inner and outer diameters of the plexiglass pipe are 50 mm and 60 mm, respectively. The axial width of the electrodes of the electrostatic sensors is 20 mm and the axial spacing between the electrodes is 40 mm. The helical surface-plate electrodes of the capacitance sensor have a axial width of 55 mm and twist one turn along the pipe surface with a pitch of $\pi \times 60$ mm. In Figure 1, the two electrostatic sensors are arranged on the same side of the capacitance sensor. If the integrated system is applied to the particle measurement in a vertical pipe, the correlation velocity from the electrostatic sensors obviously deviates from the mean velocity of particles within the capacitance sensor due to acceleration effect. This will lead to a measurement error in the solid mass flow rate. Another option is to configure the two electrostatic sensors on each side of the capacitance sensor, as shown in Figure 2. In this case, the correlation velocity accurately represents the mean velocity of the solid particles within the sensing space of the capacitance sensor. However, the axial spacing of the two electrostatic sensors then increases to 250 mm, much longer than that in Figure 1. A large axial spacing may decrease the reliability of the velocity measurement, especially under the unstable flow conditions seen at low velocity. Therefore the practical flow conditions should be considered for the configuration of the sensors in the integrated instrumentation system.

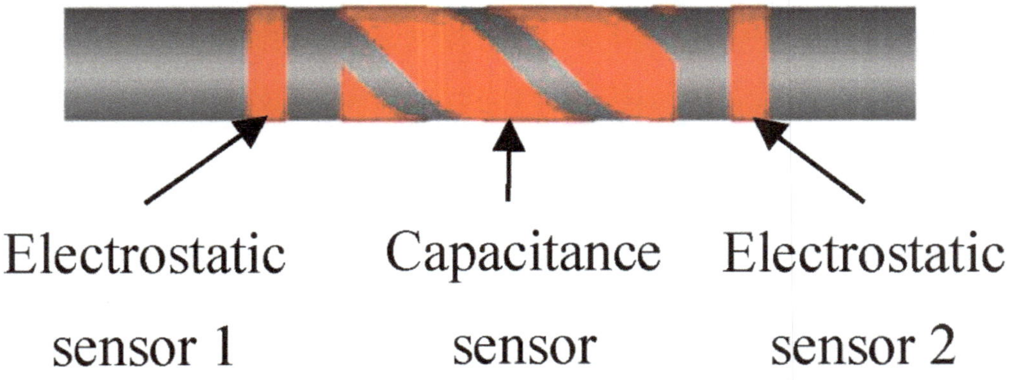

Figure 2: Sensor configuration.

The DSP-based measurement circuit mainly consists of two electrostatic charge to voltage conversion (E/V) circuits, one capacitance to voltage conversion (C/V) circuit, one multi-channel analog to digital converter (ADC), and a DSP. The E/V circuits (Figure 3) are connected to the electrostatic sensors for the electrostatic signal conditioning, while the electrostatic-immune C/V circuit (Figure 4) based on an AC-based method is used for the capacitance measurement [28,29]. The electrostatic and capacitance signals are sampled by the DSP via the ADC. Particle velocity, concentration and mass flow rate are calculated in the DSP, and then sent to PC by a serial port for storage and display. In the C/V circuit, a band-pass filter is used to eliminate the influence of particle charging on the capacitance measurement, and a differential amplifier is to balance the standing capacitance of the capacitance measurement system, only leaving the capacitance change to be further amplified and processed.

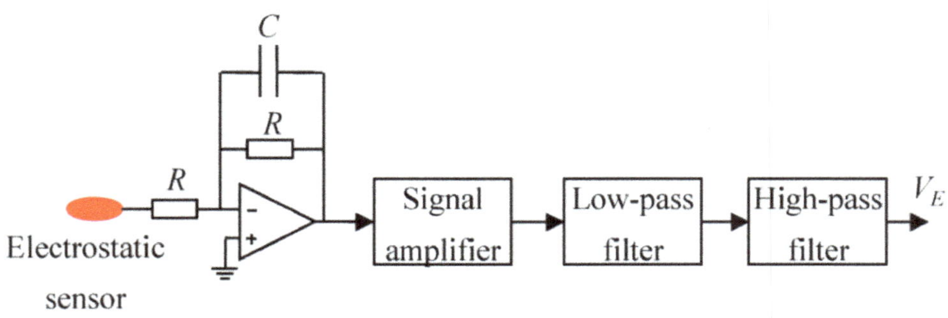

Figure 3: Charge to voltage conversion (E/V) circuit.

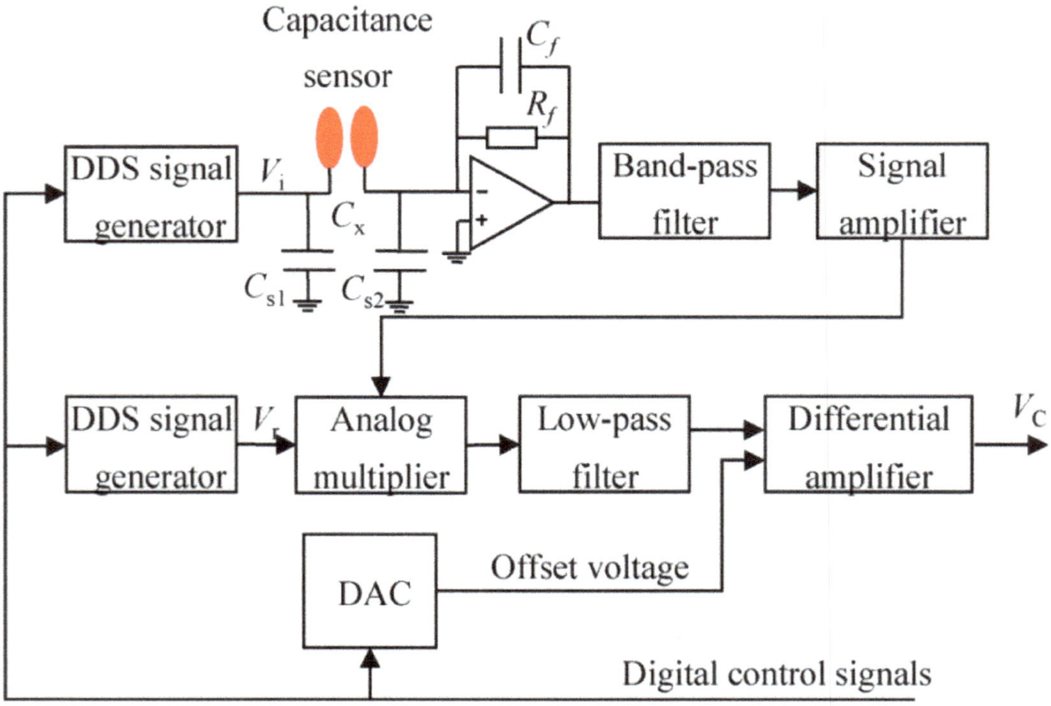

Figure 4: Capacitance to voltage conversion (C/V) circuit.

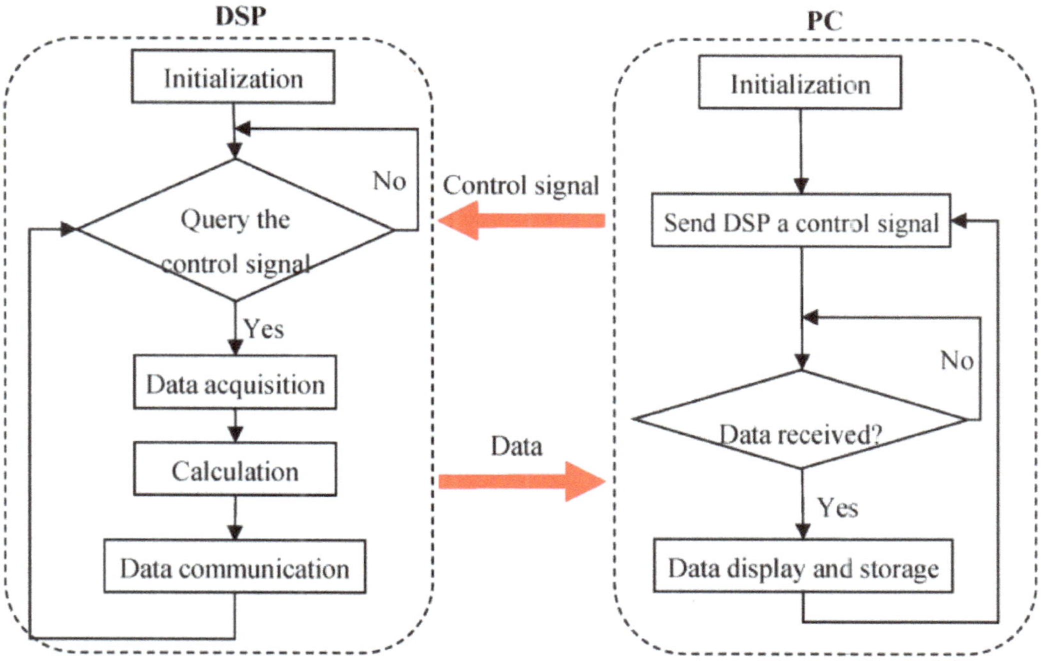

Figure 5: Flow chart of the software of the measurement system.

The flow chartof the software of the measurement system is shown in Figure 5. After the initialization of the DSP and PC, the PC sends a control signal to DSP. If the signal is accepted, the DSP will sample the electrostatic and capacitance signals at the preset sampling frequency and integration time, and then the flow parameters including particle velocity, concentration and mass flow rate are calculated. Subsequently, the DSP sends the flow parameters to the PC and waits for the next control signal. Once the PC receives the parameters, it will send the next control signal to the DSP immediately after completing the display and storage of the parameters. The number of data sampling points in the experiments is 2000, and the corresponding response time of the measurement system is about 1.5 s.

EXPERIMENTAL RESULTS AND DISCUSSION

Tests on a Pulley Rig

The instrumentation system for velocity andconcentration measurement is first tested on a V-belt pulley rig, as demonstrated in Figure 6. The belt is made of polyurethane and has a trapezoidal cross-section. Three types of belt with different sizes (Table 1) are used in the experiments. The diameter of the pulley (D) can be selected as 320 mm or 255 mm and the rotation speed can be adjusted by a variable-frequency driving motor. During the operation of the rig, the charge will be generated on the belt. Even though the motion of the belt is apparently different from the particle flow, the electrostatic induction phenomena, caused by the charge carried by belt and particles, respectively, are the same in essence. Another advantage of the rig is that the sectional area of the belt is fixed during the system operation, i.e., the solid concentration does not change with the velocity, which is beneficial to assess the performance of the C/V circuit for the concentration measurement.

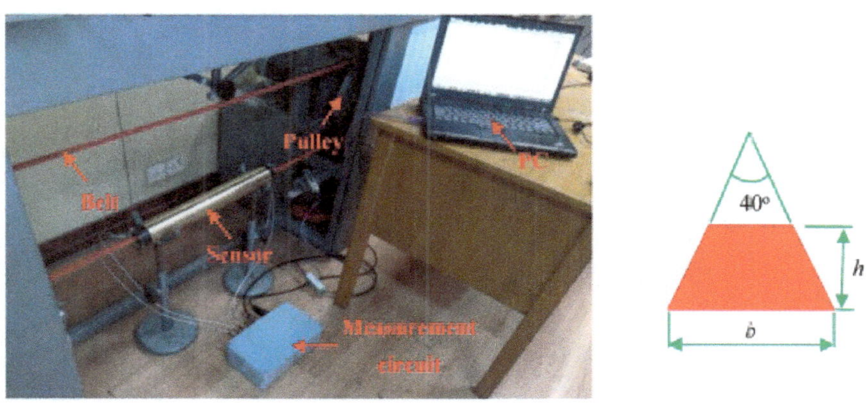

(a) (b)

Figure 6: V-belt pulley rig: (**a**) Photograph; (**b**) Cross-section of the belt.

Table 1: Belt types

Type	b	h	Area	Volume Fraction
#1	8 mm	5 mm	43.9 mm^2	2.23%
#2	13 mm	8 mm	80.7 mm^2	4.11%
#3	17 mm	11 mm	143.0 mm^2	7.28%

The rotation speed of the pulley (r) is measured by a frequency counter, and thus the velocity of the belt (v_B) can be calculated. It is regarded as the reference velocity:

$$v_B = \pi \cdot D \cdot r \tag{6}$$

The range of the rotation speed of the pulley is about 400 r/min–1300 r/min, which corresponds to a belt velocity range of 5.3 m/s–21.8 m/s. The axial spacing of the two electrostatic sensors is 40 mm. To assure that the relative measurement error in the transit time is less than ±2%, the sampling frequency is set at 20 kHz. The integration time is 0.1 s.

Figure 7 shows the typical electrostatic signals and their correlation function for belt #2 in the center of the pipe when the rotation speed of the pulley is 1164 r/min and its diameter is 320 mm. It can be seen that the two signals have a high similarity except for a time delay. An obvious peak can be observed in the correlation function and its time coordinate (2.08 ms) corresponds to the transit time, so the correlation velocity is 19.23 m/s, which is consistent with the reference velocity of 19.5 m/s. Figure 8 demonstrates the continuous measurement results of velocity and concentration for belt #2 in the center of the pipe at different rotation speeds of the pulley. The increasing rotation speed leads to a larger reference velocity, and thus larger measured velocity. The measured velocity and reference velocity are highly in agreement, even though there is a very small fluctuation due to the shaking and stretching of the belt. It also can be seen that the measured concentration is almost fixed as the ratio of the sectional area of the belt to that of the pipe is a constant for different belt velocities. Similar measurement results for the three types of belt at different rotation speeds of pulley are also obtained, which indicates that the C/V circuit has a good stability for particle concentration measurement regardless of particle velocities.

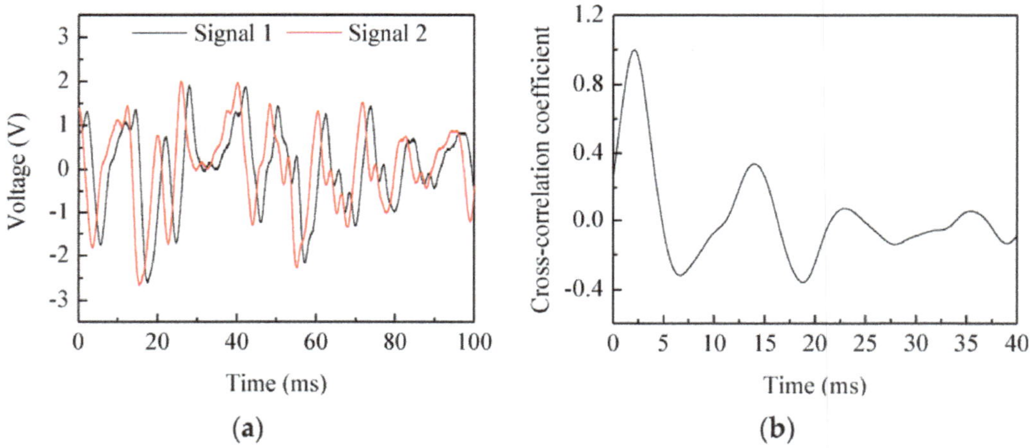

Figure 7: Typical electrostatic signals and their correlation function: (**a**) Electrostatic signals; (**b**) Correlation function.

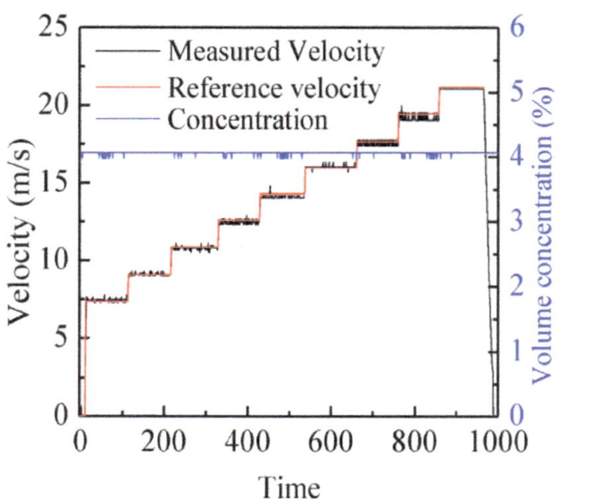

Figure 8: Continuous measurement results of velocity and concentration.

Figure 9 shows the comparisons of the measured velocities and reference velocities for the three types of belt at different rotation speeds of the pulley. It is obvious that the measured velocities are very close to the reference velocities, no matter whether the belt is in the center of the pipe or near the pipe wall (edge location). The relative error is within the range of −2%–4%, as shown in Figure 10. It verifies that the developed system can be used for accurate particle velocity measurement. Figure 11 demonstrates the measured volume concentrations for the three types of the belt. It can be seen that the measured concentrations for the location of "center" are always larger than those for the location of "edge", and the maximum relative error of the measured concentrations to the true value is about 32.1%. This is because the geometric size of the capacitance sensor is optimized based on glass, which will be introduced in Section 4.2, and the polyurethane has a different relative permittivity fromglass, so the calibrated results based on glass cannot be used directly for other materials. The results also indicate that the optimal geometric size of capacitance sensor changes and recalibration of the concentration measurement system must be conducted for different materials.

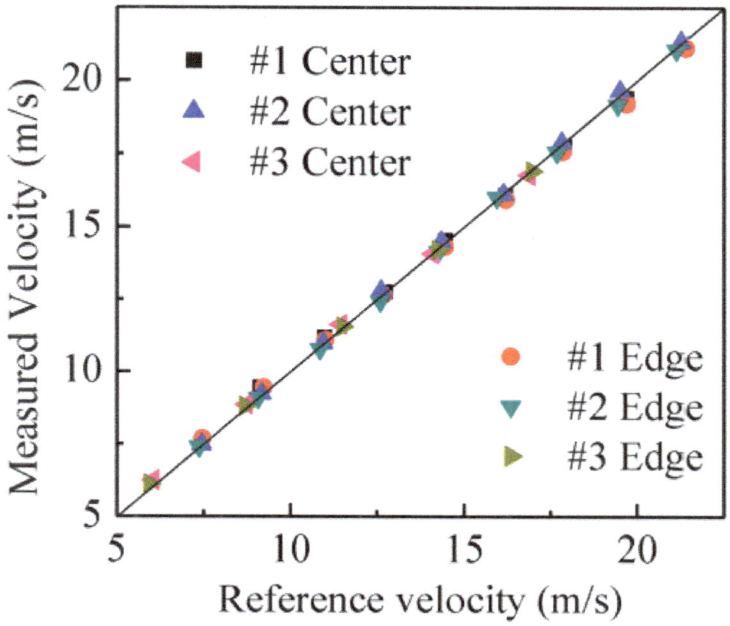

Figure 9: Comparison of measured and reference velocities.

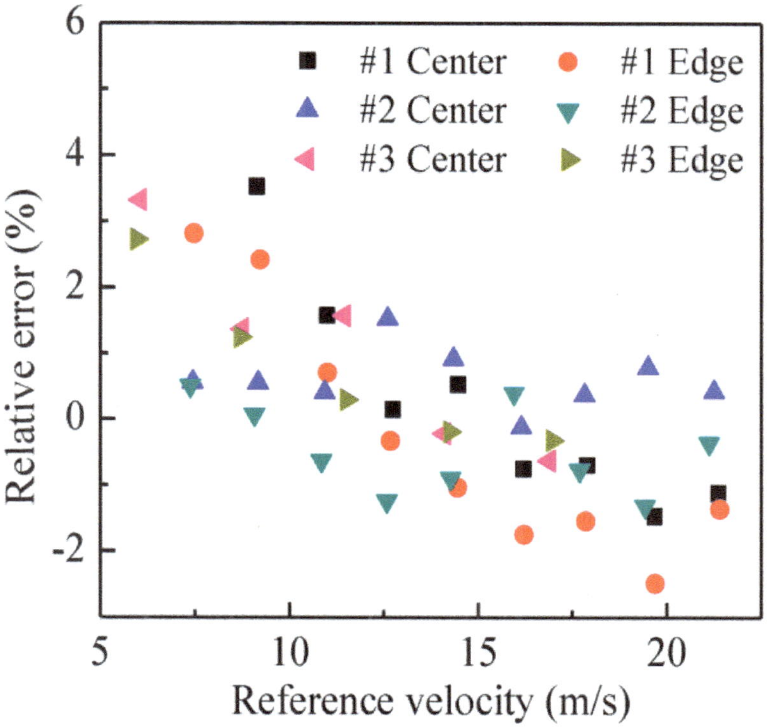

Figure 10: Relative measurement error of velocity.

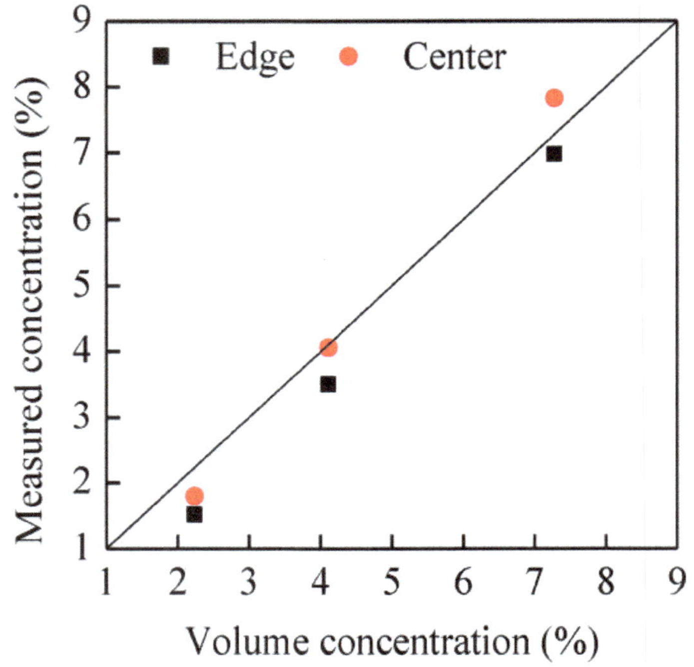

Figure 11: Measured volume concentration.

Calibration of the Capacitance Sensor

A capacitance sensor with helical electrodes has the same sensitivity on the circumference with a given radius [22]. The main factors affecting the sensitivity distribution are the pitch and axial width of the sensor electrodes. Usually, the pitch should be selected to be long enough in order to neglect the fringe effect. Then a homogeneous sensitivity distribution of the sensor can be obtained by optimizing the axial width of the electrode. In this paper, the pitch of the electrode is selected as $\pi \times 60$ mm, and the axial width is optimized by trial and error. A sequence of axial widths are tested by putting a 10 mm glass bar (parallel to the axial direction of the pipe) in the integrated system to compare the sensitivities on different locations of the pipe cross-section. As a result, the optimal axial width is 55 mm for glass. It should be noticed that the optimal axial width of the capacitance sensor varies with materials due to their different relative permittivities. When the different numbers of 10 mm glass bars are put into the pipe, the volume concentration of solid changes and its relationship with the output voltage of the C/V circuit can be calibrated. Figure 12 shows the output voltage of the C/V circuit for different volume concentration of glass. They are the average values of the output voltages when glass bars are parallel to the axis of the pipe in different optional positions over the cross-section. Figure 13 illustrates the maximum relative errors of the output voltage to the average value when glass bars are put into the pipe randomly.

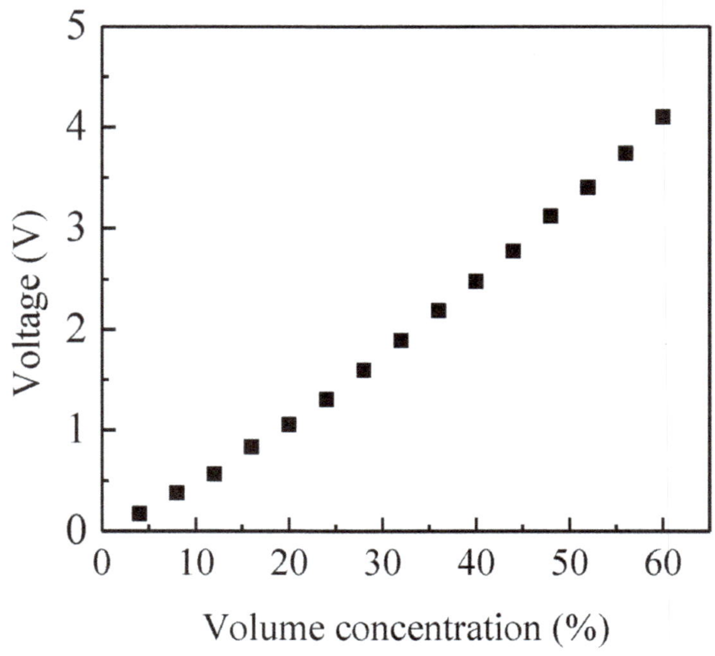

Figure 12: Average output voltage of the C/V circuit vs. concentration.

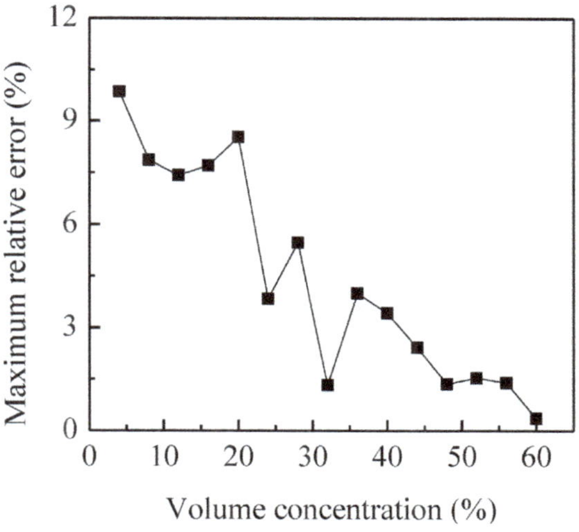

Figure 13: Maximum relative errorof the output voltageto the average value.

It can be seen that the output voltage of the measurement circuit increases with the volume concentration, but the relationship is not linear. Moreover, the maximum relative error is within ±10% anddecreases with the increasing volume concentration. By curve fitting, the relationship between the output voltage of the C/V circuit (V_C) and the volume concentration of glass can be expressed by a second-order polynomial:

$$\beta = -0.01V_C^2 + 0.194V_C \tag{7}$$

Therefore, the volume concentration can be easily calculated from the output voltage of the C/V circuit according to Equation (7).

Measurements on a Gravity-Fed Rig

Experiments are also carried out on a gravity-fed rig (Figure 14) to evaluate the effectiveness of the developed instrumentation system. The particles are glass beads with an average diameter of 3 mm, and thus Equation (7) can be directly used for the concentration measurement.

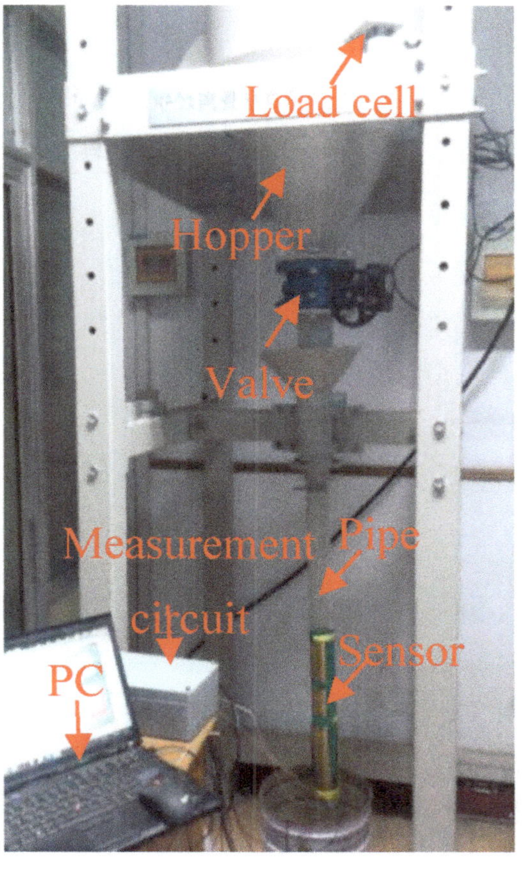

Figure 14: Gravity-fed rig.

In experiments, the mass flow rate of glass beads is adjusted by a valve at the outlet of the hopper. A load cell is installed to record the variation of the mass of the glass beads in the hopper. The average mass flow rate is calculated from its output signal and regarded as the reference value [15]. Due to the limitation of the distance between the sensors and the hopper, the particle velocity is less than 5 m/s. As for the configuration of the electrostatic and capacitance sensors in Figure 1, the sampling frequency and integration time are selected as 5 kHz and 0.4 s, respectively. Figure 15 shows the continuous measurement results of the particle velocity, concentration and mass flow rate for a typical valve opening. It can be seen that both the measured velocity and concentration have good stability, and hence the mass flow rate. Figure 16 shows the comparison between the measured mass flow rate and the estimated value from the load cells. It can be seen that the measured value is obviously larger than the estimated value. This is because the two electrostatic sensors are configured beneath the capacitance sensor. The acceleration of the particles passing through the sensors due to gravity could not be neglected. Therefore, the measured velocity is more than the mean velocity when particles flow through the capacitance sensor, which results in a larger value of the measured mass flow rate.

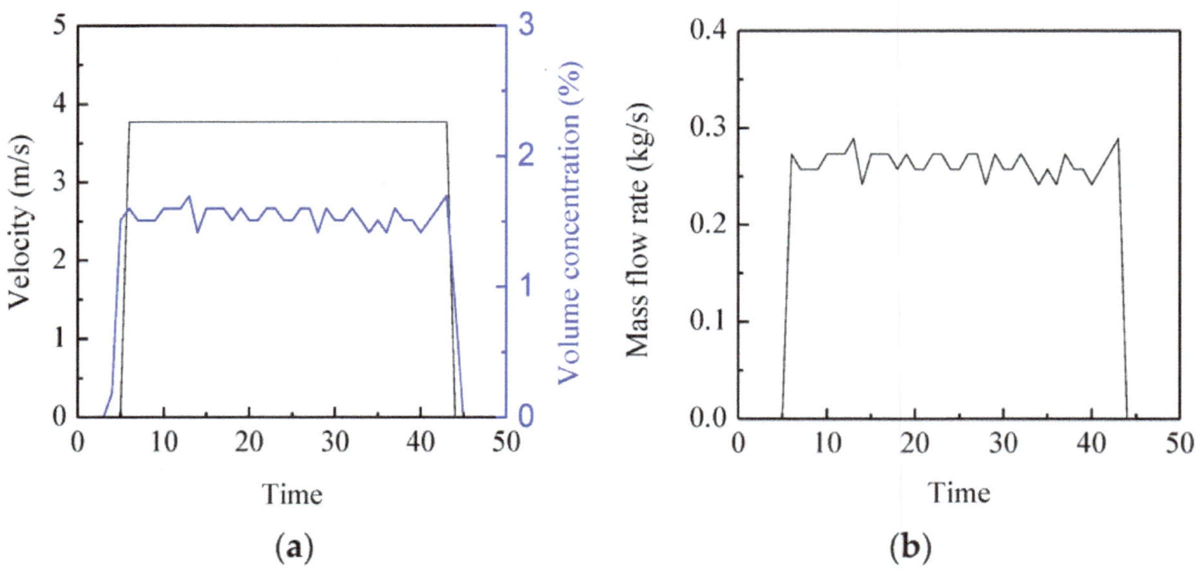

Figure 15: Continuous measurement results for a typical valve opening: (**a**) Particle velocity and concentration; (**b**) Mass flow rate.

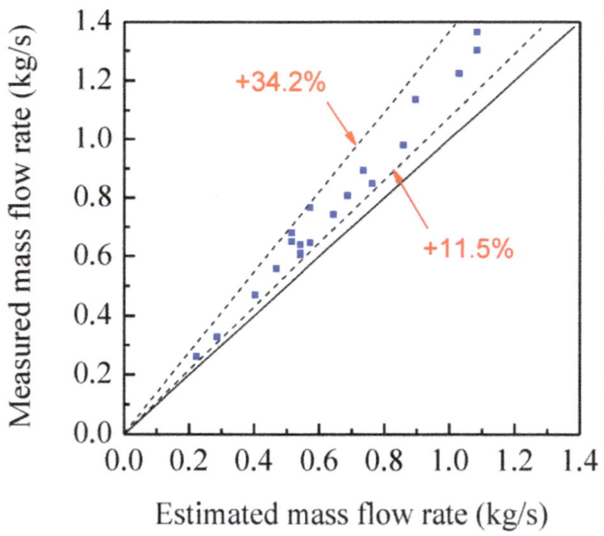

Figure 16: Comparison of the measured mass flow ratewith the estimated value.

To reduce the impact of the accelerated motion of particles, the two electrostatic sensors should be installed to each side of the capacitance sensor, as shown in Figure 2. As the axial spacing of the two electrostatic sensors is 250 mm, the sampling frequency and integration time are reselected as 1 kHz and 1 s, respectively.

Figure 17 shows the comparison of the average measured mass flow rate with the estimated value. It is can be seen that they have a good consistence. The relative error is −3%–8%, much better than that in Figure 16. As the particle concentration is withinabout 1%–8% in experiments, the measurement results are acceptable due to the relatively higher measurement error for low concentration as calibrated in Figure 13. It is expected that more accurate results can be obtained for higher concentrations of solids particles.

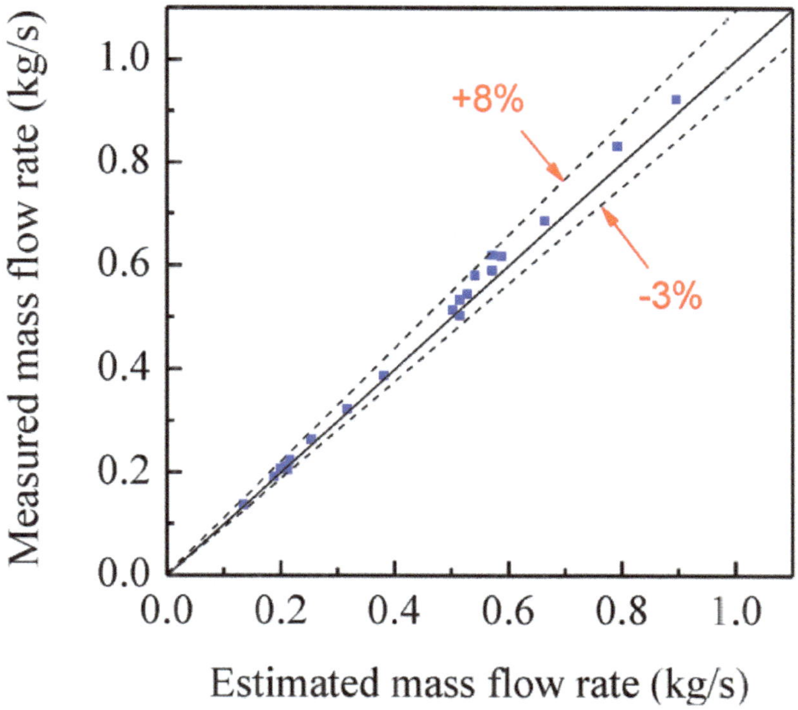

Figure 17: Comparison of the measured mass flow ratewith the estimated value for the sensor configuration in Figure 2.

CONCLUSIONS

An integrated instrumentation system based on electrostatic andcapacitance sensors has been developed in this paper. Experiments have been carried out on a V-belt pulley rig and a gravity-fed rig, respectively. The experimental results demonstrated that the measurement error of the solid mass flow rate is −3%–8% within the range of 0.13 kg/s–0.9 kg/s for glass beads. It proves that the system is capable of the online and continuous measurement of particle velocity, concentration and mass flow rate. Although the range of particle volume concentrations in the experiments is within about 1%–8%, it is expected that more accurate results can be obtained for higher concentrations of solids particles. Moreover, the instrumentation system is designed based on DSP and can be directly applied in industrial environments to provide particle flow parameters to system control units. In future work we will evaluate the performance of the instrumentation system on a dense phase pneumatic conveyor of pulverized fuel particles.

ACKNOWLEDGMENTS

The authors wish to express their gratitude to the National Natural Science Foundation of China (No. 51376049), the National Key Technology R&D Program of China (No. 2015BAA03B02) and Zhejiang Key Discipline of Instrument Science and technology (No. JL150501) for supporting the research as reported in this paper.

REFERENCES

1. Zheng, Y.; Liu, Q. Review of techniques for the mass flow rate measurement of pneumatically conveyed solids.Measurement 2011, 44, 589–604. . .
2. Cai, X.; Li, J.; Ouyang, X.; Zhao, Z.; Su, M. In-line measurement of pneumatically conveyed particles by a light transmission fluctuation method. Flow Meas. Instrum. 2005, 16, 315–320. . .
3. Nieuwland, J.J.; Meijer, R.; Kuipers, J.A.M.; van Swaaij, W.P.M. Measurements of solids concentration and axial solids velocity in gas-solid two-phase flows. Powder Technol. 1996, 87, 127–139. . .
4. Song, D.; Peng, L.; Lu, G.; Yang, S.; Yan, Y. Velocity measurement of pneumatically conveyed particles through digital imaging. Sens. Actuators A Phys. 2009, 149, 180–188. . .
5. Carter, R.M.; Yan, Y. An instrumentation system using combined sensing strategies for online mass flow rate measurement and particle sizing. IEEE Trans. Instrum. Meas. 2005, 54, 1433–1437. . .
6. Albion, K.; Briens, L.; Briens, C.; Berruti, F. Flow regime determination in horizontal pneumatic transport of fine powders using non-intrusive acoustic probes. Powd. Technol. 2007, 172, 157–166. . .
7. Tallon, S.; Davis, C.E. Use of the attenuation of acoustic pulsed waves for concentration measurement in gas-solid pipe flow. AIChE Symp. Ser. 1997, 93, 136–140. .
8. Barratt, I.R.; Yan, Y.; Byrne, B. A parallel-beam radiometric instrumentation system for the mass flow measurement of pneumatically conveyed solids. Meas. Sci. Technol. 2001, 12, 1515–1528. . .
9. Tajdari, T.; Rahmat, M.F. Review of Electrostatic Sensor Applications. J. Teknol. 2014, 70, 7–12. . .
10. Yan, Y.; Byrne, B.; Woodhead, S.; Coulthard, J. Velocity measurement of pneumatically conveyed solids using electrodynamic sensors. Meas. Sci. Technol. 1995, 6, 515–537. . .
11. Gajewski, J.B. Electrostatic nonintrusive method for measuring the electric charge, mass flow rate, and

velocity of particulates in the two-phase gas–solid pipe flows—its only or as many as 50 years of historical evolution.IEEE Trans. Ind. Appl. 2008, 44, 1418–1430. . .

12. Qian, X.; Huang, X.; Hu, Y.; Yan, Y. Pulverized coal flow metering on a full-scale power plant using electrostatic sensor arrays. Flow Meas. Instrum. 2014, 40, 185–191. . .
13. Li, J.; Fu, F.; Li, S.; Xu, C.; Wang, S. Velocity characterization of dense phase pneumatically conveyed solid particles in horizontal pipeline through an integrated electrostatic sensor. Int. J. Multiph. Flow 2015, 76, 198–211. . .
14. Yang, W.Q.; Beck, M.S. An intelligent cross correlator for pipeline flow velocity measurement. Flow Meas. Instrum. 1998, 8, 77–84. . .
15. Zhang, J.; Coulthard, J. Theoretical and experimental studies of the spatial sensitivity of an electrostatic pulverised fuel meter. J. Electrost. 2005, 63, 1133–1149. . .
16. Xu, C.; Li, J.; Wang, S. A spatial filtering velocimeter for solid particle velocity measurement based on linear electrostatic sensor array. Flow Meas. Instrum. 2012, 26, 68–78. . .
17. Li, J.; Xu, C.; Wang, S. Spatial filtering characteristics of electrostatic sensor matrix for local velocity measurement of pneumatically conveyed particles. Measurement 2014, 53, 194–205. . .
18. Elkow, K.J.; Rezkallah, K.S. Void fraction measurements in gas-liquid flows under 1-g and μ-g conditions using capacitance sensors. Int. J. Multiph. Flow 1997, 23, 815–829. . .
19. Hu, H.; Xu, T.; Hui, S.; Zhou, Q. A novel capacitive system for the concentration measurement of pneumatically conveyed pulverized fuel at power stations. Flow Meas. Instrum. 2006, 17, 87–92. . .
20. Yang, W.Q.; Liu, S. Role of tomography in gas/solids flow measurement. Flow Meas. Instrum. 2000, 11, 237–244. . .
21. Arko, A.; Waterfall, R.C.; Beck, M.S.; Dyakowski, T.; Sutcliffe, P.; Byars, M. Development of Electrical Capacitance Tomography for Solids Mass Flow Measurement and Control of Pneumatic Conveying Systems. In Proceedings of the 1st World Congress on Industrial Process Tomography, Buxton, Greater Manchester, UK, 14–17 April 1999; pp. 140–146.
22. Zhang, W.; Wang, C.; Wang, Y. Parameter selection in cross-correlation-based velocimetry using circular electrostatic sensors. IEEE Trans. Instrum. Meas. 2010, 59, 1268–1275. . .
23. Xu, L.A. Cross-Correlation Flow Measurement Technique; Tianjin University Publishing House: Tianjin, China, 1988. .
24. Jaworek, A.; Krupa, A.; Trela, M. Capacitance sensor for void fraction measurement in water/steam flows. Flow Meas. Instrum. 2004, 15, 317–324. . .
25. Li, H.; Yang, D.; Cheng, M. Sensitivity analysis of capacitance sensor with helical shaped surface plates. CIESC J. 2011, 62, 2292–2297. .
26. Abouelwafa, M.S.A.; Kendall, E.J.M. The use of capacitance sensors for phase percentage determination in multiphase pipelines. IEEE Trans. Instrum. Meas. 1980, 29, 24–27. . .
27. Tollefsen, J.; Hammer, E.A. Capacitance sensor design for reducing errors in phase concentration measurements. Flow Meas. Instrum. 1998, 9, 25–32. . .
28. Yang, W.Q.; York, T.A. New AC-based capacitance tomography system. IEE Proc. Sci. Meas. Technol. 1999, 146, 47–53. . .
29. Li, J.; Kong, M.; Xu, C.; Wang, S.; Wang, S. Influence of particle electrification on AC-based capacitance measurement and its elimination. Measurement 2015, 76, 93–103.

CHAPTER 13

A NOVEL LOW-COST INSTRUMENTATION SYSTEM FOR MEASURING THE WATER CONTENT AND APPARENT ELECTRICAL CONDUCTIVITY OF SOILS

Alan Kardek Rêgo Segundo[1], José Helvecio Martins[2], Paulo Marcos de Barros Monteiro[1], Rubens Alves de Oliveira[2], and Gustavo Medeiros Freitas[3]

[1]Department of Control and Automation Engineering and Fundamental Techniques, School of Mines, Federal University of Ouro Preto (UFOP)—campus Morro do Cruzeiro, sn, 35400-000 Ouro Preto, MG, Brazil

[2]Department of Agricultural Engineering, Federal University of Viçosa (UFV)—Avenida P.H. Rolfs, sn, 36570-900 Viçosa, MG, Brazil

[3]Vale Institute of Technology (ITV)—Avenida Juscelino Kubitschek, 31, Bauxita, 35400-000 Ouro Preto, MG, Brazil

ABSTRACT

The scarcity of drinking water affects various regions of the planet. Although climate change is responsible for the water availability, humanity plays an important role in preserving this precious natural resource. In case of negligence, the likely trend is to increase the demand and the depletion of water resources due to the increasing world population. This paper addresses the development, design and construction of a low cost system for measuring soil volumetric water content (θ), electrical conductivity (σ) and temperature (*T*), in order to optimize the use of water, energy and fertilizer in food production. Different from the existing measurement instruments commonly deployed in these applications, the proposed system uses an auto-balancing bridge circuit as measurement method. The proposed models to estimate θ and σ and correct them in function of *T* are compared to the ones reported in literature. The final prototype corresponds to a simple circuit connected to a pair of electrode probes, and presents high accuracy, high signal to noise ratio, fast response, and immunity to stray capacitance. The instrument calibration is based on salt solutions with known dielectric constant and electrical conductivity as reference. Experiments measuring clay and sandy soils demonstrate the satisfactory performance of the instrument.

INTRODUCTION

Currently the lack of drinking water has become one of the major problems faced by humanity. According to UNICEF, less than half of all people in the world have access to this resource.

Citation: Rêgo Segundo, A.K.; Martins, J.H.; Monteiro, P.M.B.; de Oliveira, R.A.; Freitas, G.M. A Novel Low-Cost Instrumentation System for Measuring the Water Content and Apparent Electrical Conductivity of Soils. Sensors 2015, 15, 25546-25563. DOI: 10.3390/s151025546

The most water-consuming activities are related to agricultural irrigation, which has to face the challenges of increasing production, while decreasing the water consumption and the environmental impacts. Because of that, it is crucial to develop and deploy technologies to allow sustainable agriculture [1,2,3].

The measurement of soil electrical properties is an alternative to estimate some of its physicochemical properties, which can aid in the management of irrigation and fertilization, making it suitable for cultivation. The dielectric constant (ε) of soil, for example, highly corralates with soil volumetric water content (θ). Therefore, technologies used to measure this parameter in real time can optimize the timing of irrigation [4]. On other hand, the conductivity (σ) of soil can be used to estimate the degree of soil salinity [5]. Furthermore, each type of crop presents optimal productivity at a given level of soil salinity [4].

The irrigation management based on the measurement of water content θ in real time allows one to restrict the amount of water applied, making its use more efficient [6,7,8,9]. Volumetric soil water content is generally regarded as an easier quantity to use than gravimetric content, particularly in irrigation scheduling and calculations of available soil water content. In addition, monitoring the conductivity σ can indicate when the soil needs correction. This parameter has long been used in the construction of spatial variability maps, which allows one to divide production areas into different management zones [5,10]. Thus, sustainable agriculture can be developed ensuring sufficient resources for future generations [4].

There are some types of sensors on the market to estimate θ and σ. However, in general, these devices are expensive, which may discourage producers from using this technology, particularly when it is necessary to import the equipment. Some researchers have proposed the development of systems for monitoring the electrical parameters of the soil [11,12,13,14,15,16,17,18,19,20].

The low-cost system developed in this study monitors water content θ, conductivity σ and temperature (T), and could be used for irrigation and fertilization control systems, providing real time measurements. Given the importance of water conservation, it is important to offer the producer various technological options to solve the problem and especially divulge the need to use instrumentation in irrigation systems.

The complete system for irrigation control is composed of: (i) a sensor to measure the soil θ, σ and T; (ii) and an automatic calibration system for θ in function of ε, for a specific soil. This paper describes the proposed instrumentation system for irrigation, consisting of a functional prototype [21], which will be deployed in field experiments for validation.

EXPERIMENTAL SETUP

Probes for Measuring θ, σ and T

Patents US2013073097 and US3882383 describe a probe capable of detecting water content in the soil [22,23]. However, the system is based only on conductivity measurements, and does not consider the dielectric constant of the soil. Probes designed to estimate the soil water content based on conductivity usually do not have good precision, because the results also depend on the soil salinity. Patent US5445178 proposes a sensor for this purpose based only on the dielectric conductivity [24], but it does not measure the soil electrical conductivity.

In this work, low-cost probes to measure θ and σ were developed based on measurement of the electrical impedance of the soil material located between two parallel stainless steel rods. In order to construct the probes, stainless steel rods (length = 130 mm, diameter = 3 mm), liquid polyester resin, a semiconductor temperature sensor (LM35), five-way cable and covers for electrical outlets plugs were used, according to the scheme shown in Figure 1a. The cover for electrical outlets plugs served as a housing and to fix the rods and the temperature sensor using polyester resin. Figure 1b illustrates the probes implementation.

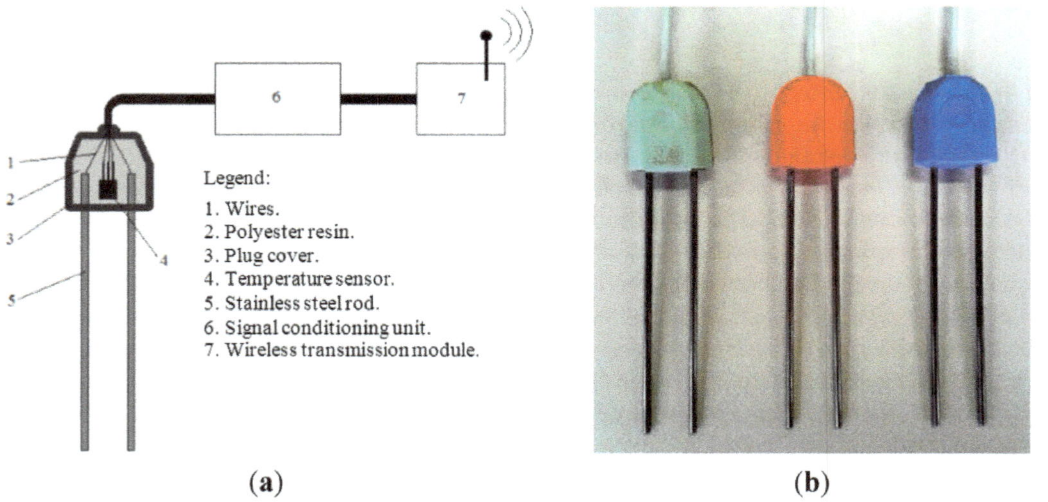

Figure 1: (**a**) Probe for measuring water content and apparent electrical conductivity of soil; (**b**) Probes for measuring temperature, electrical conductivity and relative dielectric constant.

Theory and Signal Conditioning Unit

According to Ohm's law in complex notation, the impedance (**Z**) corresponds to the ratio between the voltage (**V**) and the current (**I**) phasors. This concept assumes that the electrical properties of materials/circuits are time-invariant. The impedance establishes the relationship between module and phase of current and voltage signals in a dipole.

In some cases, it can be more convenient to use the admittance (**Y**)—inverse of impedance, for the analysis of electrical circuits. Equations (1) and (2) represent the impedance and admittance, respectively:

$$Z=R+jX \tag{1}$$

$$Y=G+jB \tag{2}$$

where $j^2 = -1$, R is the resistance and X the reactance in Ohms; G is the conductance and B the susceptance in Siemens.

The real part of Equations (1) and (2) is related to the losses by Joule effect. The imaginary part is the ability to exchange energy.

The complex relative permittivity (ε) is often used to characterize the electrical properties of materials. This parameter is related to the absolute complex permittivity (ε_a) according to Equation (3):

$$\varepsilon_a = \varepsilon_0\,\varepsilon \qquad (3)$$

where ε_0 is the permittivity of free space (=8.85 pF·m^{-1}).

The fluid parameter ε presents dielectric relaxation, where the real part decreases with increasing frequency. This phenomenon occurs in the GHz-range. In this work, the frequency range of the measurements is limited to 5 MHz. Therefore, relaxation mechanisms can be neglected, and the complex relative permittivity can be represented by Equation (4):

$$\varepsilon = \varepsilon - j\frac{\sigma}{\omega\varepsilon_0} \qquad (4)$$

where ε is the relative dielectric constant (dimensionless), σ the electric conductivity (S·m^{-1}) and ω the angular velocity (rad·s^{-1}). The imaginary part is the dielectric losses factor.

The impedance measurements of solid, liquid or gaseous substances are carried by means of a probe. Equation (5) lists the electrical properties of the substance around the electrodes, and the probe admittance:

$$\mathbf{Y} = j\omega k_g \varepsilon_0 \boldsymbol{\varepsilon} \qquad (5)$$

where k_g is the geometric constant of the probe, G=kgσ and B=kgωε0ε.

As the impedance is a complex variable, it is necessary to determine two parameters to define it: module and phase, or real and imaginary part of voltage or current. The standard method of impedance measurement consists on applying a pure sinusoidal voltage at a single frequency to the sensor electrodes and measure the phase shift and amplitude, or the real and imaginary parts, of the resulting current using either analog circuit or analog-to-digital conversion and signal processing algorithm to analyze the response [25]. In this context, different measurement techniques have being proposed based on the frequency range, required accuracy, measurement range and complexity of the system [26,27,28].

Measuring voltage and current above 100 MHz are usually difficult, and generally are not directly applicable to high-frequency devices (3–30 MHz), as in the case of the patent US5479104 [29] and US5418466 [30], which work at frequencies up to 100 MHz and 150 MHz, respectively. In this case, the determination of impedance is usually derived from the measurement of wave reflection and transmission together with distributed circuit concepts, such as Theta Probe [31], HidroSense [32], TRIME tube access probe [33]. For this purpose, it is common to employ network analyzers and time domain reflectometers (TDRs), which have higher costs when compared to low frequency measurement systems [27,34].

At lower frequencies (up to MHz range) impedance is determined by current-voltage (or I-V), bridge or resonant methods. The I-V method is quite simple, but has low accuracy. The bridge method has high accuracy, but due to the need to balance the bridge (need of balancing), it is not suitable for fast, repeated and continuous measurements. The resonance method exhibits good accuracy, but also requires adjustment of resonance, which results in the same problem of the bridge method. Moreover, all the above methods are sensitive to stray capacitances to ground, which are normally present in probes due to connecting cables or other grounded metallic parts of a probe, for example. These stray capacitances can impair the measurement impedance, which makes it necessary to use more complex circuitry to eliminate their effect [25].

Sensor 5TE (Decagon Devices Inc., Pullman, WA, USA), for example, provides a 70 MHz wave to the prongs and the bult in microprocessor measures the stored charge, which is proportional to soil ε [13]. This technique requires a third rod to measure ε and σ. Furthermore, this technique does not eliminate the effect of stray capacitances, and requires hardware capable to operate in high frequencies.

Attempting to solve some of these drawbacks, we adopt a variation of the fourth method for measuring impedance, called auto-balancing bridge or self-balancing bridge. This method has a high accuracy, fast response and simple circuit. The measurement is improved with the inclusion of an operational amplifier with high input and low output impedances (Figure 2). This setting is also known as trans-impedance amplifier or current-voltage converter. It has high signal to noise ratio and stray capacitance immunity, capable of measuring small impedances between the electrodes even in the presence of large stray capacitances to ground [25,35].

A measurement system for θ and σ using this measurement method, and neither a low cost embedded system that performs the conditioning and signal processing could not be found in the literature. This circuit allows estimating σ at low frequency (kHz) and ε at high frequency (MHz), requiring only one pair of electrodes in the probe.

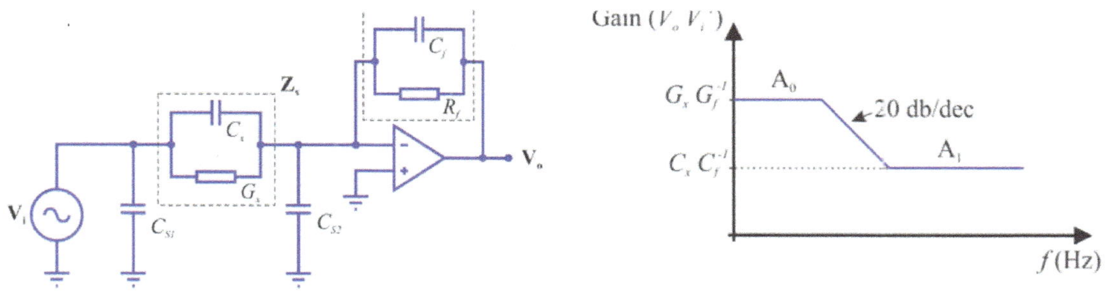

Figure 2: (**a**) Auto-balancing bridge circuit and frequency domanin response (**b**).

The basic auto-balancing bridge circuit diagram is represented in Figure 2a, where $\mathbf{V}_i$ is the input voltage, $\mathbf{Z}_x$ is the unknown impedance, $\mathbf{Z}_f$ is the impedance feedback circuit and $\mathbf{V}_o$ is the output voltage. The potential of the operational amplifier non-inverting port is connected to ground potential, which is also known as a virtual ground. Thus, the currents passing through $\mathbf{Z}_f$ and $\mathbf{Z}_x$ are balanced through the operational amplifier action, and the current through the unknown impedance is proportional to the operational amplifier output voltage.

In fact, there are stray capacitances C_{s1} and C_{s2}, caused by cables used to connect the sensor to the measuring circuit according to Figure 2b. These capacitances do not affect the measurements since C_{s1} is drained directly by the voltage source and C_{s2} is virtually grounded by the operational amplifier. This is a major advantage of auto-balancing bridge circuit [25]. The gain of the circuit shown in Figure 2b is given by Equation (6):

$$\frac{\mathbf{V}_0}{\mathbf{V}_i} = -\frac{\mathbf{Z}_f}{\mathbf{Z}_x} = -\frac{\mathbf{Y}_f}{\mathbf{Y}_x} = -\left(\frac{G_x + j\omega C_x}{G_f + j\omega C_f}\right) = -k_g\left(\frac{\sigma + j\omega\varepsilon_0\varepsilon}{G_f + j\omega C_f}\right) \qquad (6)$$

where G_x and G_f are the conductance, C_x and C_f are the capacitances of the impedances $\mathbf{Z}_x$ and $\mathbf{Z}_f$, respectively.

In Equation (6), C_x and G_x are directly connected with the determination of the sensor σ and ε. This involves the measurement of two parameters, which may be: (i) magnitude and phase in a single frequency; (ii) real and imaginary parts of $\mathbf{V}_0$ at a single frequency; or (iii) two amplitudes at different frequencies. These three possibilities are mathematically equal, but differ with respect to the complexity of the circuit [25,36]. This work proposes an electronic circuit for instrumentation that uses the third option to estimate σ and ε of the material located between the probe electrodes.

According to Equation (6), two thresholds can be identified by the ratio of $G_xG_f^{-1}$ *and* $C_xC_f^{-1}$ when frequency $f \to 0$ and $f \to \infty$, according to Figure 2b. In practice, one should choose two frequencies located on each level. The voltage gain at lower frequency (A_0) can be correlated with σ, and the voltage gain in the higher frequency (A_1) with ε of the material located between the probe electrodes.

Silva [25] presented an approach to measure ε and σ of saline solutions using auto-balancing bridge circuit. However, these experiments were performed using signal generators for circuit excitation, and data was monitored and stored in the computer through an oscilloscope and a signal-processing algorithm. Solutions ε and σ were estimated by measuring the gain of the auto-balancing bridge circuit at two different frequencies.

The present work describes the development of an embedded system in order to perform all tasks proposed in Silva [25], and to measure the soil parameters ε and σ for agriculture purposes. Furthermore, the system measures temperature to evaluate its effect on the measurements of ε and σ. Figure 3 shows the circuit diagram and the signal-conditioning unit.

In the circuit diagram shown in Figure 3a, the microcontroller provides Pulse Width Modulation (PWM) signals at different frequencies (100 kHz and 5 MHz) with 50% working cycle. These signals pass through bandpass filters to make them closer to sine functions. Therefore, they can be used as the excitation source for self-balanced bridge circuit.

The input and output of self-balanced bridge circuit are measured by the analog-to-digital converter (ADC) of the microcontroller. However, these signals are rectified before performing the reading. The multiplexers are meant to direct the power signals to the auto-balancing bridge circuit and to select which signal is read by the ADC: input signals or output. In addition, the microcontroller performs measurements of the temperature sensor located within the probe circuit and battery power level. This data is sent wirelessly to the master program installed on a computer. The measurement of load cell signal is used only in the system calibration process, which will be described later.

It is possible to use a variety of microcontrollers and amplifiers to implement the proposed circuit. For this implementation, the auto-balancing bridge uses a 680 Ω resistor, a 270 pF capacitor and an AD8065 FastFET operational amplifier. The microcontroller is a PIC16F873A. The multiplexers employ a CMOS quad bilateral switch CD4066. Both band-pass filter and rectifier are implemented with a TLE2072 operational amplifier.

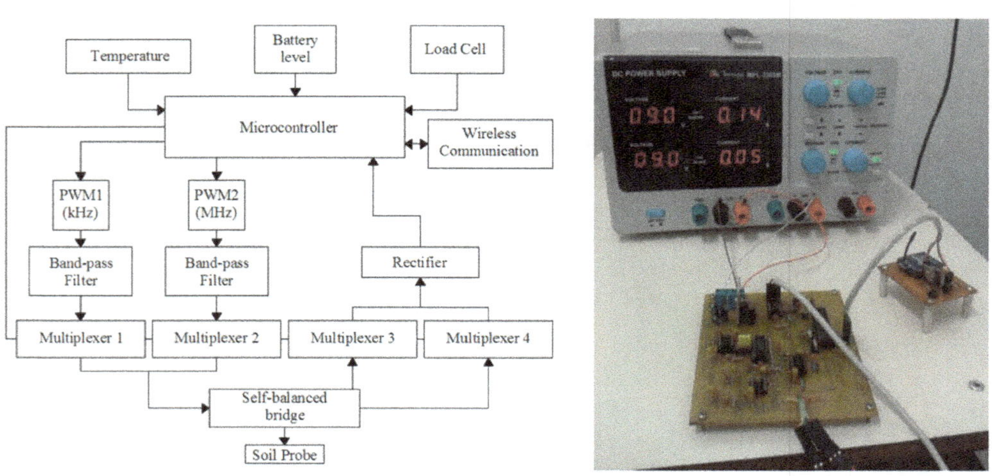

(a) (b)

Figure 3: (**a**) Signal-conditioning unit circuit diagram; (**b**) Signal-conditioning unit implemented circuit.

Calibration and Testing

Calibration Using Salts Solutions

To perform the calibration procedure seven different substances were used as references, all of which having both σ e ε known, as follows: µ

- Air (σ = 0 µS·cm−1; ε = 1);
- Deionized water. (σ = 4.5 µS·cm−1; ε = 80);
- Drinking water. (σ = 68.7 µS·cm−1; ε = 80);
- SolutionofwaterandNaCl1(σ=145.8 µS·cm−1; ε=80);
- SolutionofwaterandNaCl2(σ=348.4 µS·cm−1; ε=80);
- SolutionofwaterandNaCl3(σ=838.8 µS·cm−1; ε=80);
- Ethanol fuel (σ =7.9 µS·cm−1; ε = 24).

The constant σ of each material was measured using an apparatus with automatic temperature compensation. On the other hand, ε was obtained from the literature. These values were adopted as conventional real values, both for σ and ε of the solutions.

The tests were conducted in a climate chamber with temperature controlled at 25 ± 1 °C. Forty measurements were performed for each situation. So, there were seven treatments and 40 replicates for each of the three designed probes.

For the calibration models, circuit voltage signals from the gains A_0 and A_1 of the auto-balancing bridge at each frequency were respectively correlated with σ and ε using linear regression.

After obtaining these models, the effect of temperature was evaluated. For this purpose, measurements were taken under seven different temperature conditions described above. Subsequently, the cooling chamber was configured to vary the temperature from 5 °C to 45 °C with 5 °C steps, with each step lasting for 2.5 h. During this process, the supervisory software carried out data acquisition.

According to the solutions measurements of σ and ε accomplished at different temperatures, two correction models were obtained empirically. After, they were applied in the soil conductivity and water content measurements. Both of them have also being compared to the correction models proposed by Rhoades *et al.* [37] Equation (7) and Chanzy *et al.* [38] Equation (8). Root mean square error (RMSE) and coefficient of determination (R^2) were employed to compare the corrections difference:

$$\sigma_{25} = \left[1 - 0.20346\left(\frac{T-25}{10}\right) + 0.03822\left(\frac{T-25}{10}\right)^2 - 0.00555\left(\frac{T-25}{10}\right)^3\right]\sigma_m \quad (7)$$

where σ_{25} is the electrical conductivity adjusted to 25 °C, and σ_m is the conductivity measured according to the temperature *T*. For the complex relative permittivity, we have:

$$\varepsilon_{25} = \varepsilon_m + \alpha(25 - T) \quad (8)$$

where α is obtained empirically, based on the soil θ and σ; here we adopt α = 0.114 [38]. This model will be employed and compared to the one proposed in this work.

Tests Conducted with Soil

Two tests using clay and sandy soil were performed. These tests consisted of saturating a soil sample with water and submitting it to a drying procedure. Meanwhile, the sample mass and the signals corresponding to σ and ε were measured by the signal-conditioning unit and stored in the supervisory program database. A load cell suspending the sample measures its mass, as illustrated in Figure 4. The signal-conditioning unit is also responsible for performing these measurements.

The soil drying procedure consists on placing the structure shown in Figure 4 in the climate chamber, which was scheduled to reach 50 ± 1 °C for three hours and go back to 25 ± 1 °C for five hours, until the soil gets quite dry, and its weight practically stops changing. After this test, the soil moisture content by weight using a standard oven method for each measuring point is obtained. Finally, a correlation model between dielectric constant of the soil and the soil water content is fitted using linear regression.

Attempting to minimize the biasing influences on the sensor readings about the vertical gradient of water content from soil saturation to field capacity, the weighting container has restricted dimensions—75 mm diameter and 200 mm height.

The structure presented in Figure 4 was developed to find the calibration models for a specific soil according to a standard procedure. This procedure is necessary since the measuring circuit to obtain σ, ε and the soil temperature also detects the mass inside the weighing recipient. The acquired information is sent to the supervisory software that adjusts the calibration model. The novel calibration device greatly enhances the estimation precision of θ.

For experimental validation, we took 2608 measurements of sandy soil and 1780 measurements of clay soil, during a period of 12 and 16 days, respectively. The volumetric water content ranges for sandy and clay soil were 0.02–0.40 and 0.02–0.43. The volumetric soil water content was calculated multiplying the soil water content on weight basis by the soil bulk density ρ, which in this case is 1.27 g·cm^{-3} for sandy and 0.92 g·cm^{-3} for clay soils.

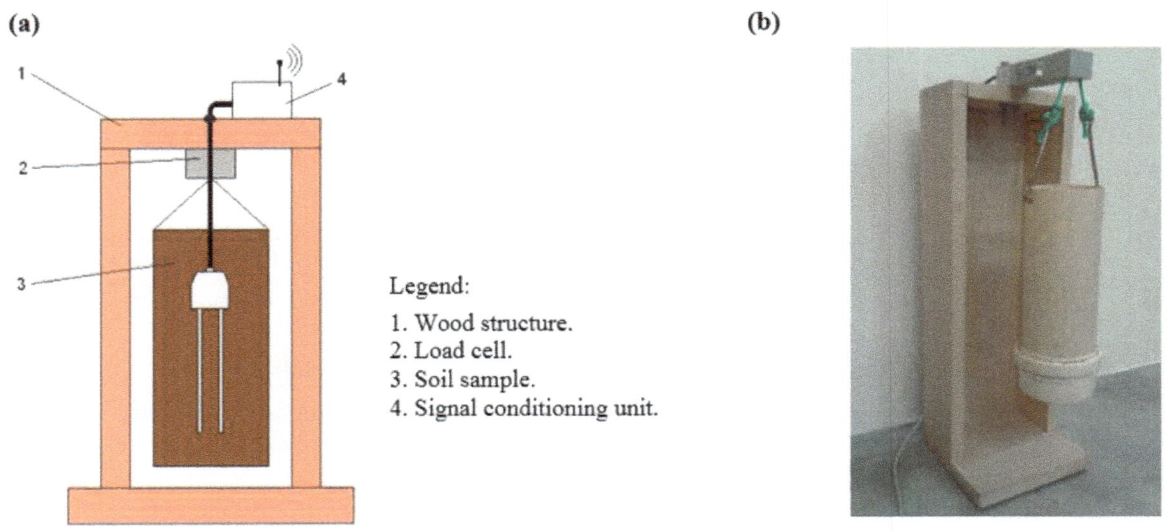

Figure 4: (**a**) Scheme support structure of the soil samples; (**b**) Structure of weighing the soil samples.

In order to validate the correlation models between the dielectric constant and the water content, we employed the K-fold cross validation method. This analysis consists on randomly dividing the original sample data in K subsamples. After that, a single subsample is retained to be used as the validation, and the remaining K − 1 subsamples are used as training data. The cross-validation process is then repeated K times (the folds), until each K subsample is employed once as the validation data.

The K results from the folds then can be averaged to produce a single estimation. The advantage of this method over repeated random sub-sampling is that all observations are used for both training and validation, and each observation is used for validation exactly once [39]. Here we applied K = 10 and 100 rounds.

The models proposed to predict θ based on ε were compared with models developed by Topp *et al.* [40], Ledieu *et al.* [41] and Malicki *et al.* [42]. These models are represented by Equations (9)–(11), respectively. To quantify the models accuracy, we analyzed the RMSE and R^2 between observed and predicted values:

$$\theta = 4.3 \cdot 10^{-6}\varepsilon^3 - 0.00055\varepsilon^2 + 0.0292\varepsilon - 0.053 \tag{9}$$

$$\theta = 0.1264\sqrt{\varepsilon} - 0.1933 \tag{10}$$

$$\theta = \frac{(\sqrt{\varepsilon} - 0.819 - 0.168\rho - 0.159\rho^2)}{7.17 + 1.18\rho} \tag{11}$$

RESULTS AND DISCUSSION

Salt Solutions

Figure 5a shows the correlation between σ measured by a conductivimeter and A_0 for a temperature of 25 ± 1 °C.Figure 5b shows the correlation between ε of the substances and A_1 for a temperature of 25 ± 1 °C.

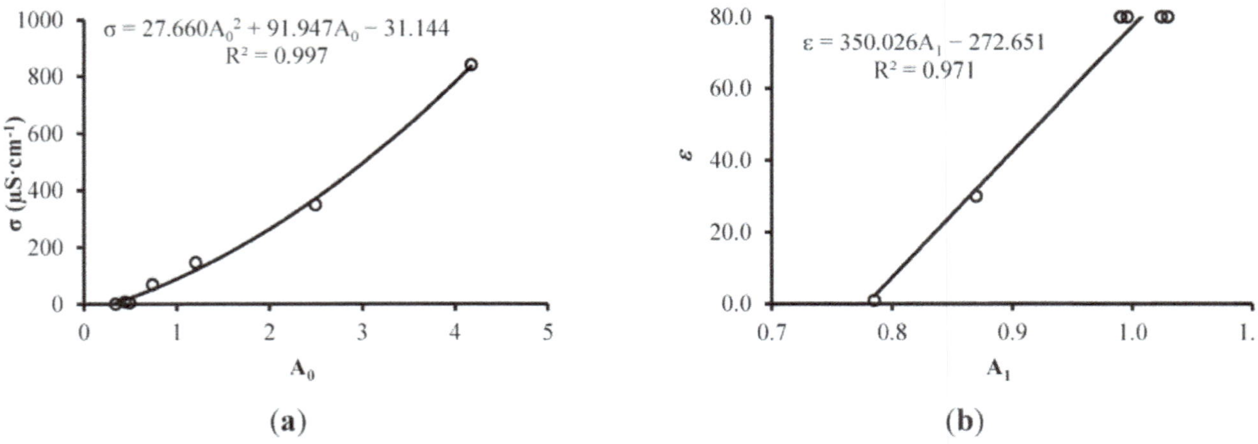

Figure 5: Relationships between σ-A_0 (**a**) and ε-A_1 (**b**).

This experiment is based on work by Silva [25], which employs air (ε = 1, σ = 0), oil (ε = 2, σ = 0), isopropanol (ε = 19, σ = 0.06 μS·cm^{-1}), glycol (ε = 37, σ = 3 μS·cm^{-1}), deionized water (ε = 79, σ = 2 μS·cm^{-1}) and water + salt (ε = 79, σ = 26 μS·cm^{-1}) to demonstrated the first order relationship between the auto-balancing bridge circuit gain at high frequency and the dielectric constant of these substances. Assuming the same first order relationship and measuring principles, we use only three reference points of ε (Figure 5b).

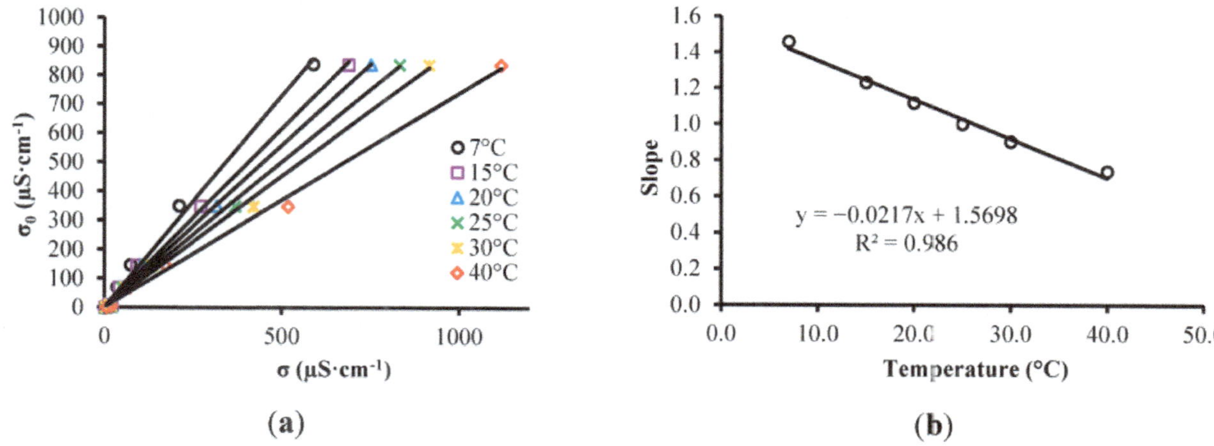

Figure 6: (**a**) σ-σ_0 relationship; (**b**) Correlation between temperature and slope of the fitted models for each temperature.

The effect of temperature on the measurement results of σ were evaluated as a function of the slopes of correlation equations between the values of electrical conductivity measured by the proposed system and electrical conductivity values measured by a conductivimeter (σ_0) at different temperatures, as shown in Figure 6a. Thus, a model of the average temperature of each solution and the slope of each model was obtained, as shown in Figure 6b. With this model, it is possible to correct the measured σ and present the measurement result corrected for 25 °C, according to Equation (12):

$$\sigma_{25} = (-0.0217T + 1.5698)\sigma_m \quad (12)$$

To evaluate the effect of temperature on the measured results of ε the same procedure described above was adopted. The slopes of correlation patterns between the measured value of ε and the conventional true value of ε (ε_0) were used for each temperature, as shown in Figure 7.

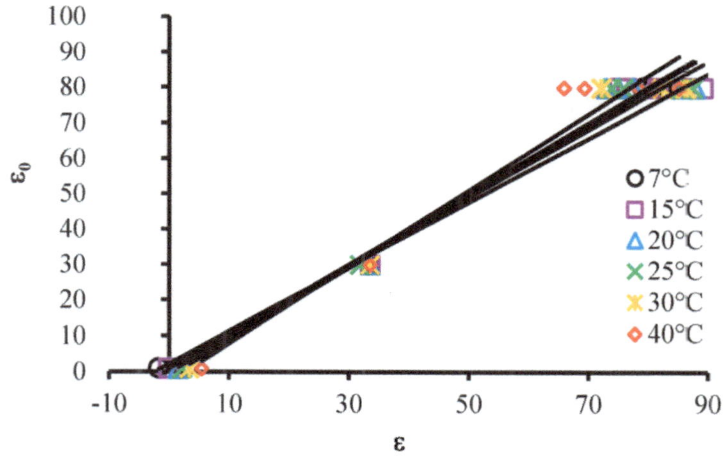

Figure 7: Models for comparison between measured values and true conventional values of ε for each different temperature.

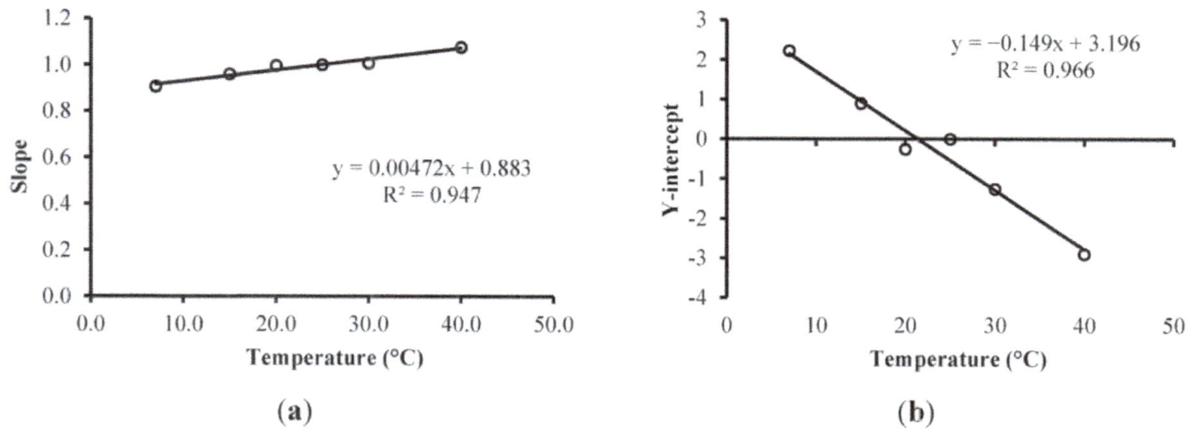

Figure 8: (**a**) Correlation between the temperature of the solutions and the slopes of the models for each temperature; (**b**) Correlation between the temperature of the solutions and the Y-intercepts of the models for each temperature.

Two models were fit between the average temperature of each solution and slopes and Y-intercepts, as shown in Figure 8. Thus, it is possible to correct the measured ε and present the measurement result at a reference temperature of 25 °C, according to Equation (13).

Equation (13) presents the correction model for ε empirically developed in this work based on the temperature:

$$\varepsilon_{25} = (0.00472T + 0.8834)\varepsilon_m - 0.1494T + 3.1957 \quad (13)$$

where ε_{25} corresponds to ε adjusted to 25 °C, and ε_m is ε measured according to the temperature *T*.

Soils

The measurement of σ is greatly influenced by temperature, as has already been observed in the tests with solutions. Figure 9 shows graphs of correlation between σ at 25 °C (σ_{25}) and σ at all temperatures (20, 25, 30, 35 and 40 °C), measured before and after correction with temperature employing the correction model presented in this work (Equation (12)) and also the model proposed by Rhoades *et al.* [37] (Equation (7)). Table 1 shows R^2 and RMSE, after and before the σ correction according to the temperature.

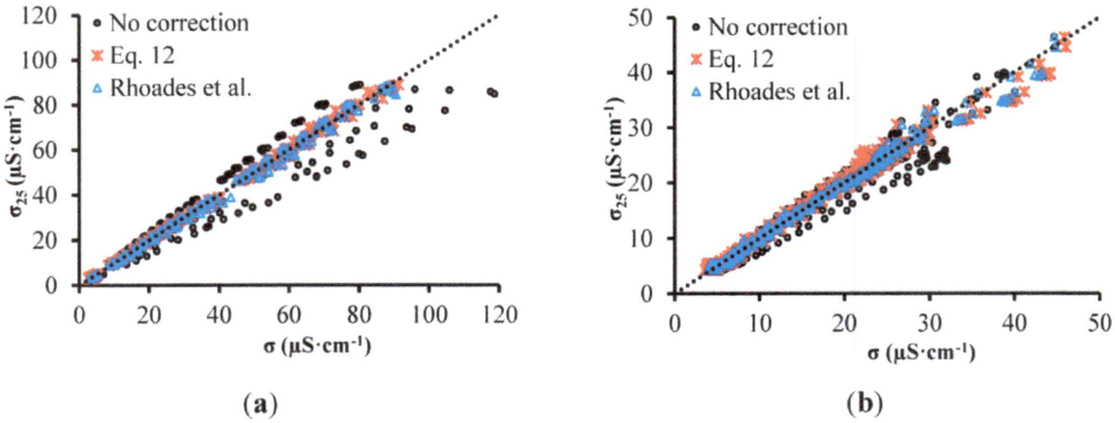

Figure 9: σ-σ_{25} relationship for sandy soil (**a**) and clay soil (**b**).

Table 1: Results from the comparison between σ correction models based on temperature

	Sandy Soil		Clay Soil	
	RMSE	R^2	RMSE	R^2
No correction	6.7727	0.9321	1.9827	0.9598
Equation 12	0.9676	0.9986	1.0620	0.9886
Rhoades *et al.*	1.3751	0.9971	0.7929	0.9936

Before correction, the adjusted models showed RMSE of approximately 6.8 μS·cm^{-1} for sandy soil and 2.0 μS·cm^{-1}for clay soil. Using the correction model presented by Rhoades *et al.* [37] Equation (7), it is possible to note that R^2increases while RMSE decreases, for both soils. Employing the model developed here, R^2 increases while RMSE decreases even more, which indicates that the correction model proposed in this work is more accurated.

In this work we propose three prediction models: the first two Equations (14) and (15) are specific for sandy and clay soils; the third model was developed for both types of soil Equation (16). Even so, similar to Malicki *et al.* [42], it also takes into account ρ:

$$\theta = 21.98 \cdot 10^{-6}\varepsilon^3 - 0.001419\varepsilon^2 + 0.03638\varepsilon - 0.1786 \quad (14)$$

$$\theta = 6.491 \cdot 10^{-6}\varepsilon^3 - 0.0004087\varepsilon^2 + 0.01180\varepsilon + 0.01907 \quad (15)$$

$$\theta = (4.518 \cdot 10^{-6}\varepsilon^3 - 0.0002746\varepsilon^2 + 0.010984\varepsilon - 0.006706)\rho \quad (16)$$

Equations (14) and (15) could be considered a recalibration of Topp's equation, estimating water content based on ε through a third-degree polynomial, whose coefficient values change from soil to soil.

Table 2 presents the results from the linear regression and cross-validation related to the models proposed in this work. Figure 10 illustrates the relation between θ predicted by each model and observed using the calibration system. Table 3 presents R^2 and RMSE for each relation.

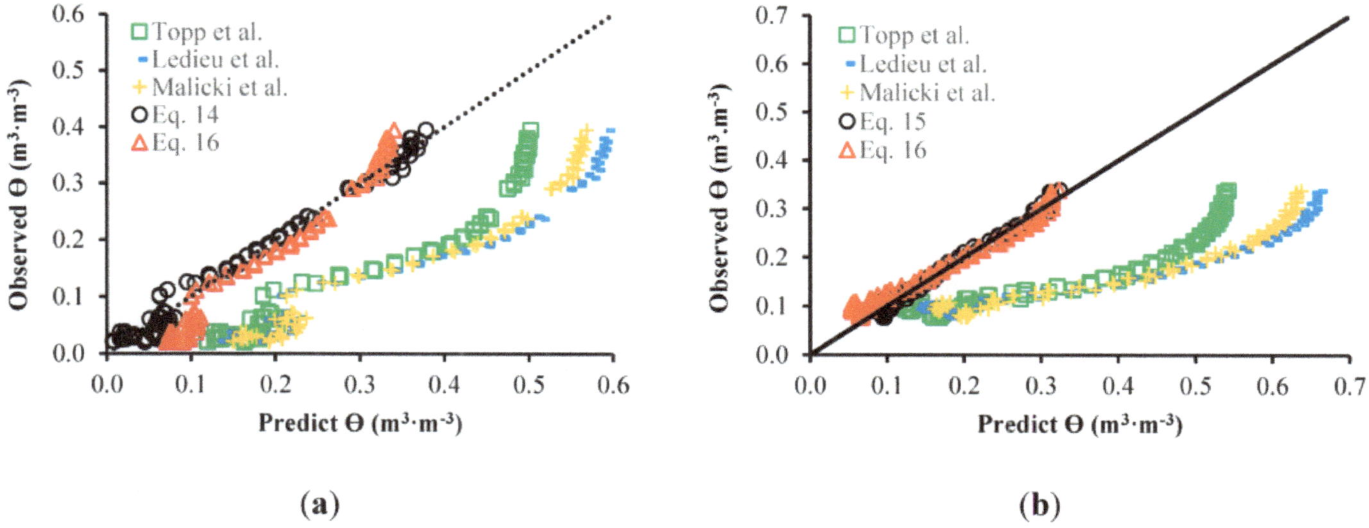

Figure 10: Relationship between predicted and observed θ for sandy (**a**) and clay soil (**b**).

Table 2: Results of linear regression and cross-validation

	Linear Regression	Cross-Validation	
	R^2	R^2	Standard Deviation
Equation (14)	0.9866	0.9856	0.0003
Equation (15)	0.9850	0.9839	0.0003
Equation (16)	0.9323	0.9298	0.0007

Table 3: Results from the comparison of models

	Sandy Soil		Clay Soil	
	RMSE	R^2	RMsSE	R^2
Topp *et al.*	0.05176	0.9319	0.02839	0.8798
Lidieu *et al.*	0.04613	0.9542	0.2374	0.9191
Malicki *et al.*	0.05114	0.9542	0.02334	0.9191
Equation (14)	0.01455	0.9866	-	-
Equation (15)	-	-	0.01001	0.9850
Equation (16)	0.03584	0.9804	0.02078	0.9797

According to Table 3, one can note that the prediction models for water content developed for each specific soil Equations (14) and (15) are more accurate due to the smaller RMSE and larger R^2. Even so, the model represented by Equation (16), employed for both soils, also presented better precision than the regular models presented in the literature (Equations (9)–(11)).

Figure 11 shows graphs of correlation between ε at 25 °C (ε_{25}) and ε at all temperatures, measured before and after correction with temperature employng the models proposed in this work Equation (13), and aslo the one presented by Chanzy *et al.* (Equation (8)). Table 4 ilustrates R^2 and RMSE, after and before the correction of ε based on the temperature of the two soils.

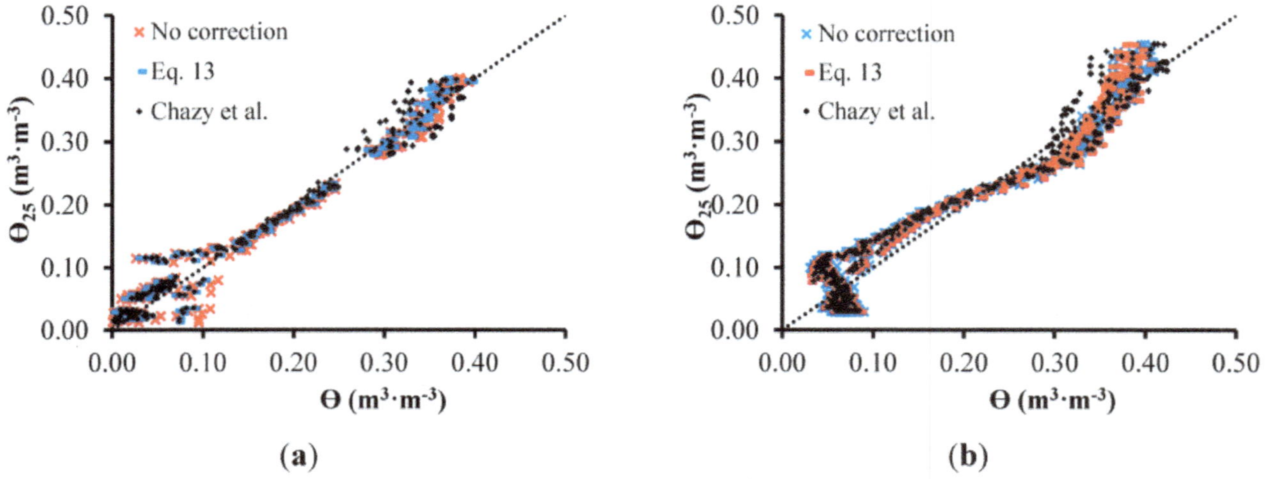

Figure 11: (**a**) θ-θ_{25} relationship for sandy soil (**a**) and clay soil (**b**).

According to Table 4, it is possible to observe that the results accuracy do not present significant changes due to the correction. The correction improves the results related to the sandy soil, but worsens the results related to the clay soil. Seyfried and Grant [43] mention a small and non linear effect for the temperature response, ranging from 5 °C to 45 °C, during the measurement of ε. According to the authors, this effect is small enough to be ignored for many applications.

Table 4: Results from the comparison between ε correction models based on temperature

	Sandy Soil		Clay Soil	
	RMSE	R^2	RMSE	R^2
No correction	0.02322	0.9713	0.03232	0.9381
Equation (9)	0.02215	0.9761	0.03250	0.9380
Chanzy *et al.*	0.02250	0.9704	0.03273	0.9367

Muñoz-Carpena [44] pointed out that the equipment that adopt the Time Domain Reflectometry (TDR), Amplitude Domain Reflectometry (ADR) and Frequency Domain Reflectometry (FDR) methods have a RMSE of 0.01 $m^3 \cdot m^{-3}$. The measurement system that uses the neutral probe is the most accurate, since it has a RMSE of 0.005 $m^3 \cdot m^{-3}$. Other inexpensive soil water content measurement systems presented RMSE of 0.02 to 0.03 $m^3 \cdot m^{-3}$ [12]. During the laboratory experiments, the measurement system developed in this work showed RMSE of 0.002 $m^3 \cdot m^{-3}$ for both clay and sandy soils, indicating good accuracy and cost-effectiveness with respect to traditional methods. Nevertheless, further experiments should validate the system accuracy under field conditions.

CONCLUSIONS

This paper presents the development, design and construction of a low-cost instrumentation system for measuring water content, apparent electrical conductivity and temperature of the soil. The measurement method is based on an auto-balancing bridge circuit. Experimental results obtained in the laboratory demonstrate the system accuracy, considered satisfactory for irrigation control. A comparison with other results in the literature also indicates the proposed system as a promising instrumentation device.

The proposed device corresponds to an efficient alternative to automatize irrigation systems, especially due to the satisfactory accuracy and low cost associated. The instrument was designed considering the final cost of the system to the farmer, which is a crucial concern while developing automation solutions for agriculture. Future work includes testing the prototype with slightly different and/or in undisturbed soils, and also manufacturing new devices for testing and field validation.

ACKNOWLEDGMENTS

The authors would like to thank the Federal University of Viçosa (UFV), the Federal University of Ouro Preto (UFOP), the Gorceix Foundation and the Vale Institute of Technology (ITV). This work was partially supported by the Coordination for the Improvement of Higher Level Personnel (CAPES), the National Council for Scientific and Technological Development (CNPq), the Research Support Foundation of the State of Minas Gerais (FAPEMIG), and the Vale Institute of Technology (ITV).

REFERENCES

1. United Nations Educational, Scientific and Cultural Organization. Launch Water Cooperation 2013. Available online: http://www.unesco.org/new/en/natural-sciences/environment/water/water-cooperation-2013/launch-water-cooperation-2013/ (accessed on 7 June 2015).
2. Corwin, D.L.; Lesch, S.M. Application of soil electrical conductivity to precision agriculture: Theory, principles, and guidelines. *Agron. J.* 2003, *95*, 455–471. .[CrossRef]

3. Ash, C.; Jasny, B.R.; Malakoff, D.A.; Sugden, A.M. Food security, feeding the future, introduction. *Science* 2010,*327*, 797. .

4. Rêgo Segundo, A.K.; Martins, J.H.; Marcos de Barros Monteiro, P.; Alves de Oliveira, R. Monitoring System of Apparent Electrical Conductivity and Moisture Content of Soil for Controlling of Irrigation Systems. In Proceedings of the International Conference of Agricultural Engineering—AgEng 2014, Zurich, Switzerland, 6–10 September 2014.

5. Corwin, D.L.; Lesch, S.M. Protocols and Guidelines for Field-scale Measurement of Soil Salinity Distribution with ECa-Directed Soil Sampling. *J. Environ. Eng. Geophys.* 2013, *18*, 1–25. .[CrossRef]

6. Hedley, C.B.; Roudier, P.; Yule, I.J.; Ekanayake, J.; Bradbury, S. Soil water status and water table depth modelling using electromagnetic surveys for precision irrigation scheduling. *Geoderma* 2013, *199*, 22–29. .[CrossRef]

7. Hedley, C.B.; Yule, I.J. Soil water status mapping and two variable-rate irrigation scenarios. *Precis. Agric.* 2009,*10*, 342–355. .[CrossRef]

8. Goumopoulos, C.; O'Flynn, B.; Kameas, A. Automated zone-specific irrigation with wireless sensor/actuator network and adaptable decision support. *Comput. Electron. Agric.* 2014, *105*, 20–33. .[CrossRef]

9. Fares, A.; Polyakov, V. Advances in crop water management using capacitive water sensors. *Adv. Agron.* 2006,*90*, 43–77. [Google Scholar]

10. Wendroth, O.; Koszinski, S.; Pena-Yewtukhiv, E. Spatial association among soil hydraulic properties, soil texture, and geoelectrical resistivity. *Vadose Zone J.* 2006, *5*, 341. .[CrossRef]

11. Rêgo Segundo, A.K.; Martins, J.H.; Monteiro, P.M.; Oliveira, R.A.; Oliveira Filho, D. Development of capacitive sensor for measuring soil water content. *Eng. Agrícola* 2011, *31*, 260–268. .[CrossRef]

12. Bitella, G.; Rossi, R.; Bochicchio, R.; Perniola, M.; Amato, M. A novel low-cost open-hardware platform for monitoring soil water content and multiple soil-air-vegetation parameters. *Sensors* 2014, *14*, 19639–19659. .

13. Scudiero, E.; Berti, A.; Teatini, P.; Morari, F. Simultaneous monitoring of soil water content and salinity with a low-cost capacitance-resistance probe. *Sensors* 2012, *12*, 17588–17607. .

14. Pelletier, M.G.; Karthikeyan, S.; Green, T.R.; Schwartz, R.C.; Wanjura, J.D.; Holt, G.A. Soil moisture sensing via swept frequency based microwave sensors. *Sensors* 2012, *12*, 753–767. .

15. Kizito, F.; Campbell, C.S.; Campbell, G.S.; Cobos, D.R.; Teare, B.L.; Carter, B.; Hopmans, J.W. Frequency, electrical conductivity and temperature analysis of a low-cost capacitance soil moisture sensor. *J. Hydrol.* 2008,*352*, 367–378. .[CrossRef]

16. Thalheimer, M. A low-cost electronic tensiometer system for continuous monitoring of soil water potential. *J. Agric. Eng.* 2013, *44*, 16. .[CrossRef]

17. Dias, P.C.; Roque, W.; Ferreira, E.C.; Siqueira Dias, J.A. A high sensitivity single-probe heat pulse soil moisture sensor based on a single npn junction transistor. *Comput. Electron. Agric.* 2013, *96*, 139–147. .[CrossRef]

18. Vellidis, G.; Tucker, M.; Perry, C.; Kvien, C.; Bednarz, C. A real-time wireless smart sensor array for scheduling irrigation. *Comput. Electron. Agric.* 2008, *61*, 44–50. .[CrossRef]

19. Gardner, C.M.K.; Dean, T.J.; Cooper, J.D. Soil Water Content Measurement with a High-Frequency Capacitance Sensor. *J. Agric. Eng. Res.* 1998, *71*, 395–403. .[CrossRef]

20. Seyfried, M.S.; Murdock, M.D. Measurement of Soil Water Content with a 50-MHz Soil Dielectric Sensor. *Soil Sci. Soc. Am. J.* 2004, *68*, 394. .[CrossRef]

21. RêgoSegundo,A.K.SISTEMADEMONITORAMENTO DO SOLO E DE CONTROLE DE IRRIGAÇÃO. BR Patent BR1020130132209, 28 May 2013. [Google Scholar]

22. Vidovich, N.V. Area Soil Moisture and Fertilization Sensor. U.S. Patent US20130073097 A1, 21 March 2013. [Google Scholar]

23. Charles, M. Soil Moisture Sensing System. U.S. Patent US3882383 A, 6 May 1975. [Google Scholar]

24. Feuer, L. Soil Moisture Sensor. U.S. Patent US5445178 A, 29 August 1995. [Google Scholar]

25. Da Silva, M.J. Impedance Sensors for Fast Multiphase Flow Measurement and Imaging. Ph.D. Thesis, Technische Universität Dresden, Dresden, Germany, 08 November 2008. [Google Scholar]

26. Kaatze, U.; Feldman, Y. Broadband dielectric spectrometry of liquids and biosystems. *Meas. Sci. Technol.* 2006,*17*, R17–R35. .[CrossRef]

27. Okada, K.; Sekino, T. *The Impedance Measurement Handbook—A Guide to Measurement Technology and Techniques*; Agilent Technologies, Inc.: Santa Clara, CA, USA, 2003. [Google Scholar]

28. Agilent Technologies. *Basics of Measuring the Dielectric Properties of Materials—Application Note*; Agilent Technologies, Inc.: Santa Clara, CA, USA, 2006. [Google Scholar]

29. Cambell, J.E. Electrical Sensor for Determining the Moisture Content of Soil. U.S. Patent US5479104, 26 December 1995. [Google Scholar]

30. Watson, K.; Gatto, R.; Weir, P.; Buss, P. Moisture and Salinity Sensor and Method of Use. U.S. Patent US5418466 A, 23 May 1995. [Google Scholar]

31. Delta-T Devices: Soil Moisture Content, Pyranometers, Data Loggers, Canopy Analysis. Available online: http://www.delta-t.co.uk/default.asp (accessed on 7 June 2015).

32. CD620: Display for the Hydrosense® System. Available online: http://www.campbellsci.com/cd620 (accessed on 17 June 2015).

33. IMKO GmbH—Precise Moisture Measurement. Available online: http://www.imko.de/ (accessed on 17 June 2015).

34. Gregory, A.P.; Clarke, R.N. A review of RF and microwave techniques for dielectric measurements on polar liquids. *IEEE Trans. Dielectr. Electr. Insul.* 2006, *13*, 727–743. .[CrossRef]

35. Yang, W.Q.; York, T.A. New AC-based capacitance tomography system. *IEE Proc. Sci. Meas. Technol.* 1999, *146*, 47–53. .[CrossRef]

36. Georgakopoulos, D.; Yang, W.; Waterfall, R. Best value design of electrical tomography systems. In Proceedings of the 3rd World Congress on Industrial Process Tomography, Banff, AB, Canada, 2003; pp. 11–19.

37. Rhoades, J.; Chanduvi, F.; Lesch, S. *Soil Salinity Assessment: Methods and Interpretation of Electrical Conductivity Measurements*; FAO: Riverside, CA, USA, 1999. [Google Scholar]

38. Chanzy, A.; Gaudu, J.-C.; Marloie, O. Correcting the temperature influence on soil capacitance sensors using diurnal temperature and water content cycles. *Sensors* 2012, *12*, 9773–9790. .

39. Rodríguez, J.D.; Pérez, A.; Lozano, J.A. Sensitivity analysis of kappa-fold cross validation in prediction error estimation. *IEEE Trans. Pattern Anal. Mach. Intell.* 2010, *32*, 569–75. .

40. Topp, G.; Davis, J.; Annan, A. Electromagnetic determination of soil water content: Measurements in coaxial transmission lines. *Water Resour. Res.* 1980, *16*, 574–582. .[CrossRef]

41. Ledieu, J.; de Ridder, P.; de Clerck, P.; Dautrebande, S. A method of measuring soil moisture by time-domain reflectometry. *J. Hydrol.* 1986, *88*, 319–328. .[CrossRef]

42. Malicki, M.; Plagge, R.; Roth, C. Improving the calibration of dielectric TDR soil moisture determination taking into account the solid soil. *Eur. J. Soil* 1996, *47*, 357–366. .[CrossRef]

43. Seyfried, M.S.; Grant, L.E. Temperature Effects on Soil Dielectric Properties Measured at 50 MHz. *Vadose Zone J.*2007, *6*, 759. .[CrossRef]

44. Munoz-Carpena, R. Field devices for monitoring soil water content. *Bull. Inst. Food Agric. Sci. Univ. Fla.* 2004,*343*, 1–16.

CHAPTER 14

CONTROL PARAMETERS OPTIMIZATION BASED ON CO-SIMULATION OF A MECHATRONIC SYSTEM FOR AN UA-BASED TWO-AXIS INERTIALLY STABILIZED PLATFORM

Xiangyang Zhou[1], Beilei Zhao[1] and Guohao Gong[2]

[1]School of Instrumentation Science and Opto-electronics Engineering, Beihang University (BUAA), Beijing 100191, China

[2]Beijing Aerospace Era Laser Navigation Technology Co., Ltd., Beijing 100094, China

ABSTRACT

This paper presents a method based on co-simulation of a mechatronic system to optimize the control parameters of a two-axis inertially stabilized platform system (ISP) applied in an unmanned airship (UA), by which high control performance and reliability of the ISP system are achieved. First, a three-dimensional structural model of the ISP is built by using the three-dimensional parametric CAD software SOLIDWORKS®; then, to analyze the system's kinematic and dynamic characteristics under operating conditions, dynamics modeling is conducted by using the multi-body dynamics software ADAMS™, thus the main dynamic parameters such as displacement, velocity, acceleration and reaction curve are obtained, respectively, through simulation analysis. Then, those dynamic parameters were input into the established MATLAB® SIMULINK®controller to simulate and test the performance of the control system. By these means, the ISP control parameters are optimized. To verify the methods, experiments were carried out by applying the optimized parameters to the control system of a two-axis ISP. The results show that the co-simulation by using virtual prototyping (VP) is effective to obtain optimized ISP control parameters, eventually leading to high ISP control performance.

INTRODUCTION

An inertially stabilized platform (ISP), which is used to support and stabilize the imaging loads so that the sensor's line of sight (LOS) can track the target accurately in real-time, plays an important role in aerial remote sensing. In recent years, due to their potential applications in different fields, great interest in the utilization of unmanned aerial vehicles (UAVs) has arisen [1,2,3,4,5,6]. The reason why unmanned airships (UAs) have become the ideal platform is mainly because they are less intrusive, noiseless, capable of hovering, less energy consuming, cost efficient and able to stay in the air for longer periods [2,6]. Although robotic airships have some advantages over other UAVs at low speeds and low altitude applications, they present a challenging control problem too. Furthermore, their kinematic and dynamic models are highly nonlinear and coupled [4]. Similarly, there are some disadvantages in stabilization when UA is used as aviation platform in an aerial remote sensing system, such as easily pitch maneuvering, low leveling stabilization, heading sideways swinging and so on.

Citation: Zhou, X.; Zhao, B.; Gong, G. Control Parameters Optimization Based on Co-Simulation of a Mechatronic System for an UA-Based Two-Axis Inertially Stabilized Platform. Sensors 2015, 15, 20169-20192. DOI: 10.3390/s150820169

Therefore, higher requirements are put forward for the control system when an ISP is supported by an UA platform.

In order to achieve the tracking accuracy and stabilization required by an UA-based aerial remote sensing system, the ISP control system needs to have more accurate and reliable system parameters under the condition of supporting large loads and being installed on an unsteady suspension pod. Generally, to optimize the control system parameters, many complicated, repeated and rigorous experiments have to be carried out on the ground by simulating the aircraft environment. Moreover, in order to ensure the robust adaptability of the control system in a real environment, the electrical and mechanical cooperation debugging process is crucial. In the conventional design process, the structural system modeling is usually simplified, which easily decreases the fidelity of the simulation. In addition, since much time is required by the parameter optimization and reliability testing process of the control system, this can potentially damage the ISP. As a result, it is hard to obtain good control performance when traditional methods are employed. Therefore, to improve the performance of an ISP's control system, control parameter optimization based on co-simulation using virtual prototyping is indispensable.

In order to significantly reduce the amount of physical testing that is required, virtual prototyping (VP) can be used, especially in the design stage [7]. VP is a process used for shortening the time to market and reducing the product cost. It makes use of a digital model for testing and evaluating the specific characteristics of a product and for simulating the manufacturing processes in a computational environment [8]. Physical prototyping, although is more desirable, may prove to be expensive and time consuming [7]. Repeated, efficient, and extensive use of prototypes is a vital activity that can make the difference between successful and unsuccessful entry of new products into the competitive world market [9]. The value of VP is rapidly being recognized for a wide range of engineering applications. These applications range from illustrating the potential of a system early in the technology assessment activity or early developmental phase to detailed analysis of mature designs in the advanced engineering phase [10]. ADAMS™ is a successful software for VP analysis [11].

The co-simulation technique based on ADAMS™ and SIMULINK® cooperation can be useful tool for improving the development cycle, which is suitable for the design of a mechatronic system with a complex mechanical structure and dynamic behavior with a control system [12,13]. MATLAB® and SIMULINK® focus on different areas and, if used in isolation, are unable to provide satisfactory answers to all the questions involved in the global simulation of mechatronic systems [14]. The control module is generated in ADAMS™, which is the interface between ADAMS™ and MATLAB®, and contacts the model in ADAMS™ with control system in MATLAB® [11,15]. The method of co-simulation using ADAMS™ and MATLAB® provides a new method for studying the dynamics of complex systems, which can simplify the simulation process and make the simulation results more accurate and increase the design reliability [16]. Co-simulation is not only widely used for verification of the control strategy of complicated dynamic systems, but also used for controller designing for dynamic systems [17]. In [18], it was shown that the use of co-simulation for robot design was more efficient than without co-simulation. The efficiency of co-simulation techniques for the development of mechatronic systems has been proven use [12]. Interactive co-simulation of ADAMS™ and MATLAB® could help designers consider the two prime parts of mechanics and control, and consequentially result in higher design efficiency [19].

In this paper, to meet the high precision and high stability requirements of an UA-based two-axis ISP for remote sensing, control parameter optimization based on co-simulation of the mechatronic system is carried out using ADAMS™ and SIMULINK® cooperation. The purpose is to improve the control performance of the ISP by utilizing the significant characteristics of co-simulation and virtual prototyping. The dynamics prototype is established in the ADAMS™ environment instead of using the simplified model of traditional design, whose dynamic parameters are then used as inputs for a SIMULINK® controller to optimize the control system parameters. As a result, a control system with high precision and high stability characteristics can be achieved. To verify the methods, experiments are carried out by applying the optimized parameters to the real control system of a two-axis UA-based ISP.

BACKGROUND ANALYSIS

Aerial Remote Sensing System

Figure 1 shows the schematic diagram of an aerial remote sensing system. Generally, an aerial remote sensing system consists of four main components [20,21], *i.e.*, a multi-axis ISP, an imaging sensor, a position and orientation system (POS) and the aircraft vehicle. When applied, the multi-axis ISP is mounted on the aviation platform, and the imaging sensor and POS are fixed on the inner of the ISP's gimbals. When the aviation platform rotates or jitters, the control system of multi-axis ISP gets the high-precision attitude reference information measured by the POS and then routinely controls the LOS of the imaging sensor to achieve accurate pointing and stabilization relative to the ground level and flight track.

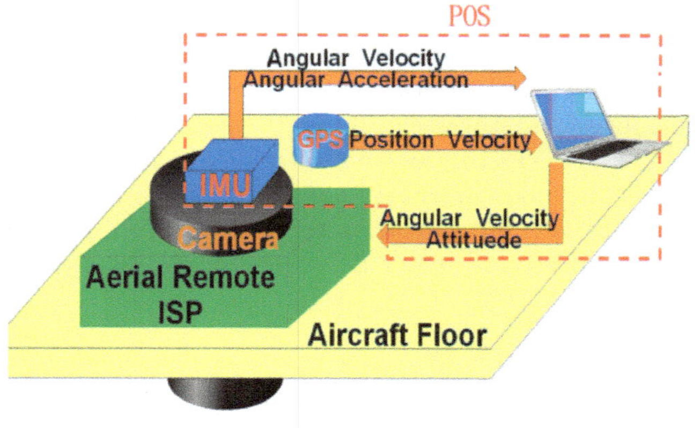

Figure 1: Schematic diagram of an aerial remote sensing system.

Working Principle of Two-Axis ISP

Figure 2 shows the schematic diagram of the two-axis ISP's working principle. We can see that the ISP consists of two gimbals, which are an azimuth gimbal (A-gimbal) and a pitch gimbal (P-gimbal) [20]. The P-gimbal is assembled on the A-gimbal and can rotate around the Y_p axis. The A-gimbal is assembled on the base of the aviation platform and can rotate around the Z_a axis. G_p and G_a stand for the rate gyros that measure the inertial angular rate of the P-gimbal and A-gimbal, respectively. E_p and E_a respectively stand for the photoelectric encoders installed on the P-gimbal and A-gimbal, which are used for measuring the relative angles between gimbals. M_z and M_y respectively stand for the gimbal servo motors which drive the A-gimbal and P-gimbal to keep them steady in inertial space. Ay represents the accelerometer installed on the P-gimbal that is used to measure the gimbal's rotary angular acceleration.

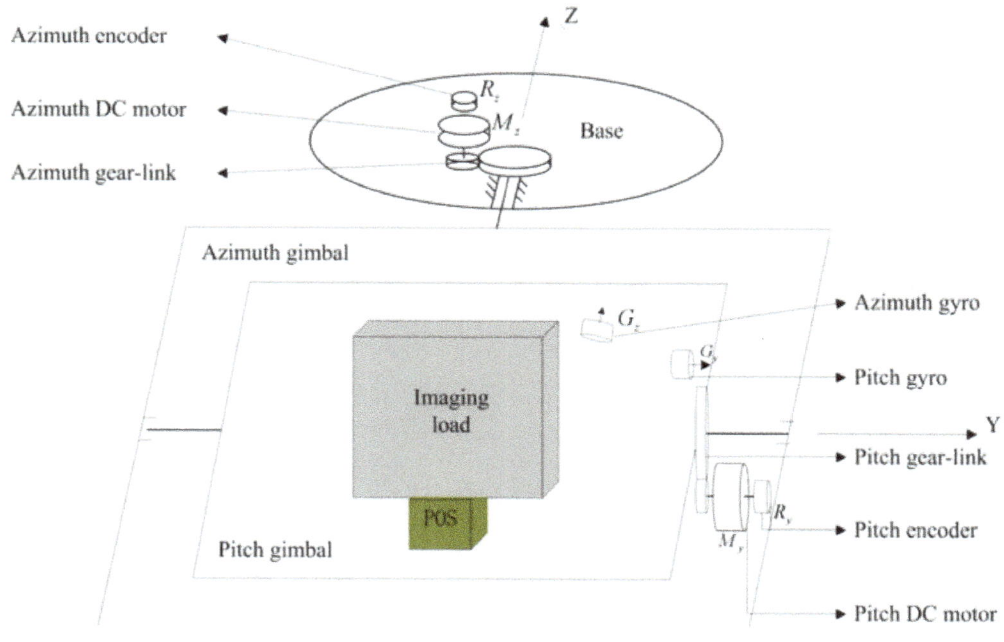

Figure 2: Schematic diagram of a two-axis ISP's working principle.

Three Closed-Loop Compound Control Scheme

Figure 3 shows the block diagram of a three-loop control system for an ISP. Conventional stabilization techniques employ rate gyros, rate integrating gyros, or rate sensors to detect angular rate disturbances of the LOS [22]. The blocks of G-pos, G-spe and G-cur separately represent the controllers in the position loop, speed loop and current loop; the PWM block represents the power amplification used for amplifying the current to drive the torque motor; L represents the inductance of a torque motor and R represents the resistance; K_t represents the torque coefficient of the motor and N is the transition ratio from the torque motor to the gimbals; J_m represents the moment of inertia of the motor and J_l represents the moment of inertia of the gimbals along the rotation axis.

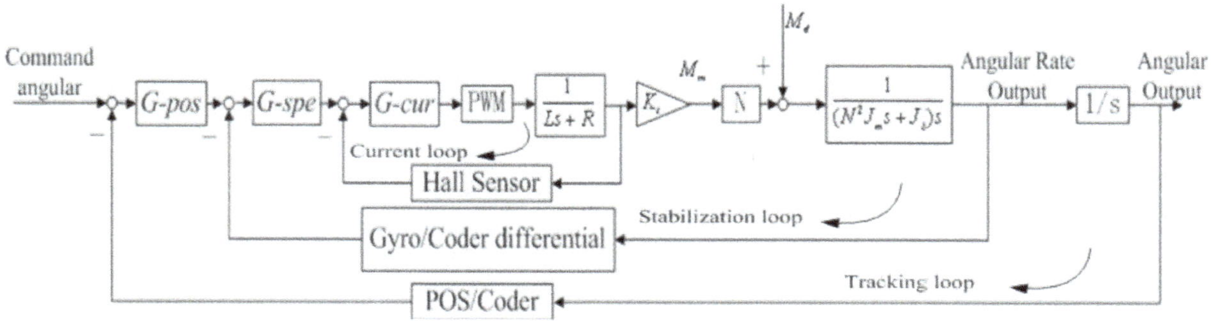

Figure 3: A block diagram of a traditional three-loop control system for an ISP.

Since the maximum required velocity of an ISP is usually low, and rarely exceeds 100 degrees per second, gearing can be considered in an attempt to reduce the size and weight of the actuator, particularly when the torque requirements are demanding [23]. Thus, to meet the requirements of high driving torque, heavy load and small size,*etc.*, the gimbals are often designed to be rotated by indirect drive motors linked to each gimbal through gear trains. To satisfy the high quality imaging requirements, the ISP needs to realize high pointing precision without jitter under lower tracking speed. However, since most gearing arrangements inevitably introduce additional friction and torsional resonances in the system, the reaction torque from a geared actuator constitutes an equivalent torque disturbance that can degrade the stabilization performance [23].

Although requirements for ISPs vary widely depending on the application, they all have a common goal, which is to hold or control the line of sight (LOS) of one object relative to another object or inertial space [23]. It should be noted that in this paper, the design of an ISP system is subjected to experimental requirements in both pointing and stabilization accuracy of the whole remote sensing system. To meet the requirements of high driving torque capacity, gear trains are employed in a two gimbals drive system. Therefore, for the designed system, it is difficult to obtain a higher accuracy level compared to a direct drive system since there will be errors derived from the gear train.

THREE-DIMENSIONAL CAD MODELING

The Structure of ISP System

Figure 4 shows a three-dimensional CAD structural model of a two-axis ISP built using the three-dimensional parametric CAD software SOLIDWORKS®. The model represents a compromise between a series of conflicting requirements, such as small size and heavy load capacity, light weight and high stiffness, *etc.*, which has to undergo appropriate modifications to realize optimization by adjusting parameters. It can been seen that the two-axis ISP is mainly made up of five sub-systems, *i.e.*, shaft supporting load system, gimbal structure system, control system, inertial measurement system and drive-transmission system, by which the main function of disturbance rejection can be realized in real-time. As a result, the LOS are kept tracking the target in the directions of both the azimuth and pitch axes all along.

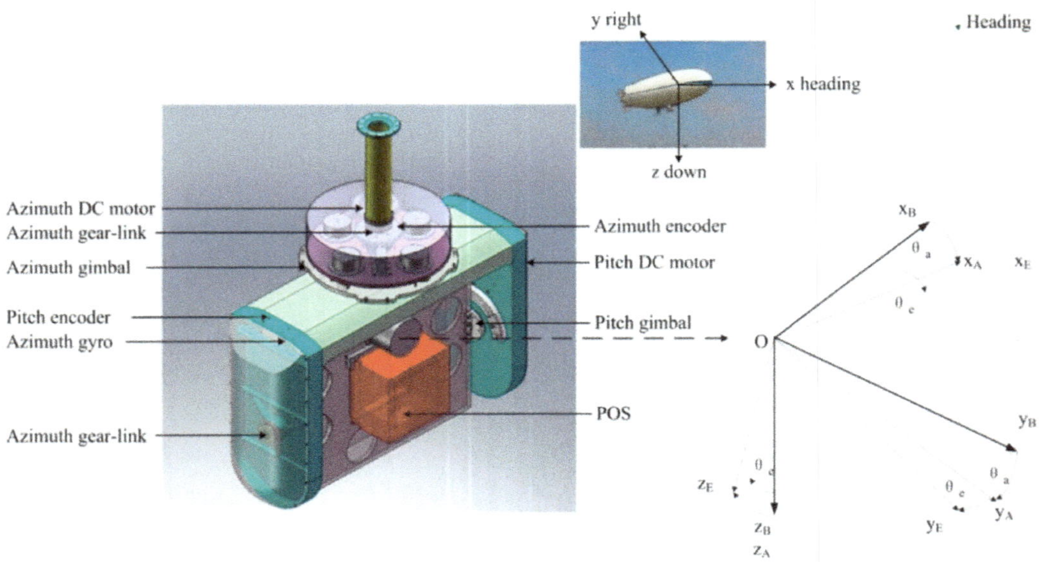

Figure 4: Three-dimensional CAD structural model of a two-axis ISP built using SOLIDWORKS®.

FEM Analysis

Static Analysis Results

To validate the static and dynamic performance, finite element analysis (FEA) is performed by the ANSYS Workbench Platform (AWP). The FEA model was directly imported from the established SOLIDWORKS® model which is then meshed by using the highly-automated AWP meshing function. Figure 5 shows the FEA static results that satisfy the strength design requirements.

Dynamic Analysis Results

Modal analysis can provide a bidirectional connection with all major CAD systems, enabling more efficient simulation-driven design decisions. A structural mode can be thought of as a shape and a frequency at which a structural shape resonates [23]. Modal analysis has become a major technology in the quest for determining, improving and optimizing the dynamic characteristics of engineering structures [24]. The fundamental principles of modal FEM analysis can be simply generalized as follows [25,26]:

The modal representation of a mechanical structure can be determined analytically if a lumped mass-spring system is concerned. In the general case of a continuous structure, a numerical approximation by means of a Finite Element Model (FEM) is made, discretizing the structure in a finite number of physical coordinates. The equations of motion describing this approximated system in the time and Laplace domain are given by:

$$[M]\{\ddot{x}(t)\}+[C]\{\dot{x}(t)\}+[K]\{x(t)\}=\{f(t)\} \quad (1)$$

$$s^2[M]+s[C]+[K]\{X(s)\}=\{f(s)\} \quad (2)$$

where [M], [C] and [K] respectively represent the mass, damping and stiffness matrix.

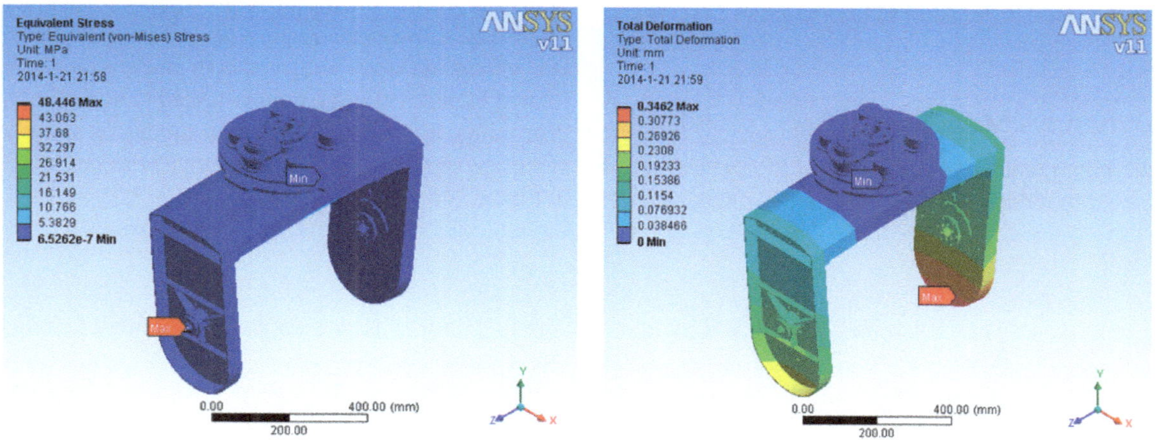

(a) (b)

Figure 5: FEA static results: (**a**) Displacement distribution and (**b**) Stress distribution.

The solution of these equations leads to an eigenvalue problem that can be solved in terms of the modal parameters. Eventually, the natural frequency and shape of each order for a system with N Degrees of Freedom (DOFs) can be obtained. Figure 6 and Table 1 shows the FEA dynamic results that satisfy the dynamic design requirements.

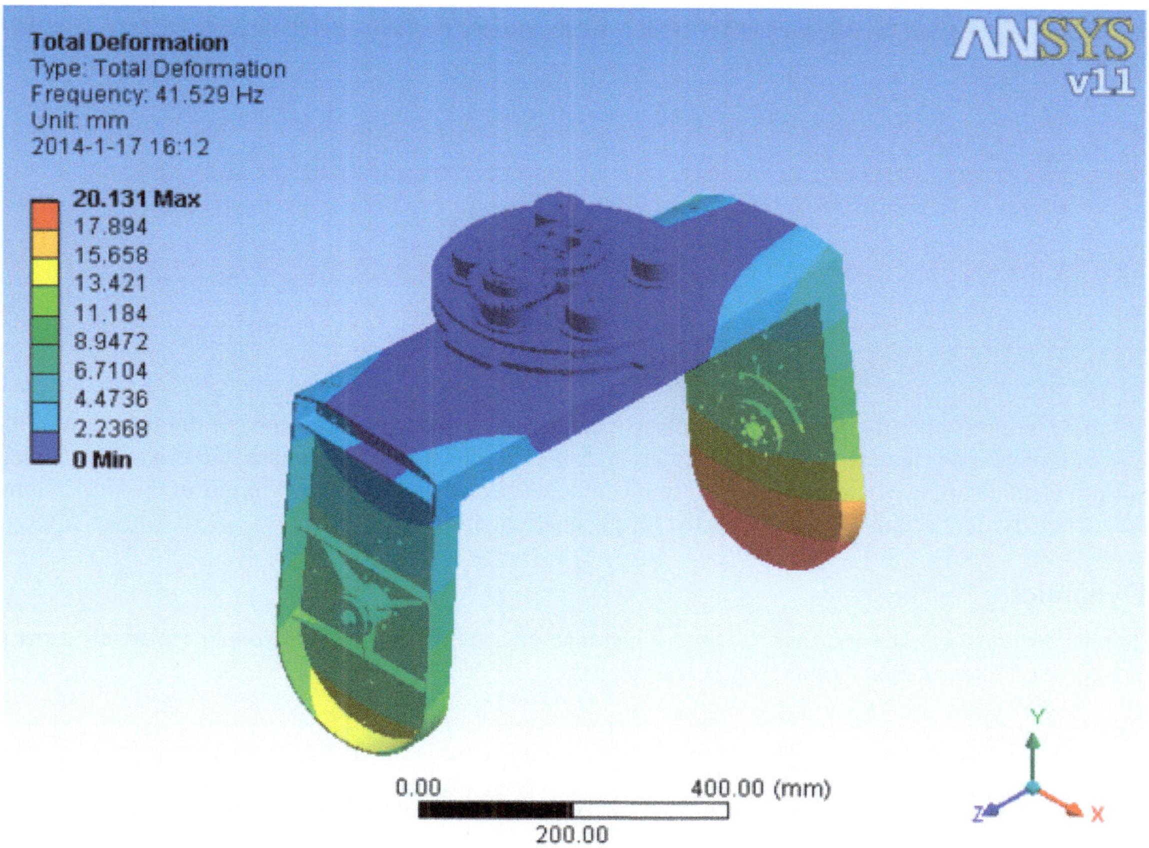

Figure 6: The first order mode of ISP structure.

Table 1: The FEA results of the prior six modes at natural frequency

Models	1	2	3	4	5	6
Frequency/Hz	41.5	45.2	55.1	74.8	121.8	185.2

SYSTEM DYNAMICS MODELING AND ANALYSIS

System Dynamics Modeling

Since the maximum required velocity of an ISP is usually low, gearing are used to reduce the size and weight of the actuator, particularly when the torque requirements are demanding [23]. Therefore, to meet the requirements of highly driving torque, heavy load and small size, *etc.*, both gimbals are designed to be rotated by indirect drive motors linked to each gimbal through gear-trains. The gear drive system model, as shown in Figure 4, can be simplified with a fixed transmission ratio, as shown in Figure 7.

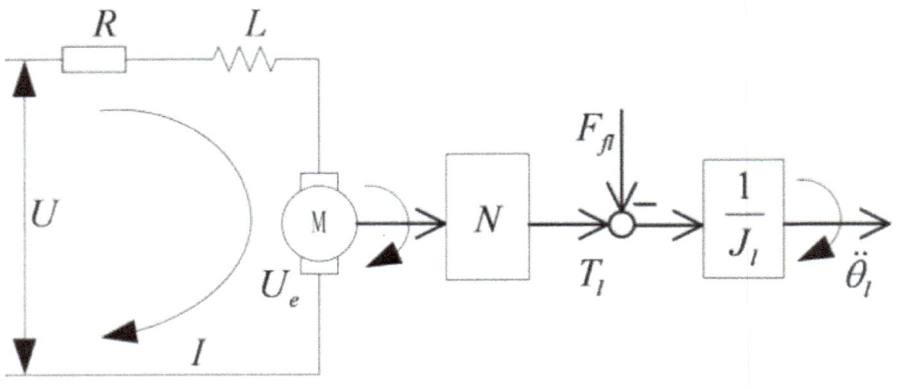

Figure 7: Block diagram for a simplified gear drive system with fixed transmission ratio.

According to the current balance equation and the torque balance equation, at the side of the motor output axis, the following equations are satisfied:

$$U = IR + L\frac{dI}{dt} + k_b\omega_m \tag{3}$$

$$J_m\ddot{\theta}_m + b_m\omega_m = T_m - \tau_l - F_{fm} \tag{4}$$

$$T_m = k_T I \tag{5}$$

where *U*, *I*, *R* and *L* are respectively the armature voltage, current, resistance and inductance of torque motor; k_b is the back EMF constant of motor, ωm is angular velocity of motor; J_m is the rotational inertia of motor, θm is angular displacement of motor, b_m is equivalent damping of motor, T_m is electromagnetic torque of motor, $_l$ is load torque at the side of motor output axis, F_{fm} is the inner friction torque of motor; k_T is the torque coefficient of motor.

Gimbal Dynamics

According to references [20,21], the traditional Newton-Euler method is used to model the two-axis ISP in this paper. The basic Newton-Euler equation for a rotational rigid body is described by

$$\sum \boldsymbol{M} = \boldsymbol{J}\dot{\boldsymbol{\omega}} + \boldsymbol{\omega} \times \boldsymbol{J}\boldsymbol{\omega} \tag{6}$$

where ∑M is the resultant moment of force applied to a rigid body, J is the inertia moment. ω and ω˙ are the absolute angular rate and angular acceleration, respectively. Assuming that the two gimbals are symmetrical, then the dynamic models of the two gimbals can be described as below.

Dynamic Model of the Pitch Gimbal

Based on the coordinate definition shown in Figure 4, we assume that the resultant moment of force applied to the pitch gimbal is $\sum M_E$, dynamic model of the pitch gimbal is given by:

$$\sum \boldsymbol{M}_E = \boldsymbol{J}_E\dot{\boldsymbol{\omega}}_{IE}^E + \dot{\boldsymbol{\omega}}_{IE}^E \times \boldsymbol{J}_E\dot{\boldsymbol{\omega}}_{IE}^E \tag{7}$$

Because the pitch gimbal only has rotational freedom in the direction of x_E, its dynamic model around axis x_E is given by:

$$M_{mEx} - M_{dEx} = J_{Ex}\dot{\omega}_{IEx}^{E} + (J_{Ez} - J_{Ey})\omega_{IEy}^{E}\omega_{IEz}^{E} \tag{8}$$

Dynamic Model of the Azimuth Gimbal

Assume that the resultant moment of force applied to the pitch gimbal is $\sum M_A$, dynamic model of the azimuth gimbal is given by:

$$\sum \boldsymbol{M}_A = \boldsymbol{J}_A \dot{\boldsymbol{\omega}}_{IA}^{A} + \boldsymbol{\omega}_{IA}^{A} \times \boldsymbol{J}_A \boldsymbol{\omega}_{IA}^{A} \tag{9}$$

Then dynamic model of the azimuth gimbal around axis z_A is given by:

$$M_{mAz} - M_{dAz} + M_{E/Az} = J_{Az}\dot{\omega}_{IAz}^{A} + (J_{Ay} - J_{Ax})\omega_{IAx}^{A}\omega_{IAy}^{A} \tag{10}$$

where $M_{E/Az}$ refers to the coupling torque between the pitch gimbal and the azimuth gimbal expressed in the direction of z_A, and it is given by:

$$M_{E/Az} = \sin\theta_e[J_{Ey}\dot{\omega}_{IEy}^{E} + (J_{Ex} - J_{Ez})\omega_{IEx}^{E}\omega_{IEz}^{E}] + \cos\theta_e[J_{Ez}\dot{\omega}_{IEz}^{E} + (J_{Ey} - J_{Ex})\omega_{IEx}^{E}\omega_{IEy}^{E}] \tag{11}$$

Thus, dynamic model of the two-axis ISP can be described as:

$$\begin{cases} \dot{\omega}_{IEx}^{E} = \dfrac{M_{mEx} - M_{dEx} - (J_{Ez} - J_{Ey})\omega_{IEy}^{E}\omega_{IEz}^{E}}{J_{Ex}} \\ \dot{\omega}_{IAz}^{A} = \dfrac{\begin{Bmatrix} M_{mAz} - M_{dAz} - (J_{Ay} - J_{Ax})\omega_{IAx}^{A}\omega_{IAy}^{A} - \sin\theta_e[J_{Ey}\dot{\omega}_{IEy}^{E} + (J_{Ex} - J_{Ez})\omega_{IEx}^{E}\omega_{IEz}^{E}] \\ +\cos\theta_e[J_{Ez}\dot{\omega}_{IEz}^{E} + (J_{Ey} - J_{Ex})\omega_{IEx}^{E}\omega_{IEy}^{E}] \end{Bmatrix}}{J_{Az}} \end{cases} \tag{12}$$

CO-MODELING OF THE MECHATRONIC SYSTEM

The Principle of Electromechanical Co-Simulation

The co-modeling of the mechatronic system involves two aspects: mechanical system modeling and control system modeling. When the mechanical models are established, the dynamic performance of the system can be improved by mode simulation analysis and structural optimization, which is the basis of the control system design. When the control system parameters are optimized, the high-fidelity VP based on ADAMS™ and SIMULINK® cooperation can be used for the performance testing, visualization simulation of the electromechanical system, control system parameter optimization, and reliability prediction, *etc.* Eventually, high-fidelity parameters for the control system can be obtained so that a control system with high accuracy and stabilization is achieved. In conclusion, the ADAMS™ model can be used for a control design that is developed in the SIMULINK® environment. State variables of the ADAMS™ model are connected with the control system in SIMULINK® and the whole VP is tested [12]. Figure 8shows the flow chart of electromechanical co-simulation.

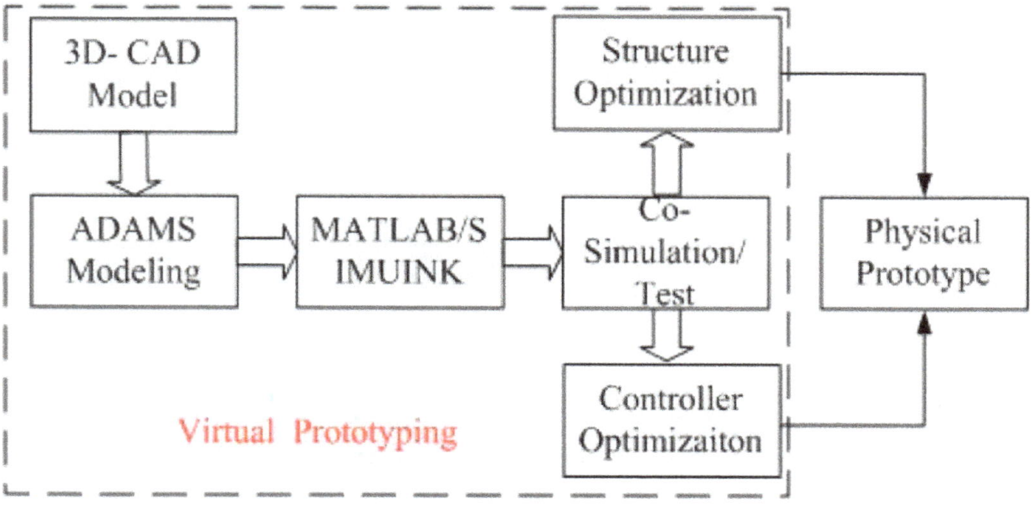

Figure 8: Flow chart of electromechanical co-simulation.

Mechanical System Modeling

ADAMS™ Environment Analysis

The co-simulation of the mechatronic system is an important part of VP technology. In the traditional modeling and simulation, the mechanical system is generally assumed to be an ideal model, in which many factors are not taken into good account, such as the installation error, mass unbalance, and so on. In particular, the coupling between the base angular motion and the mass center bias cannot be included in the modeling.

The ADAMS™ environment is able to provide a multi-body model for static, kinematic and dynamic analysis [12]. Moreover, ADAMS™ provides a way to analyze complex electromechanical systems by co-simulation. ADAMS™/Controls is a plug-in for ADAMS™ that helps one add sophisticated controls to the ADAMS™ model.

ADAMS™-Based Mechanical System Modeling

First, a three-dimensional structural model of the ISP established in SOLIDWORKS® is input into the ADAMS™ environment. Then, an ADAMS™-based mechanical system model can be obtained, whose procedures are as follows:

- Setting Parameters. Add the necessary parameters into the ADAMS™ model, such as the gravitational acceleration, material properties, *etc.* Furthermore, adjust the center of mass to minimize the mass unbalance, which is helpful to reduce the structural coupling effects.
- Adding constraints, driving moments and loading. The different movement constraints need to be defined to reflect the system movement. Through constraints, the components are associated with each other and the relative movement is limited [17]. The ISP contains five fixed pairs and five transmission pairs. Two gear pairs are respectively set for azimuth and pitch transmission links to realize the connection from the rotation of the DC motor to the movement of two gimbals. Figure 9 shows the constraints addition of ISP under the ADAMS™ environment.
- Defining state variables. In order to exchange information between the mechanical and control systems, we need to define the state variables and build up the interfaces between input and output. State variables are the key links of the internal information inflow and outflow in the ADAMS™ model. State variables for the system output are the angular velocities and rotational angles of the azimuth gimbal and pitch gimbal.
- Establishing interfaces. The principle of the co-simulation between ADAMS™ and SIMULINK®-Driveline is that digital signals which are produced in one step in ADAMS™ and SIMULINK® are respectively transferred through interfaces. This process continues until the simulation ends [17].
- Obtaining the co-simulation interface module. The ADAMS™ model is fed under the SIMULINK® environment to be analyzed. The model is used in the control scheme to predict the behavior of the system using a PID controller [27]. Figure 10 shows the co-simulation interface of the ISP.

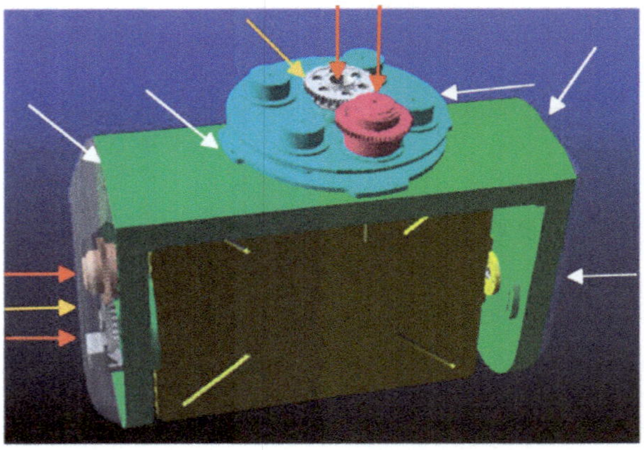

Figure 9: Constraints addition of ISP under ADAMS.

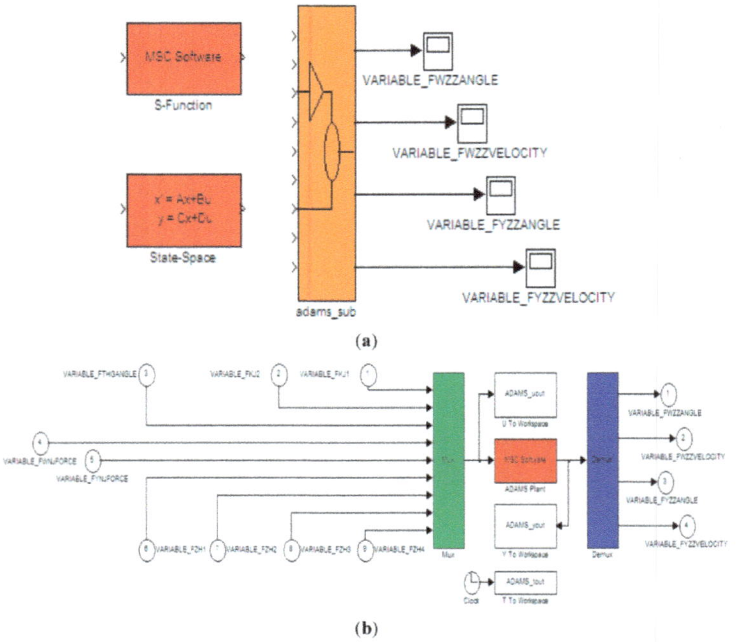

Figure 10: Co-simulation interface module of the ISP: (**a**) connection between ADAMS™ and SIMULINK® models; and (**b**) ADAMS™ module in co-simulation model.

Control System Modeling

SIMULINK® Environment Analysis

In order to execute a co-simulation between ADAMS™ and MATLAB®, definition of an acceptable format for the inputs and outputs of each program is required. The objective of co-simulation is to make a connection so that any change in one of the programs affects the other one [28]. The response speed and stabilization performance of an ISP can be influenced seriously by the mechanical system. MATLAB® is used in the model construction of the control system, and the mechanical model of ADAMS™ is connected into the MATLAB® control model. Thus, co-modeling between the mechanical dynamics and the control system is performed. Based on electromechanical co-modeling, the ISP control parameters can be optimized by co-simulation of the mechatronic system.

SIMULINK®-Based Control System Modeling

(1) Control system model

The transfer function analysis in control system is as follows: the drive motor can be equivalent to LR circuit, which is composed of inductors and resistors in series. The transfer function is:

$$G_m = \frac{K_m}{T_e s + 1} \tag{13}$$

where Te is the electromagnetic time constant, Te=LR; Km is the motor current coefficient, Km=1R.

The equivalent transfer function of motor BEMF is wk_e, where w is the motor speed, ke is the BEMF coefficient of the motor.

The PWM module of a drive motor can be equivalent to a first-order inertia link:

$$G_{pwm} = \frac{U}{T_{pwm} s + 1} \tag{14}$$

where U represents the supply voltage of PWM module, and T_{pwm} is the period of carrier wave of PWM output. The current measurement module is taken as the proportion of links K_f, in which the Hall sensor is employed.

Rate gyro can be taken as a first-order inertia link, $\frac{1}{T_G s+1}$, where T_G is the sampling period. Based on the analysis above, we can obtain the three closed-loop compound PID control scheme of two-axis ISP, as shown in Figure 11. In Figure 11, the mechanical part is outlined with a red dashed line.

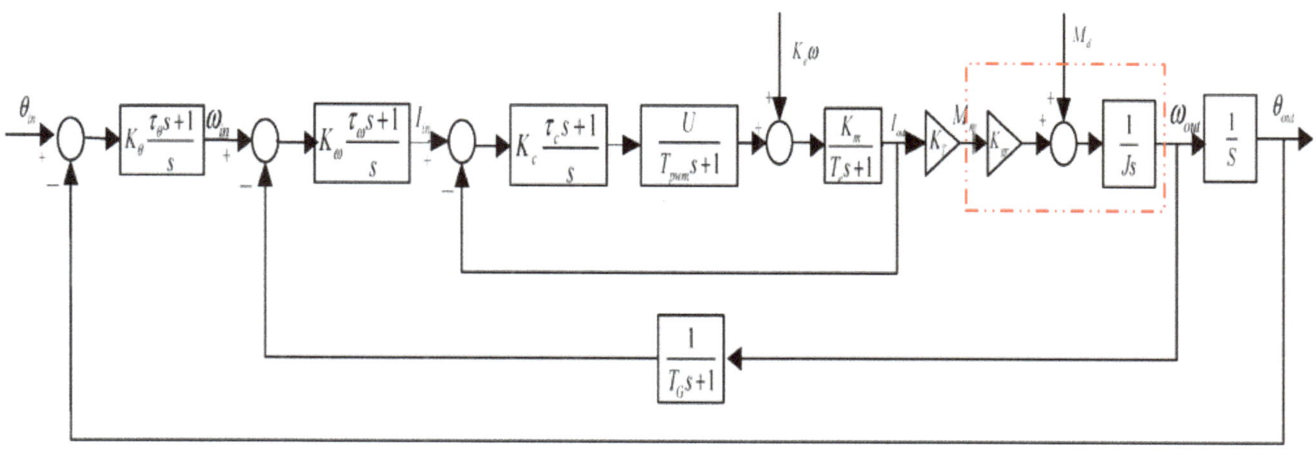

Figure 11: Block diagram of three closed-loop compound PID control scheme of two-axis ISP.

(2) SIMULINK®-based modeling

From Figure 11, we can first analyze the mechanical parts of the three closed-loop compound PID control scheme of the two-axis ISP: for the "Adams-sub" module established in ADAMS™, its input part is the motor torque, and its output part is the angular position and speed of gimbals resolved by the module automatically in real-time. For the system model established by MATLAB®, its input part is the output part of "Adams-sub", *i.e.*, the angular position and speed of gimbals resolved by the module automatically in real-time under ADAMS™.

Figure 12 is the ISP model produced in SIMULINK®. By inputting the "Adams-sub" block in the SIMULINK®environment, the ISP model is suitable for control and motion simulation as a defined system in MATLAB®. When the co-simulation model is established, it is necessary to simplify the transfer function models of the electrical part. For example, the function of motor BEMF, the time delay of the power-drive, and the time delay of the gyro signal can be ignored since their effects are too weak to be considered.

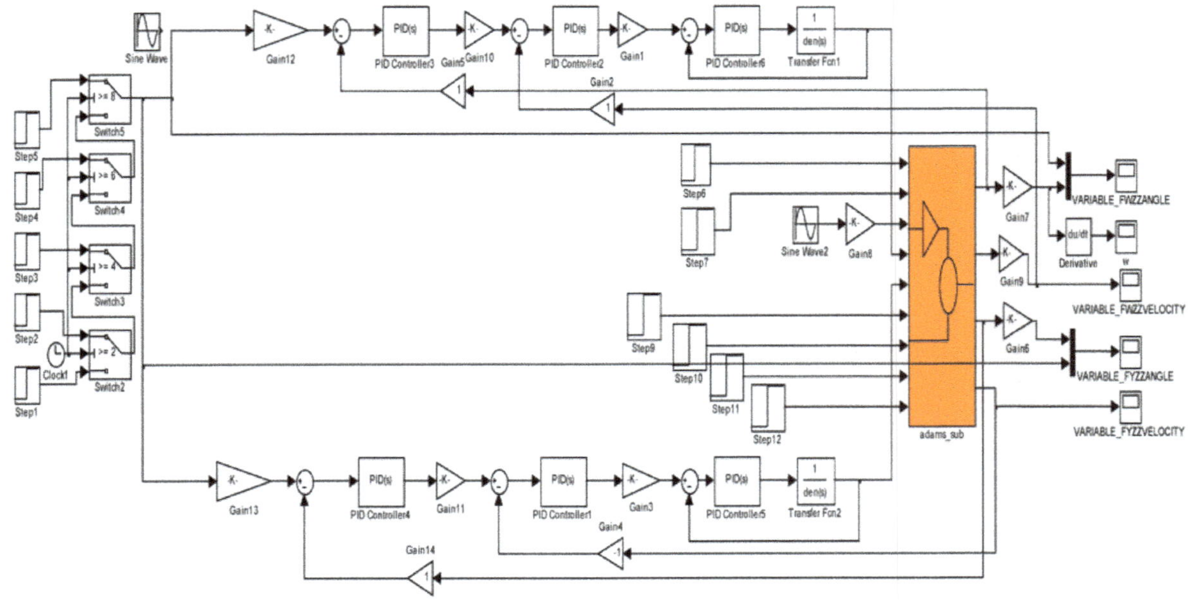

Figure 12: ISP model in SIMULINK®.

CO-SIMULATION AND TESTING

Dynamic Performance and Steady-State Accuracy Testing

Step Response

Figure 13 shows the step response of the azimuth gimbal and the pitch gimbal under co-simulation based on ADAMS™ and SIMULINK® cooperation.

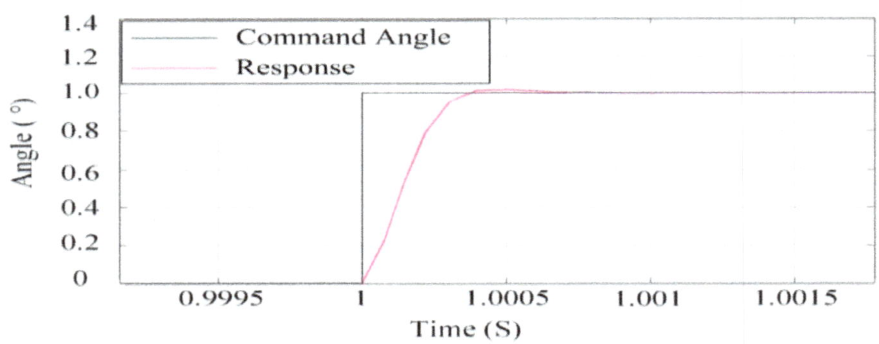

(a)

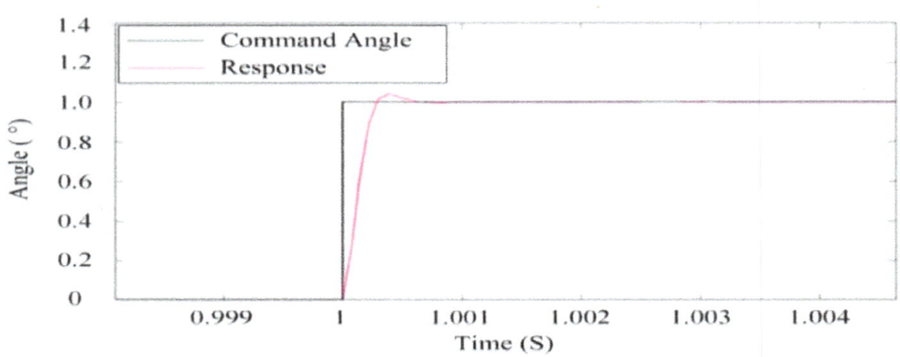

(b)

Figure 13: Step response of co-simulation: (**a**) the azimuth gimbal; (**b**) the pitch gimbal.

The black curve represents the set angular position and the purple curve is the response of the ISP. The interaction time interval is 0.0001 s. It can be seen that with help of the control parameters of the electromechanical simulation model, the ISP can realize high response speed and provide steady-state accuracy with mechatronic system co-simulation.

Sinusoidal Response

Figure 14 shows the response curve of the azimuth and pitch gimbals for the sinusoidal angular input command under co-simulation based on ADAMS™ and SIMULINK® cooperation. Likewise, the black curve represents the set angular position and the purple curve is the response of the ISP. It can be seen that with the help of the control parameters of the electromechanical simulation model, the ISP can track the command angle rapidly and accurately, by which the high stabilization performance can be attained.

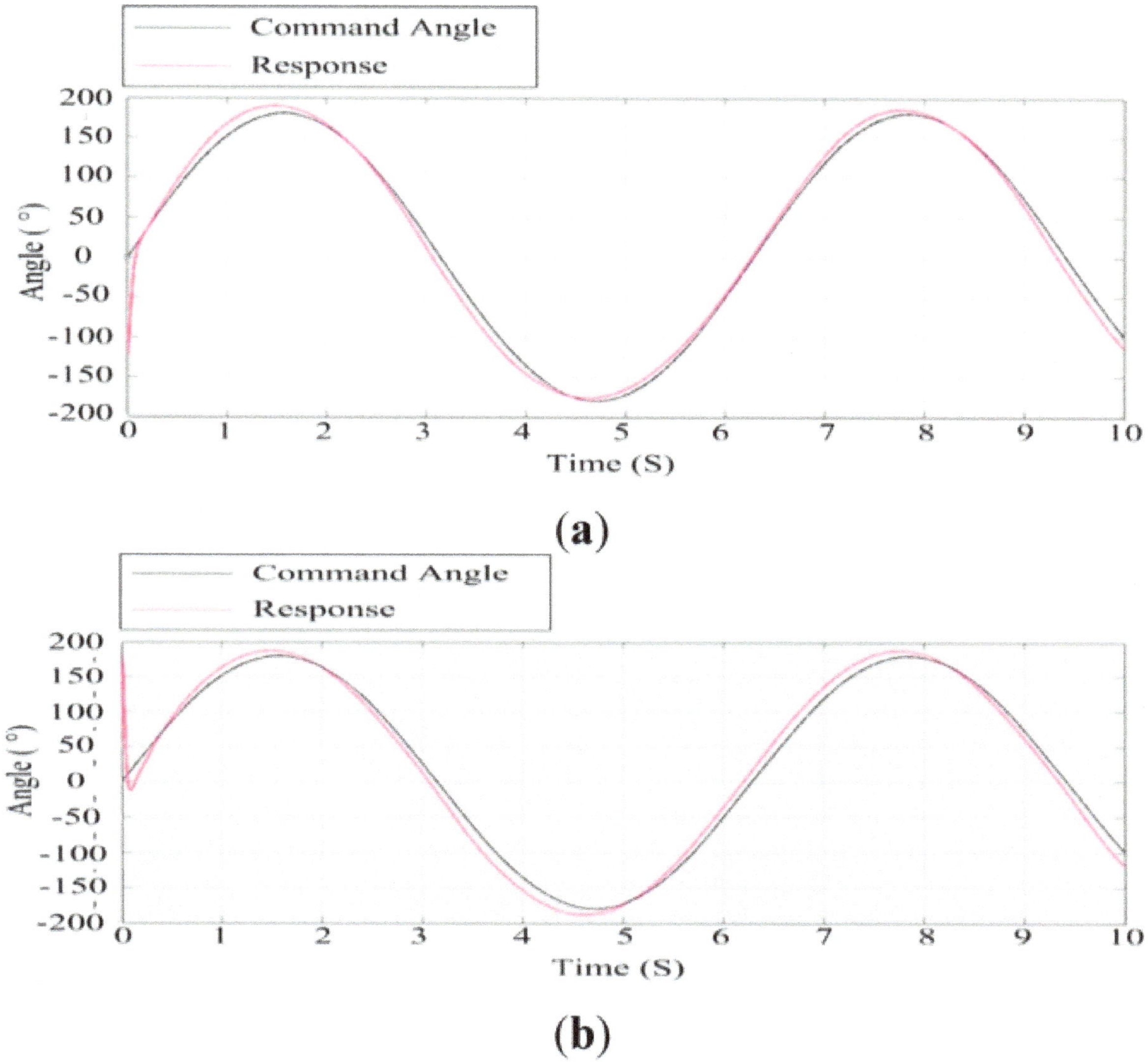

Figure 14: The response curve to a sinusoidal angular command in the electromechanical co-simulation model: (**a**) the azimuth gimbal; (**b**) the pitch gimbal.

Simulation Testing under Movable Base Conditions

Due to fact the pitch axis of the ISP coincides with the roll axis of the UA, the roll swing of the UA corresponds to the input of the ISP pitch disturbance. Therefore, it is necessary to test the control performance of the ISP under movable base conditions by simulating the roll swing of the UA at various angles. Figure 15 shows the response curve for a 10° sinusoidal angle roll swing of the UA under co-simulation. It can be seen that under this movable base condition, the ISP has good tracking accuracy whether for azimuth gimbal or pitch gimbal.

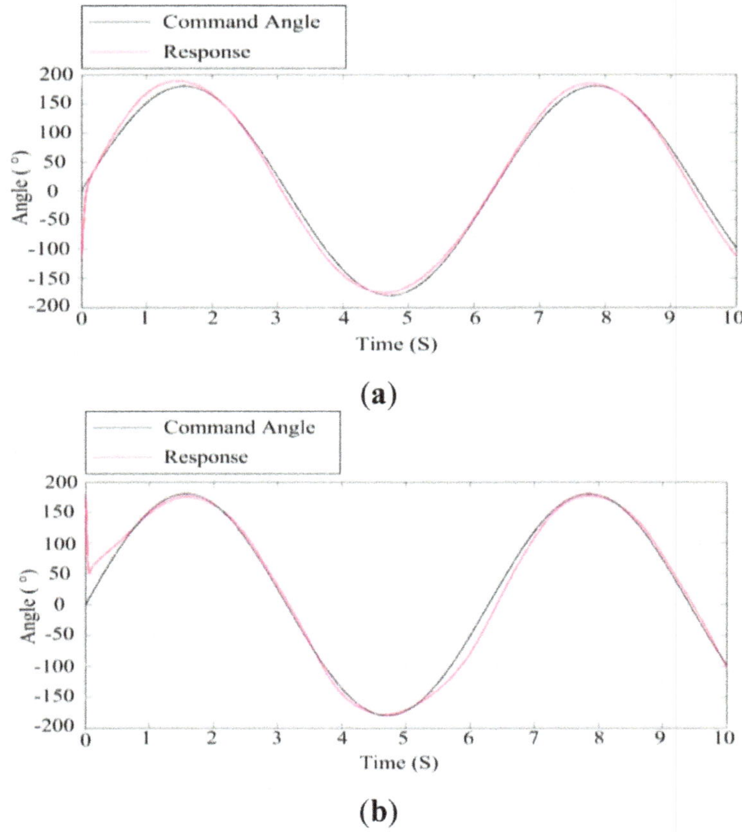

Figure 15: The response curve under roll swing of 10° sinusoidal angle of the UA: (**a**) the azimuth gimbal; (**b**) the pitch gimbal.

Figure 16 shows the response curve for a 30° sinusoidal angle roll swing of the UA under co-simulation. It can be seen that under this movable base condition, the azimuth gimbal still has good tracking accuracy, but the tracking accuracy of the pitch gimbal begins to decrease since there is an obvious deviation between the command and its response. This means that when the roll swing of the UA is up to about 30°, the ISP no longer has enough capability to isolate the UA disturbance with high accuracy.

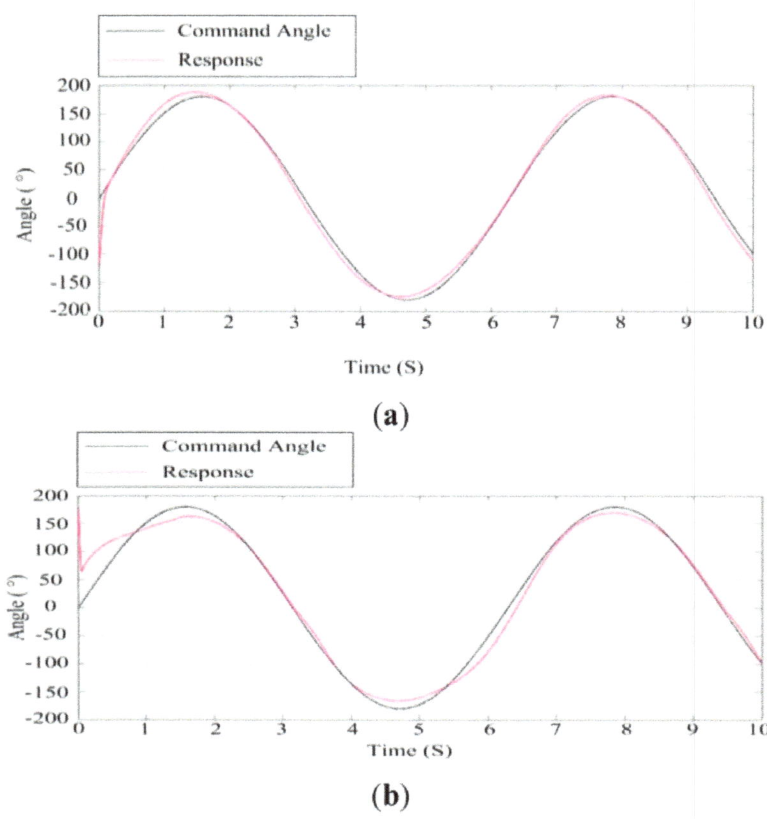

Figure 16: The response curve under 30° sinusoidal angle roll swing of the UA: (**a**) the azimuth gimbal; (**b**) the pitch gimbal.

Analysis of the Effect of Wind Speed

The wind force is an important factor influencing the performance of an ISP. Therefore, performance testing of the ISP under the wind force is simulated. According to the Bernoulli wind-pressure equation, we get:

$$F = \frac{v^2}{1.6} S \tag{15}$$

where F is the wind force on the object, v is the wind speed, S is swept area of the object. The total swept area of the ISP is approximately 0.5 m^2 and the wind speed is 10 m/s, so we get F is about 31 N.

Figure 17 and Figure 18 show the results of simulation tests under different wind speeds by applying a wind force disturbance to the ISP. Comparing Figure 17 with Figure 18, we can see that when the wind speed is 10 m/s, the azimuth gimbal and the pitch gimbal of the ISP have better tracking performance than that when the wind speed is 14 m/s. This means when the wind speed is larger, the tracking performance of the ISP will decrease, particularly for the pitch gimbal. Therefore, we can determine the wind speed reliability range to guarantee the stabilization performance of ISP.

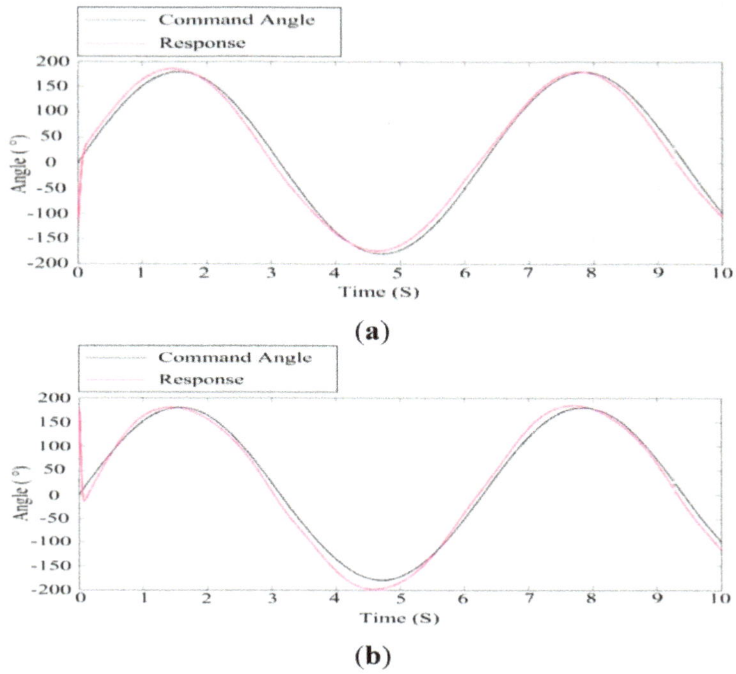

Figure 17: Response curves of ISP at the wind speed of 10 m/s: (**a**) the azimuth gimbal; (**b**) the pitch gimbal.

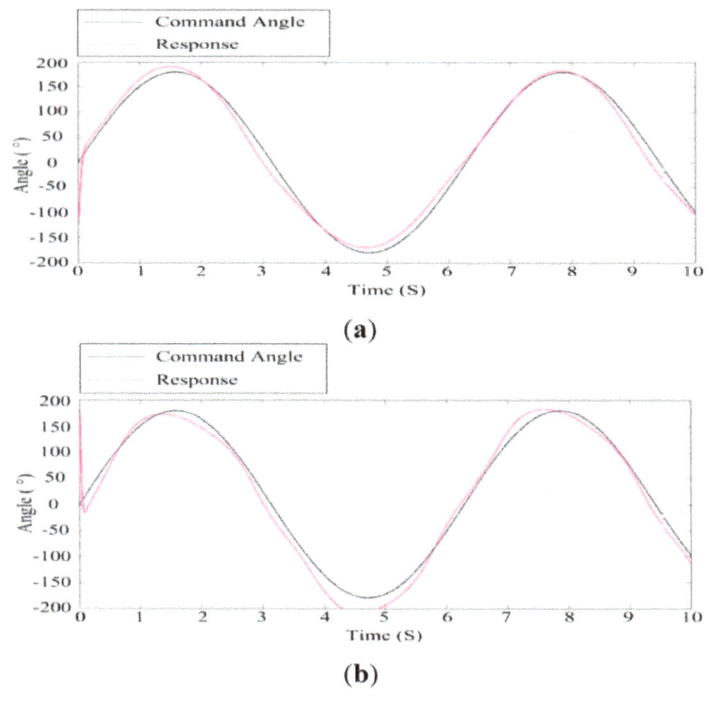

Figure 18: Response curves of the ISP at the speed of 14 m/s: (**a**) the azimuth gimbal; (**b**) the pitch gimbal.

EXPERIMENTAL VERIFICATION

Testing Under Movable Vehicle Conditions

Figure 19 shows a picture of our real experimental two-axis ISP system and the movable vehicle. The ISP originally works with the control parameters optimized by co-simulation. Then, the parameters are further optimized and improved based on actual electromechanical tests. The eventual control parameters are based on those of co-simulation and only differ very little from the simulation results. This illustrates that the co-simulation is an important way to design control systems with high performance.

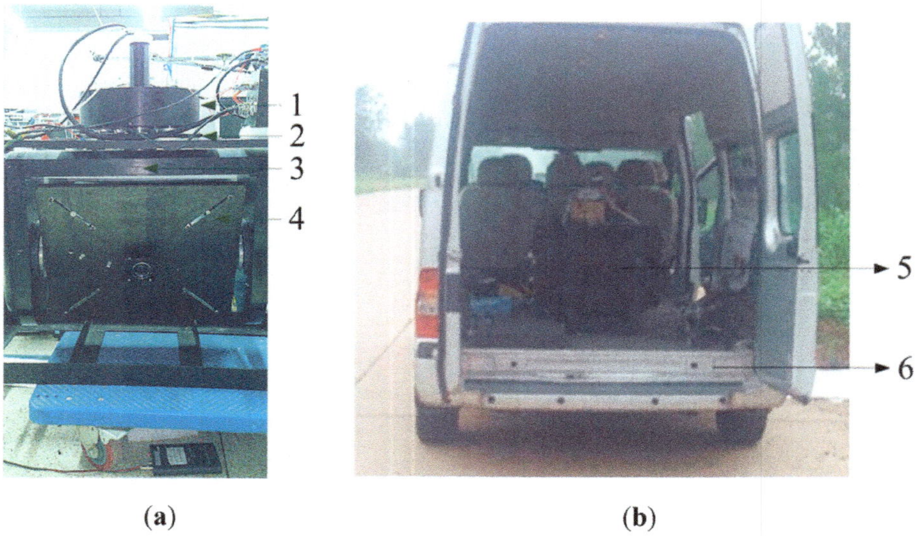

Figure 19: Picture of the ISP during experiments under movable base conditions: (**a**) ISP and (**b**) movable vehicle. 1-Loads; 2-Base; 3-Azimuth gimbal; 4-Pitch gimbal; 5-ISP; 6-Movable vehicle.

Figure 20 shows the tracking results of the ISP in a real movable vehicle experiment. It can be seen that the ISP has high tracking accuracy to the instruction command that makes it track stably to the target. The steady pointing accuracy is less than 0.5° (RMS) under the conditions where the tracking angular speed is less than 40°/s.

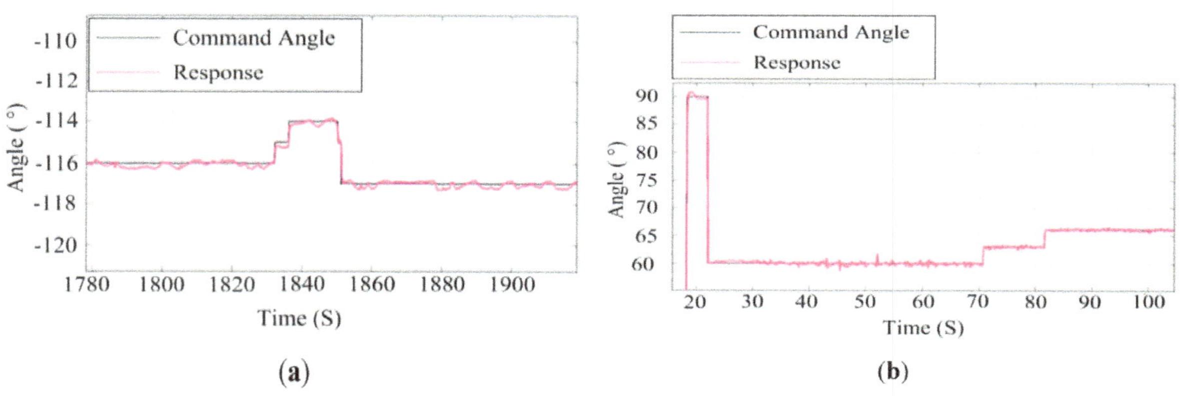

Figure 20: The tracking accuracy of the ISP in a real movable vehicle experiment: (**a**) the azimuth gimbal; (**b**) the pitch gimbal.

Testing the UA in Air

In ISP design, the wind force interaction is an important disturbance. If an ISP and payload are exposed to the vehicle's aerodynamic wind stream, direct buffeting occurs [29]. However, the torque due to wind is modeled using a moment coefficient which is quite difficult to estimate, and the model of the air movement in the flight environment is so complex that many theories have been formed, such as the Von Karmen vortices. Therefore, to simplify the design, the operating disturbances including wind force are generally modeled as equivalent torque disturbances [29]. In this paper, thus, we use a simple wind force model to simulate the ISP performance when the airship is subjected to a certain wind speed. The air experiments were carried out under good weather conditions with only a little wind speed.

Figure 21 shows a picture of an experiment with a two-axis ISP system on an UA in air. The control parameters are those determined in the movable vehicle experiments. Figure 22 shows the ISP tracking results of the experiments on a real UA in the air. It can be seen that under the real conditions of the UA in the air, the ISP still has high tracking accuracy with the instruction command, which is similar to the results in the movable vehicle experiments. The steady-state accuracy of the system under UA in air is 0.6° (RMS), and the velocity under the tracking angular is 20 °/s.

Figure 21: The experiments with a two-axis ISP system on a UA in the air.

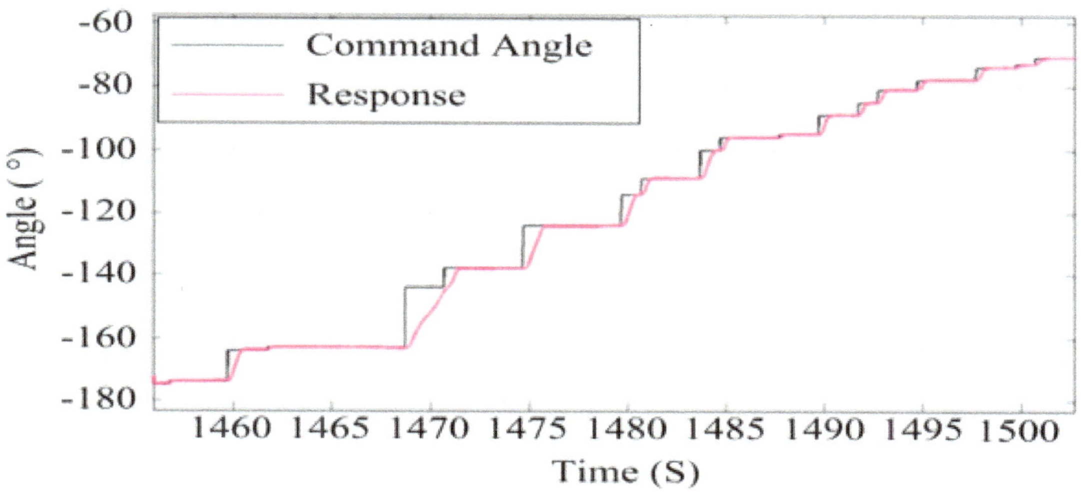

(a)

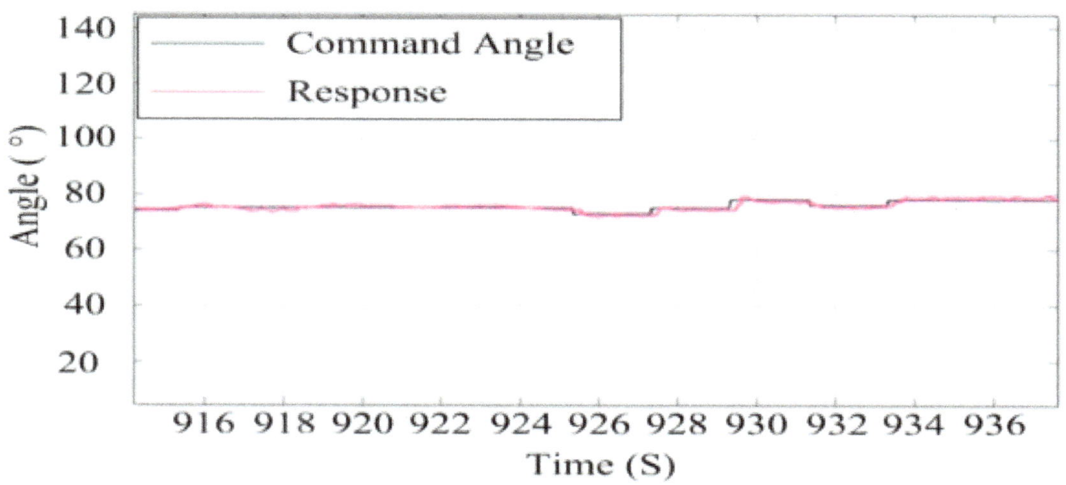

(b)

Figure 22: The tracking results of the ISP in the experiments on a real UA in the air: (**a**) the azimuth gimbal; (**b**) the pitch gimbal.

CONCLUSIONS

In this paper, to realize high control performance of a two-axis ISP applied in UAs, a method based on co-simulation of the mechatronic system is proposed to optimize the control parameters. Depending on the built three-dimensional CAD structural model, the dynamics model is established under an ADAMS™ environment, by which the main dynamic parameters are obtained through simulation analysis. Then, the dynamic parameters are input into the SIMULINK® controller to simulate and test the performance of the control system to optimize the control parameters. Experiments were carried out to verify the method. The results show that the proposed method is effective in that it can obviously improve the accuracy and reliability of the ISP.

ACKNOWLEDGMENTS

This work was supported in part by the National Natural Science Foundation of China (No. 51375036, No. 51205019), the China Scholarship Council (CSC), the National High Technology Research and Development Program of China (No. 2012AA120601) and the National Basic Research Program of China (No. 2009CB724001). We acknowledge sincerely the valuable supports and kind helps of Liwei Lin during the first author studied as a visiting scholar in Department of Mechanical Engineering, Berkeley Sensor and Actuator Center, University of California, Berkeley, CA 94720, USA.

REFERENCES

1. Lee, S.J.; Kim, S.P.; Kim, T.S.; Kim, H.K.; Lee, H.C. Development of autonomous flight control system for 50 m unmanned airship. In Proceedings of the IEEE International Conference on Intelligent Sensor, Sensor Networks and Information Processing (ISSNIP), Berkeley, CA, USA, 14–17 December 2004; pp. 457–461.
2. Azinheira, J.R.; Carneiro de Paiva, E.; Ramos, J.G., Jr.; Beuno, S.S. Mission path following for an autonomous unmanned airship. In Proceedings of the IEEE International Conference on Robotics and Automation (ICRA), San Francisco, CA, USA, 24–28 April 2000; Volume 2, pp. 1239–1275.
3. Beniovenqo, F.; Paiva, E.; Bueno, S. Sliding mode control approaches for an autonomous unmanned airship. In Proceedings of the 18th AIAA Lighter-Than-Air Systems Technology Conference, Seattle, Washington, USA, 4–7 May 2009.
4. Repoulias, F.; Papadopoulos, E. Dynamically feasible trajectory and open-loop control design for unmanned airships. In Proceedings of the Mediterranean Conference on Control & Automation, Athens, Greece, 27–29 June 2007; pp. 1–6.
5. Cappadona, A.; Lecca, R.; Vazzola, M.; Gili, P.; Farina, P.; Surace, C. Innovative unmanned airship structural analysis: Dual-hull and exoskeletal configurations. *J. Phys.: Conf. Ser.* **2009**, *181*, 012097. . .
6. Xu, T.; Zhang, X.D.; Lu, Y.J. Onboard controlling system design of unmanned airship. In Proceedings of the International Conference on Electronic and Mechanical Engineering and Information Technology (EMEIT), Harbin, China, 12–14 August 2011; pp. 3028–3031.
7. Malik, S.M.; Lin, J.; Goldenberg, A.A. Virtual prototyping for conceptual design of a tracked mobile robot. In Proceedings of the Canadian Conference on Electrical and Computer Engineering (CCECE), Ottawa, ON, Canada, 7–10 May 2006; pp. 2349–2352.
8. Raffaeli, R. Knowledge-based approach to flexible part design. *J. Eng. Des.* **2010**, *21*, 7–29. . .
9. Patil, M.J.; Satish, S.K.; Patil, M.J. Virtual prototyping techniques for product development and realization. In Proceedings of the Total Engineering, Analysis and Manufacturing Technologies, Bangalore, Karnatka, India, 22–24 September 2008.
10. Schaaf, J.C.; Thompson, F.L. System concept development with virtual prototyping. In Proceedings of the Winter Simulation Conference, Atlanta, GA, USA, 7–10 December 1997; pp. 941–947.
11. Chen, J.; Han, D. The control of tendon-driven dexterous hands with joint simulation. *Sensors* **2014**, *14*, 1723–1739. . . [PubMed]
12. Brezina, T.; Hadas, Z.; Vetiska, J. Using of co-simulation ADAMS-SIMULINK for development of mechatronic systems. In Proceedings of the 14th International Symposium, on MECHATRONIKA, Trencianske Teplice, Slovakia, 1–3 June 2011; pp. 59–64.
13. Spiryagin, M.; Simson, S.; Cole, C.C.; Persson, I. Co-simulation of a mechatronic system using Gensys and Simulink. *Veh. Syst. Dyn.: Int. J. Veh. Mech. Mobil.* **2012**, *50*, 495–507. . .
14. Wagner, M.; Schneider, P.; Hinnerichs, A.; Bruder, T.; Stork, A. Functional DMU: Co-simulation of mechatronic systems in a virtual environment. In Proceedings of the ASME World Conference on Innovative Virtual Reality, Milan, Italy, 27–29 June 2011; pp. 193–198.
15. Chen, C.Z.; Nie, H.; Chen, J.B.; Wang, X.T. A Velocity-based impedance control system for a low impact docking mechanism (LIDM). *Sensors* **2014**, *14*, 22998–23016. . . [PubMed]
16. Zhu, D.L.; Qin, L.Y.; Zhang, H.; Xia, M.M. Research on Co-simulation using ADAMS and MATLAB for active vibration isolation system. In Proceedings of the International Conference on Intelligent Computation Technology and Automation (ICICTA), Changsha, China, 11–12 May 2010; pp. 1126–1129.
17. Liu, Y.L.; Xiang, C.L.; Wang, W.D.; Huang, K. Co-simulation of a dual-mode electro-mechanical transmission in adams and MATLAB. In Proceedings of the IEEE International Conference on Mechatronics and Automation (ICMA), Tianjin, China, 3–6 August 2014; pp. 1096–1101.
18. Broenink, J.F.; Ni, Y.Y.; Groothuis, M.A. Model-driven robot-software design using integrated models and co-simulation. In Proceedings of the International Conference on Embedded Computer Systems: Architectures, Modeling and Simulation, Samos, Greece, 16–19 July 2012; pp. 339–344.

19. Deng, Y.B.; Zhang, H.; Ma, G.H. A co-simulation of seam tracking system based on ADAMS and SIMULINK. In Proceedings of the Second International Conference on Mechanic Automation and Control Engineering (MACE), Hohhot, China, 15–17 July 2011; pp. 8–11.

20. Zhou, X.Y.; Gong, G.H.; Li, J.P. Decoupling control for a three-axis inertially stabilized platform used for aerial remote sensing. *Trans. Inst. Meas. Control* **2014**. . .

21. Zhou, X.Y.; Zhang, H.Y.; Yu, R.X. Decoupling control for two-axis inertially stabilized platform based on an inverse system and internal model control. *Mechatronics* **2014**, *24*, 1203–1213. . .

22. Kennedy, P.J.; Kennedy, R.L. Direct *versus* indirect line of sight (LOS) stabilization. *IEEE Trans. Control Syst. Technol.* **2003**, *11*, 3–15. . .

23. Hilkert, J.M. Inertially stabilized platform technology: Concepts and principles. *IEEE Control Syst. Mag.* **2008**,*28*, 26–46. . .

24. Bai, Z.F.; Zhao, Y.; Ma, W.L. Modal analysis for small satellite system with finite element method. In Proceedings of the Systems and Control in Aerospace and Astronautics (ISSCAA), Shenzhen, China, 10–12 December 2008; pp. 1–5.

25. Zhou, X.Y.; Yu, R.X.; Li, J.P.; Li, D.P. Structure optimal design of roll gimbal for an aerial three-axis ISP based on FEM modal analysis. In Proceedings of the IEEE ICMTMA, Shanghai, China, 6–7 January 2011; Volume 3, pp. 373–376.

26. Van der Auweraer, H. Structural dynamics modeling using modal analysis: Applications, trends and challenges. In Proceedings of the 18th IEEE Instrumentation and Measurement Technology Conference, Budapest, Hungary, 21–23 May 2001; pp. 1502–1509.

27. Affi, Z.; Romdhane, L. ADAMS/Simulink interface for dynamic modeling and control of closed loop mechanisms. In Proceedings of the 7th WSEAS International Conference on Automatic Control, Modeling and Simulation, Prague, Czech Republic, 13–15 March 2005; pp. 353–356.

28. Cheraghpour, F.; Vaezi, M.; Jazeh, H.E.S.; Moosavian, S.A.A. Dynamic Modeling and Kinematic Simulation of Staubli© TX40 Robot Using MATLAB/ADAMS Co-simulation. In Proceedings of the 2011 IEEE International Conference on Mechatronics (ICM), Istanbul, Turkey, 13–15 April 2011; pp. 386–391.

29. Masten, M.K. Inertially stabilized platforms for optical imaging systems: Tracking dynamic targets with mobile sensors. *IEEE Control Syst. Mag.* **2008**, *28*, 47–64.

CHAPTER 15

CURRENT ADVANCES IN SYNCHROTRON RADIATION INSTRUMENTATION FOR MX EXPERIMENTS

Robin L. Owen[a], Jordi Juanhuix[b], Martin Fuchs[c]

[a]Diamond Light Source, Harwell Science and Innovation Campus, Didcot, OX11 0DE, UK

[b]Alba Synchrotron, Carrer de la llum 2-26, Cerdanyola, 08192, Spain

[c]National Synchrotron Light Source II, Brookhaven National Lab, Upton, NY, 11973, USA

ABSTRACT

Following pioneering work 40 years ago, synchrotron beamlines dedicated to macromolecular crystallography (MX) have improved in almost every aspect as instrumentation has evolved. Beam sizes and crystal dimensions are now on the single micron scale while data can be collected from proteins with molecular weights over 10 MDa and from crystals with unit cell dimensions over 1000 Å. Furthermore it is possible to collect a complete data set in seconds, and obtain the resulting structure in minutes. The impact of MX synchrotron beamlines and their evolution is reflected in their scientific output, and MX is now the method of choice for a variety of aims from ligand binding to structure determination of membrane proteins, viruses and ribosomes, resulting in a much deeper understanding of the machinery of life. A main driving force of beamline evolution have been advances in almost every aspect of the instrumentation comprising a synchrotron beamline. In this review we aim to provide an overview of the current status of instrumentation at modern MX experiments. The most critical optical components are discussed, as are aspects of endstation design, sample delivery, visualisation and positioning, the sample environment, beam shaping, detectors and data acquisition and processing.

INTRODUCTION

The first successful proof-of-principle of macromolecular crystallography (MX) experiments at synchrotron accelerators were performed in the mid 1970s [124][67] and [117]. These structures have allowed the visualisation of proteins and viruses on an atomic level, and led to a much greater understanding of the function of the building blocks of life. The relevance of this knowledge has been recognized through the award of a large number of Nobel prizes (summarised in Ref. [21]) Over these 40–50 years, synchrotron sources, beamlines and experimental approaches have constantly evolved, dramatically changing what is considered possible, or can be obtained, from an MX experiment. Recent reviews, such as those by Refs. [38] and [46] give an excellent overview of the history of MX and synchrotron radiation.

As an illustration of the impact of synchrotron radiation on MX and the rate at which the field is expanding, the

Citation: Robin L. Owen, Jordi Juanhuix, Martin Fuchs, Current advances in synchrotron radiation instrumentation for MX experiments, Archives of Biochemistry and Biophysics, Volume 602, 15 July 2016, Pages 21-31, ISSN 0003-9861, http://dx.doi.org/10.1016/j.abb.2016.03.021.

number of structures held in the Protein Data Bank (PDB) at the time of these reviews and now can be compared. In 2010, the PDB comprised ~60,000 structures of which ~73% were determined at a synchrotron. At the time of writing, the number of structures has increased to more than 112,000 with synchrotrons now accounting for ~77% of the total number of depositions and approximately 90% of all new depositions. This rate of increase is reflected in challenging sub-areas of the field such as membrane protein crystallography where the number of unique structures has increased from 280 to 518 over the last five years. The striking increase in both the number and rate of depositions is, in no small part, due to continuing advances in beamline instrumentation, experimental approaches and tools to facilitate MX at X-ray sources.

There are now over 100 beamlines dedicated to MX at more than 20 synchrotrons worldwide[1] (Fig. 1) and the instrumentation in all areas of these beamlines has continued to evolve. This evolution has driven scientific discoveries, much in the way put by Sydney Brenner in 1980 – "Progress in science depends on new techniques, new discoveries, and new ideas, probably in that order" [122]. Crystals of poor diffraction quality that previously would have required spending another year in the crystallization lab for optimization, can now provide complete datasets and lead to successful structure determination. Generally, beamline scientists attempt to set up their data acquisition interface with default settings to maximize the data quality users can obtain. However, it is often successful communication between the structural biologist, to describe the challenges of the sample, and the beamline scientist, to describe new advances in instrumentation and software, that really makes the difference in meeting experimental goals. New techniques or instrumentation often originate from pioneering experiments or developments at 'standard' MX beamlines. The success of the technology developed for MX beamlines also inspires other synchrotron fields such as powder diffraction and small angle scattering, which are adopting instruments such as the robotic sample manipulators or the experiment control software.

Figure 1: Current worldwide distribution of MX beamlines. Area of the outer circle is proportional to the number of MX beamlines currently operating at the facility.

This review therefore, while attempting to touch on historical developments, focuses on outlining the current state of the art of MX beamline instrumentation, and tries to point out the ongoing and new trends and developments such as an increasing degree of automation, increasing numbers of experiments performed with remote access, cross fertilization with free electron laser experiments, micro- and nano-crystallography, native SAD phasing, in-situ room temperature data collection and multi-crystal and serial crystallography.

X-RAY SOURCE, OPTICS AND DIAGNOSTICS

X-ray Sources

The first "parasitic" protein crystallography data collections, at synchrotron accelerators primarily dedicated to high-energy physics, exposed the potential of purpose-built facilities for the field of MX [67], [117] and [124]. The first dedicated synchrotron sources (DORIS, Hamburg-Germany; SRS, Daresbury-UK; NSLS, Brookhaven-USA; Photon Factory, Tsukuba-Japan) started operation in the early 1980s with bending magnet beamlines dedicated to MX. Synchrotron sources made another step forward in the 1990s with the advent of 3rd generation sources which featured a lower emittance and the extended use of insertion devices (ID) installed in straight sections as a photon source for beamlines. The photon flux, beam size and beam divergence of these photon sources match the weak diffracting power, small crystal size, and the large unit cell parameters of typical macromolecular crystals.

Insertion devices (IDs) are periodic arrays of magnets which bend the path of the electrons as they circulate the storage ring. Changes in the electron beam trajectory at each magnet results in the emission of radiation. Undulators are IDs that exploit constructive interference between the radiation emitted at different periods and are the brightest photon source at current synchrotrons [3]. This constructive interference results in sharp peaks in the spectral profile where the intensity of radiation produced is proportional to the square of the number of poles. Undulators can emit radiation at the wavelengths typically used in MX (around 1 Å) only in storage rings with electron energy above ~2.25 GeV. Below this energy, radiation of around 1 Å must be emitted by wigglers, where the interference is no longer preserved and the flux is proportional to the number of poles: a significant reduction.

New developments to greatly reduce the emittance and increase the brilliance of storage rings, and hence improve the X-ray beam properties achievable at the sample position, are now being realised [69] and [147]. By using damping wigglers and increasing the circumference, the NSLS-II currently achieves the smallest sub-nm rad horizontal emittance in regular operation [149]. In a new fourth generation of storage rings, the multibend achromat magnetic lattices [48] and [49] allow the construction of diffraction-limited storage rings [10], typically of circumference 500–1500 m and possessing an emittance below 0.4 nm rad. The first fourth generation storage ring due to become operational is MAX IV in Lund, Sweden, which will begin operation in 2016 [138]. Such is the attraction of these storage rings that several other such sources are under construction, and the lattice of a number of existing sources is being redesigned. The X-ray beams produced by these new sources will represent an increase by a factor of 20 in brilliance and ~1.5 in flux in comparison to third generation sources.

Aided by the smaller emittance of the new storage rings, a new generation of undulators using superconducting magnets [58] or, especially, cryogenic permanent magnets [66]aims at reducing the magnetic period. A shorter undulator period is of interest, as it allows for a reduction of the electron energy required for the generation of a specific photon wavelength, a larger number of periods per length (resulting in higher flux), an increased spectral tunability, and an increase of the radiation hardness of the undulator.

Monochromators

The beamline monochromator selects X-rays of a single wavelength from the broad energy spectrum emitted by the bending magnet or insertion device, and also absorbs the power of the disregarded energies. A historical overview of the development of monochromators at synchrotron sources is given by Ref. [12]. Briefly, the photon wavelength is selected using Bragg's law and can be varied by changing the angle of crystals with respect to the incoming beam. The means of crystal mounting is chosen according to the design priorities of the beamline. Currently the most popular type of monochromator is the double-crystal monochromator (DCM), in which the second crystal surface is adjusted to maintain a fixed beam path after the monochromator at any photon wavelength. By sagittally bending one of the monochromator crystals, the x-ray beam can be focused horizontally. Also, two identical DCMs placed oppositely can be used to maintain the beam path after the monochromator at any selected wavelength [8]. An alternative to a DCM is the channel-cut monochromator (CCM), in which the two crystal surfaces are part of the same silicon block. CCMs are less prone to introduce beam vibrations and drifts, but change the path of the output beam when the beamline energy is changed, unless two identical channel-cut crystals are used as in the case of the Bartels monochromator. For long microfocus beamlines, the required high stability of the monochromator crystals can indicate a design with a vertical rotation axis [57], which minimizes the projection of gravity onto the Bragg axis and greatly reduces the construction's height above the supporting granite structure. Moreover, since the source size is still larger in the horizontal than in the vertical direction, a horizontal diffraction plane has less influence on the beam stability. Due to its axis being perpendicular to the x-ray polarization, this sacrifices some flux at long wavelengths and energy resolution for a more stable construction.

Silicon crystals are very well suited for selecting the photon wavelength for MX experiments and high quality crystals can be obtained at relatively modest cost. Moreover, silicon has excellent physical properties at cryogenic temperatures: the coefficient of thermal expansion at liquid nitrogen temperatures is close to zero and the thermal conductivity is largely increased with respect to room temperature. Monochromator silicon crystals are usually cut such that the crystal surface is parallel to a Bragg plane (symmetric cut), for ease of manufacturing and to preserve both beam size and divergence when changing the photon energy.

The most commonly used silicon reflection is (111) as it delivers a strong diffracted beam due to its large structure factor. Moreover, the natural width of the (111) reflection is similar to the typical vertical photon beam divergence. This similarity maximizes the flux while keeping an energy bandpass of $\Delta E/E \sim 2 \cdot 10^{-4}$ low enough to allow energy-dependent experiments. Other advantages of silicon are that the lattice parameter ($d = 3.1356$ Å) allows a large deflection angle of $2\theta \sim 18.5$ deg at a wavelength of 1 Å, and the diamond cubic symmetry of the crystal structure results in the absence of a (222) reflection and thus the possible presence of X-rays of twice the desired energy at the sample position. Other reflections, notably (220) and (311), have been used to improve the energy resolution at the expense of a lower photon flux at the sample. Nonetheless the use of these reflections is limited as only a small subset of experiments greatly benefit from this increased energy resolution. Multilayer monochromators with a wide energy bandpass, and hence large photon flux throughput, are currently being developed and installed in order to exploit high frame rate detectors and new fast data collection strategies.

Focusing Optics

The role of beamline optics is to focus the cone of X-rays emitted by the photon source to a size comparable to, or smaller than, the sample. Concurrently, the beam divergence, which increases by the same factor as the size is reduced, must be minimised to avoid increasing the apparent crystal mosaicity seen on the diffraction pattern. The variety of source properties, and scientific cases at MX beamlines has led to a number of optical layouts. Some representative examples are shown in Fig. 2.

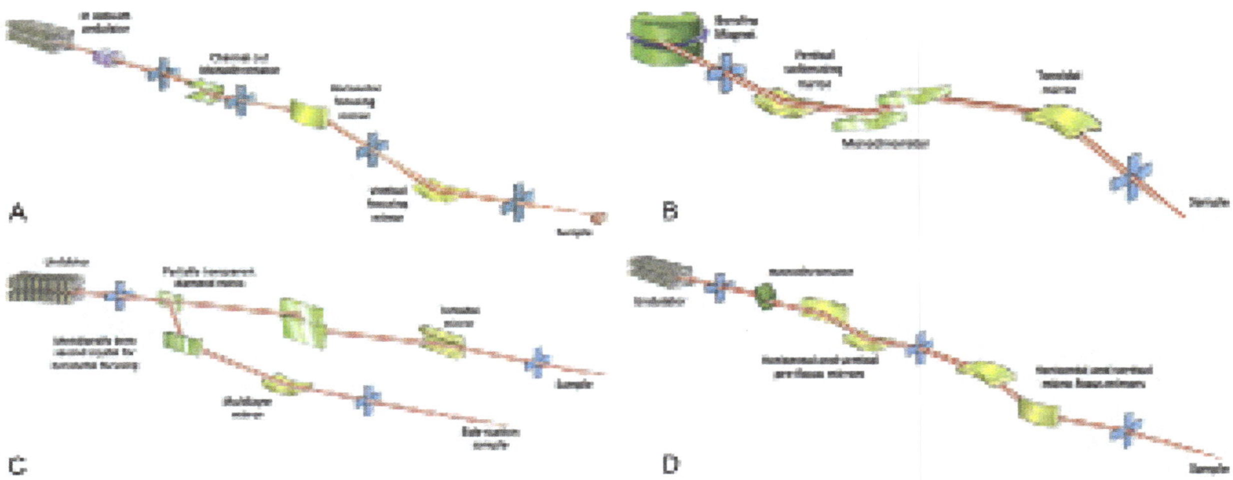

Figure 2: Schematics of different optical arrangements used at MX beamlines. In an effort to include a variety of optical elements some schemes incorporate aspects of multiple beamlines, and so may not represent an actual beamline. A. Undulator beamline with a double crystal monochromator and focusing provided by two mirrors in a KB arrangement. B. Bending magnet beamline showing the X-ray beam collimated in the vertical by a collimating mirror placed before the double crystal monochromator. C. Undulator beamline with a fixed-wavelength sid-estation. A fraction of the beam from the undulator is deflected by a partially transparent monochromator crystal. The beam passing through is incident on a double crystal monochromator and then focused using a toroidal mirror. D. Two-stage focusing arrangement for obtaining a micro-focused beam at the sample position.

Curved reflective mirrors with lengths between 0.2 and 1 m are the most common method for focusing the X-ray beam onto the sample or the detector position, or to create a secondary source in a 2-step focusing scheme. Aberration-free focussing can be achieved through use of an ellipsoidal surface although toroidal and spherical shapes are used as well. Mirrors made of silicon, fused silica or ceramic and coated with one or several high-Z metal stripes (usually Rh, Pd, Pt, Ir) are placed at typical grazing incidence angles of ~0.3 deg to ensure total external reflection. As reflectivity of these coatings falls dramatically above an element-dependent cut-off energy, the mirror effectively removes any higher harmonics passing the monochromator.

One of the mirror configurations most commonly used is the Kirkpatrick-Baez (KB) arrangement, consisting of a pair of plane-elliptical mirrors mounted orthogonally, so that the beam can be adjusted independently in horizontal and vertical dimensions. Meridional optics (curved along the beam direction) such as these have a long radius of curvature (0.2–10 km) and this is often achieved by bending a plane surface or through use of bimorph mirrors [154], in which a piezo-actuator is attached every ~20 mm along the length of the mirror. Adaptive mirror optics based on mechanical actuators are also under study.

Very strict tolerances on the mirror slope errors are required to maintain the source brightness in low-emittance synchrotrons and Free Electron Lasers (FELs). Roughness of the mirrors has the effect of broadening the spot size, while longer period slope errors are additionally the major source of beam heterogeneities, especially when the beam is defocused to match the crystal dimensions. Several approaches have been used to improve the surface quality in mirrors. Classical mechanical polishing using massive planetary polishing systems is proven to achieve flat mirror surfaces to 0.2 μrad RMS. New innovative approaches are Ion Beam Figuring [155] and, in particular, Elastic Emission Machining (EEM) [156]. The EEM technique, offered by J-Tec (Osaka, Japan), can achieve a RMS slope error of less than 50 nrad and a figure accuracy of <1 nm in a 350 mm-long mirror and is the current state-of-the-art. Advances on mirror polishing have been only possible with accompanying new optical metrology instruments, notably the Nanometer Optical Metrology (‹NOM›) long-trace linear profilometers [157]. Monitoring of the surface quality is possible with in-situ, at-wavelength methods such as the pencil beam method or the shearing interferometry [148]. Overall, these advances have led to a 5-fold improvement of the mirror surface quality in the last 10 years.

An alternative to reflective optics are refractive lenses which exploit the slightly different refractive index between light-Z materials (Be, Al, Si) and vacuum to focus the X-ray beam [158]. Transfocators are arrays of refractive lenses that can focus the beam in one or two dimensions depending on the lens shape [141]. The photon energy and the number of identical lenses brought to the beam path determine the focusing distance. Although transfocators deliver less flux due to the limited transmission through the lenses, the beam is very stable and the change of the beam size (i.e. the focus) or the energy which imply changing the number of lenses, take only a few minutes. The use of transfocators is currently expanding worldwide, especially in Europe, being particularly interesting for beamlines offering small beam sizes and moderate to high photon energies, or concerned to reduce the wavefront distortions.

With the improvement of the source and focusing technologies, there are now over 20 microfocus beamlines worldwide with beam sizes below 20 μm [133] with beamlines under construction to achieve beam sizes even below 1 μm.

Diagnostics

To maximize the signal-to-noise and overall data quality of macromolecular crystallography data sets, the size and divergence of the X-ray beam should be matched to the crystal properties [34] and [107]. It is the beamline diagnostics system that provides feedback on the current value and stability over time of these beam parameters.

Beyond interactive beam optimization, beam diagnostics are critical for automatic beamline operation, and for the logging of experiment parameters to image headers or other metadata information to inform data processing. Beam diagnostic information is essential for data collection strategy planning. With the beam profile available, the 3D dose distribution within a crystal can be visualised and integrated in the experiment planning by RADDOSE-3D [152]. Last but not least, diagnostics are a means for the user›s verification of the beamline›s proper operation – is a lack of diffraction due to the sample or is it the beamline?

Beam monitors can provide intensity, position, and shape or profile information, with different types providing different combinations of these three key parameters. The main performance specifications of beam monitors are beam transmission, sensitivity, radiation hardness, time resolution, dynamic range, size, and photon energy dependence. Invasive monitors such as fluorescent screens that completely block the beam are often used for commissioning the beamline. Diagnostics in use during data collection have to be transmissive, such as the monitor types discussed below.

In addition to controlling the beam parameters, ensuring the beam›s stability [120] is crucial for optimising data quality. In case it is impossible or impractical to identify and remove them, beam-drifts, and to some extent also vibrations, can be stabilized through position feedback schemes. Typically, the measured beam position is used as a corrective input to beam steering optical elements such as X-ray mirrors or monochromator crystals. The speed of such feedback systems depends on the mechanical properties of the optical elements, which place a practical limit around 10 Hz. Beam stabilization can also be achieved by steering the electron beam orbit via the storage ring magnets [26], also at higher frequencies [106]. Moreover, feedback on monochromator crystal angular settings can be used to stabilize the X-ray beam energy and intensity [15], [85] and [140], and in principle the stabilization of the full beam profile is possible.

Monitors for white beams require special solutions that can handle the high heat load. Blade beam position monitors measure photo currents from tungsten or diamond blades exposed to the X-ray beam fringes [14], [73], [76] and [132]. Quadrant electrode monitors obtain photocurrents from a thin diamond membrane [103]. For both monitor types, a position is derived from the differential signal coming from the sensors located in opposite sides of the beam.

For monochromatic X-ray beams, a variety of position monitors are used: CVD diamond quadrant electrode photo current monitors with a polycrystalline [131] and [130], (Dectris Rigi) or a single crystalline membrane [42], [92] and [103], quadrant diode monitors detecting backscatter fluorescence from a metal foil [2] and split electrode ion chambers[97], [128] and [129]. All of these can double as intensity monitors. An intensity monitor combined with an aperture that cuts into the beam can also provide dynamic positional information such as the frequency spectrum of vibrations, or – given the beam profile – relative position data.

Pixelated CVD diamcnd quadrant electrode monitors can provide beam profile data [153]. The NanoBPM profile monitor [81] commercialised by FMB Oxford is essentially a pinhole camera observing the fluorescence generated by the X-ray beam footprint in a thin membrane.

ENDSTATION AND EXPERIMENT

Endstation

In designing a macromolecular crystallography beamline's experimental station, several partially opposing requirements have to be considered: Performance, versatility, flexibility, ease of use, size, and automated operation. These conflicting requirements together with other constraints such as source and optical parameters, and even the available space and resources, mean that many different approaches have been adopted to design the core functionality of an MX endstation. Typically this functionality encompasses sample movement and positioning, sample visualisation, temperature control, and automated sample exchange.

Several key boundary conditions have greatly evolved throughout the years, thereby greatly changing the requirements for experimental stations. Beam sizes and average sample sizes shrunk from millimetres to a few microns and data collection times have diminished from days to minutes and now seconds due to the development of faster detectors and brighter X-ray beams. To follow these advances, endstation designers had to improve goniometer precision, microscope resolution, incorporate sample automation, and greatly improve the overall stability of the apparatus.

An example of a highly integrated, high-precision and widely implemented design for an experimental station is the EMBL/ESRF MD2 microdiffractometer [114] and its successor, the MD3, both of which have been commercialised.[2] At the other extreme, the use of a six axis robotic arm as a basis for a diffractometer in the G-Rob [87] or the ESRF RoboDiff [102] is an implementation designed for highest possible flexibility and throughput. An example for extreme versatility in terms of the experiments possible at a single beamline is the PETRA III P11 beamline combining macromolecular crystallography, imaging, and time-resolved techniques into one single experimental station for bio-imaging and diffraction [96]. Endstation layouts at MX beamlines worldwide have been developed to make optimal use of the locally available spatial conditions, such as so-called mini-hutches that provide user-access to the sample area through a small window, rather than requiring the opening of the hutch-door [90]. Several experimental stations have added specialized features, such as advanced visualisation techniques like fluorescence imaging [63] or SONICC detection [91], or concurrent optical spectroscopies and data collection from crystallization plates (both discussed below). Beamlines that aim to provide extreme figures of merit require dedicated solutions, such as the optimization for smallest beam and crystal sizes for micro- and nano-crystallography [121], or for longest wavelengths with helium-flooded [71] or even in-vacuum sample environments [105].

The experimental stations at the authors› beamlines may serve as illustrative examples of the variation of designs laid out above (see Fig. 3). The experimental station of the XALOC beamline at the ALBA synchrotron has at its core the microdiffractometer MD2[80]. The D3 diffractometer at the Swiss Light Source [56] combines a microcrystallography goniometer with an in-situ on-axis microspectrophotometer for concurrent UV/Vis absorption, fluorescence and Raman spectroscopy. The I24 endstation at Diamond provides two separate goniometers for rapidly switching between micro-cryocrystallography and high precision room-temperature crystallography in plates (initial design overview given in Ref. [65]).

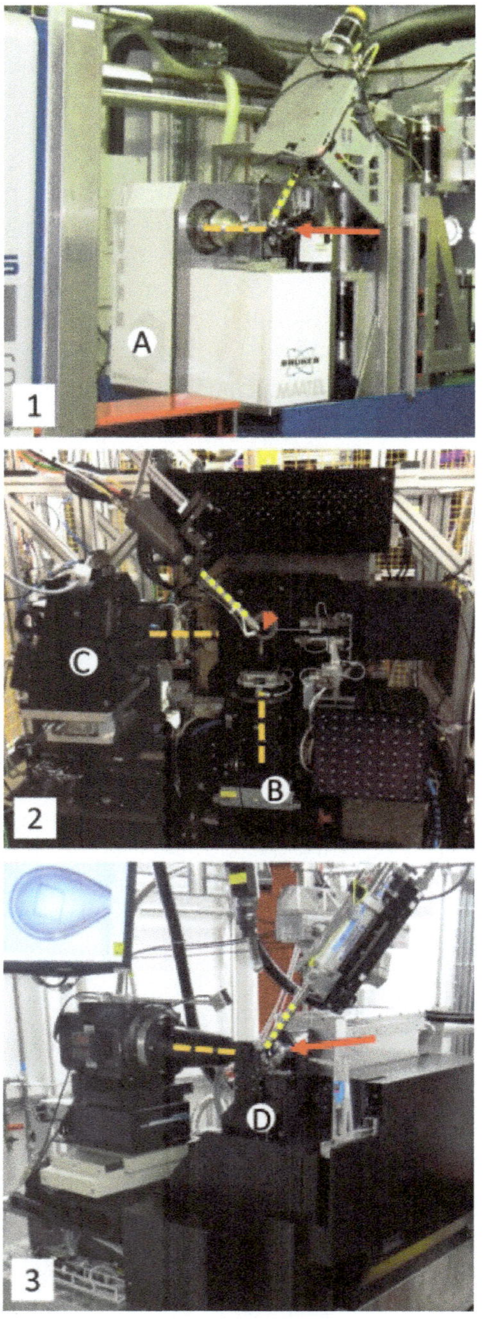

Figure 3: Experimental stations (ES) of the beamlines XALOC at ALBA (1), I24 at Diamond Light Source (2) and X10SA at the Swiss Light Source (3). The beam direction is indicated by the red arrow, goniometer spindles by the orange dashed lines and the cryo cooler nozzle by the yellow dotted lines. The XALOC ES is built around the microdiffractometer MD2 (Arinax, France) (A). The I24 ES features two goniometers, the main goniometer with a vertical spindle (B) and a dedicated crystallization plate goniometer with a horizontal spindle (C). The X10SA diffractometer D3 features a microspectrophotometer for concurrent diffraction and optical spectroscopy experiments (D).

Sample Positioning

The goniometer is central in the oscillation crystallographic method used in almost every MX experiment, in which the crystal rotates while recording diffraction images. Goniometer technology has seen an evolution from the use of mechanical bearings to air bearings, and from the use of large circles carrying proportional detectors to small goniometer spindles rotating the crystal while the diffraction pattern is recorded simultaneously with an area detector. Arguably the key performance parameter of a goniometer is its sphere of confusion – the smallest sphere containing the goniometer's sample mount position while rotating through its complete parameter space. The requirements on a goniometer's sphere of confusion are dictated by the size of crystals measured, the size of the beam, and the Bragg angle range over which crystals are rotated during a data collection. Assuming the experimenter has carefully matched the beam size to the crystal size, and wants to collect a 180° wedge, the sphere of confusion should typically be less than 10% of the beam and crystal size. Reductions in both of these parameters to the single micron scale have driven forward goniometer technology.

Modern microcrystal diffractometers have been able to achieve sub-micron spheres of confusion (see for example the D3 diffractometer [56], the Kohzu QKSU-1 goniometer[71] and the MD3 microdiffractometer, though many others now exist at, for example, beamlines such as I24 at Diamond, GM/CA at APS, and P11 at PETRA III). By measuring the reproducible error of a goniometer, the angular dependence of a goniometer's gravitational sag can be determined and stored in a lookup table, to be used in as part of an active correction scheme to reduce the goniometer's sphere of confusion. With their rotational axis aligned to the gravitational force, vertical axis goniometers typically can achieve smaller spheres of confusion of around or even below 100 nm[84] and [121]. In principle, for longer wavelengths the horizontal polarization of the synchrotron radiation may limit the diffraction resolution attained by vertical axes in the oscillation method for crystallographic data collection as used in MX [4]. Nevertheless, to the authors› best knowledge, for the resolutions and data qualities typical in MX, current experiments are yet to encounter this limitation.

Multi-axis goniometers, providing three rotational degrees of freedom, open the possibility for further optimization of data collection strategies [20]. These strategies include the reorientation such that Friedel pairs are recorded on the same image to aid anomalous phasing, or the alignment of the crystal with a long crystal axis along the rotation axis to minimise diffraction spot overlap. The orientation of the unit cell axes is typically determined via software such as XOalign[3] following collection of a small number of indexing frames.

Due to geometrical constraints and their comparatively large sphere of confusion, multi-axis goniometry has not been extensively used in the past decade. However, such goniometry mounted on the omega axis has met increasing interest: first with the minikappa device in combination with the MD2 and MD3 diffractometers [20] and more recently with the PRIGo [145].

While the classical oscillation method consisted of centering a crystal in the beam and then rotating it while recording diffraction images, the use of line, or helical, scanning [159]has become an increasingly common means of collecting crystallographic data. By translating a crystal in a synchronized movement with the omega rotation, a helical path through/along a crystal is traced. Translation during exposure to X-rays means that the absorbed dose can be evenly distributed over the crystal, to make optimal use of the diffracting volume and avoid discontinuities in the scaling correction that would result from collecting data from discrete positions. In addition to the beam trajectory, an optimal data collection strategy has to include a suitable beam shape – for a needle-shaped crystal, this could be a vertical line focus swept along the needle – and the distribution of the X-ray dose, see below.

Beam Shaping and Sample Environment

Optimization of data collection includes careful tailoring of the X-ray beam properties to the crystal. By adjusting the beam focus to the size of the crystal, the background scattering from the buffer surrounding the crystal can be minimized and signal to noise maximised [51]. Where the beamline's native focus is larger than the sample, the beam size can be adjusted at the expense of flux through the use of slits or pinhole apertures (as demonstrated for example at GM/CA, APS [53] or offered as part of the functionality of the MD2 microdiffractometer). For crystal morphologies such as plates, for optimal data collection the beam shape should be changed as the crystal rotates such that the beamsize matches the projected dimensions of the crystal along the beam axis [68] – while this ideal approach is not commonly implemented yet, a beamsize adjustment to the average projection size is good common practice. Alternatively, the beam is sometimes expanded at the sample position by defocusing the mirrors to uniformly illuminate the crystal or to focus the beam at the detector.

For all experiments, care must be taken to minimise the contribution of air scatter to the data collected. Most frequently this is addressed through the use of a collimating tube that encloses the beam to within a few mm of the sample. For micro-crystal data collection, and data collection at long wavelengths, absorption and scattering by air becomes a significant limitation and additional care is required. Air scatter can be reduced through the use of a helium path between the beam stop and detector surface, switching the sample cooling from gaseous nitrogen to helium, or even by encasing the complete endstation environment and flushing it with He gas [71]. Further gains – especially crucial for measurements at long wavelengths – can be made through use of an in vacuum endstation as, for example, at beamline I23 at Diamond Light Source [105].

Sample cooling using a nitrogen gas flow at cryogenic temperature is an essential part of all 'standard' MX experiments at synchrotron sources (radiation damage aspects of cryo-cooling are discussed below). Further control of the sample cooling can be achieved through the use of a cryo-shutter that allows controlled sample annealing. Under certain circumstances, this can improve the diffraction quality of a crystal, though the chances of success can be hard to predict. For room temperature measurements, the use of humidity control devices [127] can prevent samples from drying out. By monitoring the diffraction properties of the crystal as humidity conditions vary, improvements in crystal quality can be found in some cases [18]. A key consideration in the endstation design is temperature stability: the effects of cold nitrogen gas can be mitigated through the use of a cryo gas extraction tube opposite the cryo nozzle, which will also reduce the nitrogen flow turbulence. The use of on-axis cryo-cooling, with a heating shield between the sample pin and goniometer to stabilize its temperature, greatly reduces the sample pin›s projection to the gas flow [11]. Automation also helps improve temperature stability as the need to open the experimental hutch regularly is removed.

Data collection from crystals held in crystallographic plates, or *in situ* data collection, is routinely used for both crystal screening and the collection of datasets. As crystallization trays were not initially designed with X-ray data collection in mind, an early use case was to obtain rapid feedback on the crystal identity and on the quality of hits in crystallization trials. As the data collected at room temperature is free of artefacts from handling, cryo-protectants and from the cooling process itself, *in situ* data collection can provide a rapid means of optimising crystallization conditions. The first implementations used a six-axis robotic arm with a gripper for SBS format crystallization plates to position and rotate the plates in the X-ray beam [77], while more recent implementations have incorporated dedicated goniometry into the endstation [13], [87] and [7]. Crystallization plates have also evolved to facilitate X-ray data collection, with plates designed to minimise background scatter and maximize the accessible rotation range becoming commercially available. *In situ* plates compatible with lipidic cubic phase crystallization, a critical part of membrane protein crystallography, have also been developed allowing datasets to be collected from membrane proteins without the need for harvesting from the drop [75]. The wells of CrystalDirect plates [28] are designed to be laser cut and mounted on special pins to avoid crystal fishing, sidestepping some of the problems associated with *in situ*crystallography.

Complementary Spectroscopic Techniques

Several beamlines offer the acquisition of optical spectroscopic data concurrent with the collection of diffraction data sets [22], [30], [39], [50], [111], [113], [118], [125],[126] and [136]. Key applications of this complementary information are the identification of kinetic reaction intermediates in structural enzymology, as well as the quantitative monitoring of changing oxidation states of metal centers due to X-ray induced photoelectrons. The spectroscopic methods of choice are UV/Vis absorption, fluorescence, and Raman spectroscopy. The design of the micro-spectrophotometers differs considerably between the different beamlines. The range of methods offered varies from absorption spectroscopy, to combined absorption and fluorescence spectrometers, to the combination of all three methods listed above. While some implementations require a reconfiguration of the endstation, and therefore are typically operated in dedicated spectroscopy shifts, others are permanent installations. A further defining characteristic is the alignment of the optical axes with respect to the X-ray beam – on-axis systems provide an immediate view of the relative beam overlap, while off-axis systems require rotational realignments between diffraction and spectroscopic measurements. Since the acquisition of spectroscopic data can be considerably more time consuming than the diffraction experiment, several beamlines provide accompanying off-line laboratories to perform a spectroscopic pre-characterization of the crystals before taking them to the beamline.

Sample Visualisation

High quality visualisation of samples is essential for accurate alignment of crystals in the X-ray beam. In its most simple implementation this takes the form of a standard microscope pointing at the sample position, though as beam and crystal sizes have reduced this approach has proved insufficient to accurately and reliably align crystals in the X-ray beam. On-axis video microscopes (OAV), such as that developed for the MD2 diffractometer [114], with a drilled objective and 45° mirror provide a view of the sample as seen from the source, and this arrangement is now considered standard at MX beamlines. OAV images and recognition algorithms are used to identify loops and automatically centre these in the X-ray beam [82], [86] and [119]. It can be challenging, however, to apply such routines based on an optical microscope to small crystals or to crystals whose mother liquor becomes opaque in the cryo-cooling process. Errors of only a few microns are sufficient to move the crystal out of the X-ray beam during data collection.

In order to address some of these challenges, the optical properties of protein crystals can also be exploited to facilitate crystal identification and alignment. UV-induced fluorescence can be used to identify crystals [24], [63] and [142] aiding automated alignment of crystals within loops. UV light provided by a laser or LED induces fluorescence from aromatic amino acids between 300 and 450 nm, which is readily detected by a standard microscope giving high contrast between crystals and the surrounding mother liquor. Second-order nonlinear optical imaging of chiral crystals (SONICC) has also recently been explored for imaging protein crystals [29], [43] and [83]. This allows identification of protein crystals within opaque amorphous material such as lipidic cubic phase (LCP).

Where optical recognition is not possible, and for very small crystals, grid scanning can be used. The diffraction grid scan rasters samples through the X-ray beam resulting in a 2D map of diffraction images [1], [17], [25] and [135]. Automatic scoring of each image allows rapid visualisation of crystal location, or in the case of crystals larger than the beamsize, variations in diffracting power across a crystal. Sequential collection of grids at multiple angles is sufficient to ensure centering of a sample so it remains in the X-ray beam during data collection. Grid scanning is particularly applicable to crystals grown or cryocooled in opaque material. In the case of extremely small crystals, from which it is not possible to collect a large wedge of diffraction data, sequential grids often take the form of a coarse grid to locate crystals followed by a closely spaced grid to fine-tune the crystal position for data collection.

X-ray imaging, or tomography, can also be used to identify and characterise crystals[19] and [146]. The contrast in X-ray beam properties required for imaging means this approach has not become widely implemented, but it may prove of particular use at long wavelength beamlines where increased absorption of X-rays should be taken into account for optimal data processing.

Detectors

Macromolecular crystallography has benefited greatly from the development of large area detectors. In the early days of crystallography these took the form of X-ray sensitive film (1980s) and image plates (mid-1990s) before the introduction of CCD detectors to synchrotron sources (late 1990s). The increasing area of detectors allows data to be collected with improved signal to noise as the detector can be positioned further away from the sample when collecting data to a given resolution, especially when the beam is focused after the sample. This reduces the contribution of diffuse X-ray scatter to the recorded pattern, and increases the inter-spot separation. This and other aspects of optimal data collection are discussed by Ref. [37].

Until recently, the vast majority of MX beamlines were equipped with a CCD detector possessing a readout time on the order of a few seconds. At synchrotron sources this permitted a frame rate of one image every 2–3 s, with the X-ray shutter closing while the detector was read-out. During this time the goniometer would be stopped, wound back and then set rotating again ready for the next image in the dataset adding a time overhead and stringent requirements on synchronisation of beamline hardware. The advent of Pixel Array Detectors (PADs) for MX has revolutionised how data are collected, with an increasing number of MX beamlines now equipped with a PAD.

A major advantage of PADs and the newest generation of CCD detectors over ‘classical’ detectors is their short readout time. Readout times of less than a millisecond can be achieved as all pixels on the detector are read-out in parallel. Pioneered for MX at the Swiss Light Source [47], and commercialised by Dectris, this allows ‘continuous’ data collection, where the crystal is continually rotated and the X-ray shutter remains open during data collection. This approach greatly simplifies shuttering and synchronisation of beamline hardware, and dramatically reduces the duration of the experiment.

The large increase in achievable frame-rate and absence of readout noise means that experimental protocols such as fine-phi slicing, which improve data quality[115] and [101], but would previously have been prohibitively time consuming are now considered routine.

The dramatic reduction in the data collection times that can be achieved from both advances in detector technology and beamline hardware has proven a driving force for automated sample exchange. The combination of these two developments means it can be considered routine to collect many hundreds of datasets in a single visit to a beamline.

The second type of detector present at all beamlines that support anomalous data collection is an energy dispersive X-ray Fluorescence (XRF) detector. By monitoring the element specific fluorescence line of a heavy atom scatterer while scanning the monochromator across its absorption edge, the exact position of the peak and inflection points of the edge can be determined using a program like CHOOCH [52], to optimize the anomalous signal. The most common form of detector in use is a silicon drift detector[137] which permits input count rates of up to 10^6 photons/s. To avoid detector saturation the beam intensity must therefore be carefully optimized to keep the detector dead time below values around 10–15%.

Sample Delivery

Automated sample exchange has greatly benefited synchrotron MX allowing the high throughput promised by shorter exposure times and fast detector readout to be achieved. Increased throughput allows large numbers of crystals to be screened during a single visit which can prove to be essential when only a small fraction of crystals diffract well, and (or) when crystals are unsuitable for a home source. Automated sample exchange also permits remote access, which can facilitate large numbers of research groups accessing a beamline for short visits in a short period and regular access to facilities remote from the home laboratory.

Industrial multi-axis robots have been often used for sample mounting (for a list of systems, see references in Refs. [95] and [116] This approach has been sidestepped at some sources in favor of automounters with a limited range of movement (most notably the SC3 at the ESRF and the Berkeley automounter [134]). The strict requirements imposed by automated sample exchangers have driven the development of standardised sample mounts. The Structural Proteomics In Europe (SPINE) project developed a standard sample holder for general use [27] based on the Hampton CrystalCap, and the SPINE pin remains the standard at the majority of synchrotrons. Standardised containers for the transport of SPINE pins between sites, and mounting in sample changers have also been developed, with multiple standards (for example the Unipuck and ESRF/EMBL basket) co-existing. However, the increasing rate of crystal testing and collection is driving the development of even more compact pin designs such as the miniSpine and the NewPin standard,[4] that are currently being introduced.

The strict requirements on sample holder uniformity, the need to synchronise robots with other beamline equipment, and the necessary combination of moving components and liquid nitrogen has, coupled with the high success rates required for successful, ice-free, sample exchange, resulted in automation being a challenging area since the first auto-mounters were developed in the early 2000s [31] and [99].

To date, the majority of data collections at synchrotrons are carried out on single crystals individually mounted by hand on a holder. While in some cases, the number of crystals may be increased to double-digits, for example in the case of polyhedral crystals mounted on micromeshes [6], the number of crystals per mount nonetheless remains low. Experiments at Free Electron Lasers (FELs) have driven new modes of sample delivery, driven by the impact a single FEL pulse has on a protein crystal: with $> 10^{12}$ photons in a pulse of 5–50 fs, only a single still image can be collected before the sample is destroyed. Thousands of crystals are needed, driving the need for new modes of sample delivery.

A number of approaches have been developed to enable serial crystallography [23]. The most widely used has been the liquid jet or gas dynamic virtual nozzle [41] which streams a narrow jet of liquid containing crystals through the X-ray beam. High Viscosity Extrusion jets (or "toothpaste jets") with their slower flow speeds have since also been used at synchrotron beamlines [108] and [16]. Approaches more akin to those used at synchrotron sources have been developed as well. These are generally based on a goniometer 'borrowed' from a synchrotron beamline [32] and [72] and successes with this approach have led to the emergence of serial crystallography at synchrotrons [16],[33] and [62]. On-going work in new sample mounts, such as grid-based supports [9],[75], [100], [104] and [123] and microfluidic flow cells [44], [70] and [89], may see sample delivery techniques optimal for use at both FELs and synchrotrons.

Sample Lifetime

The loop mounting and cryo-cooling of protein crystals prior to data collection has had a dramatic effect on MX. Once cryo-cooled, crystals effectively become immortal (prior to exposure to X-rays), and sample transport and automatic mounting is greatly simplified. The development of techniques to optimize the cryo-cooling process [61] means that in a large fraction of cases, crystal quality is not adversely affected upon cooling to 100 K, though notable exceptions to this exist, for example in the field of virus crystallography. Cryo-cooling also provides a large benefit in terms of radiation damage, increasing the lifetime of crystals by a factor of ~70 [60].

Even the protection conferred by cryo-cooling does not allow radiation damage to be ignored however, and the finite lifetime of crystals is now frequently the primary reason for choosing the start angle for data collection with care. Site-specific damage means that in addition to a global decay in diffracting power during data collection the relative intensities of reflections change [112] and [35]. The magnitude of changes to individual reflections resulting from site-specific damage is comparable to the changes induced by the introduction of heavy atoms so the best possible anomalous data are collected with Friedel pairs recorded close together in time (dose). An elegant way of achieving this is through the use of a multi-axis goniometer.

In recent years there has been a resurgence in room temperature (RT) crystallography at synchrotron sources with facilities for *in situ* data collection developed at a number of sources [7], [13] and [78]. Data collection from crystals held in a crystallization tray means the potentially damaging steps of crystal harvesting and the cryo-cooling process can be sidestepped. In some cases, most frequently in virus crystallography, successful cryo-cooling can prove impossible, meaning RT data collection is the only option for structure solution. In such cases *in situ* data collection eliminates the need to transfer crystals to capillaries. Even while cryo-cooling appears to be successful: the cooling process can introduce artefacts in side-chain interactions [79] or hide conformation diversity [55]making a RT structure highly desirable. Although the rapid onset of radiation damage means that it may be possible to collect only a few degrees of data from each crystal, strategies have been developed to exploit new instrumentation such as fast PADs and maximize the amount of data that can be collected from each crystal [110] and [109]. Approaches for identifying the similarity of different crystals and how to optimally merge data from these have also been developed [54] and [88]. The concomitant development of hardware and software optimised for RT data collection and processing means the approach can be applied to structure solution and phasing of challenging targets such as membrane proteins [75] and [5].

Data Acquisition and Processing

Key to the wide uptake of MX at synchrotron sources has been the development of control software to allow non-expert users carry out crystallography experiments. Control of the experiment usually relies on a high-level graphical software layer on top of the low-level beamline control system such as EPICS [36], TANGO [64] or MADOCA [93]. Some widely implemented systems are Blu-Ice [94], MxCuBE [59], GDA[5] and BSS [139]. To ensure the portability to different synchrotrons, some systems such as MXCuBE are encompassing a further dissection between the graphical interface (common to all facilities) and the functionality logic (particular to each beamline). The software layer can embrace all aspects of the experiment from sample loading and sample alignment to data collection and analysis. A high-level software layer allows complex data collections to become routine. Examples of this include automated grid X-ray scans, helical data collection, automatic adjustment of kappa goniometry, and interweaved (inverse beam) data collections. A reliable, comprehensive software layer is required for automatic beamline operation for hours through sample queueing, as well as for remote operation of the beamline.

The speed of data collection and throughput has dramatically increased in recent years. This is illustrated by 2 MAD experiments made 15 years apart, both taking approximately a little less than 25 min [144] and [45]. In 1999, a total of 240 images, 3 s exposure per image, were taken on one crystal at three wavelengths. In 2014, ~43,000 frames, 0.02 s exposure per frame, were taken on four crystals at four wavelengths each. The larger*quantity* of collected data has increased the productivity of the beamlines, as shown by the progressive increment of the maximum annual PDB releases achieved by the beamlines. It also favours the variety of successful projects, as more collection strategies are available (e.g. fine phi-slicing, multi-crystal data clustering).

Increases in throughput and speed mean that control systems face much tighter requirements on synchronisation and must be able to cope with large volumes of data. The amount of data generated boosts the need for automated data processing. This ranges from preliminary analysis of raw images, to automated phasing and structure solution. Pipelines such as *ELVES* [74], *xia2* [150], autoPROC [143], EDNA [98] and*MeshAndCollect* [151] aim to perform the automatic steps of the data processing, as well as to provide quick feedback on data already collected to optimize the following collections.

As an umbrella to the experiments performed at the beamlines, Laboratory Information Management Systems (LIMS) such as ISPyB [40] allow large volumes of both raw data and the results of auto-processing to be tracked and managed also from home laboratory. In the future LIMS will likely be coupled to user and safety office databases.

CONCLUSIONS

In recent years, significant progress has been made in all aspects of MX experiments, from photon sources in storage rings to the management of processed data. Taken together, these developments have transformed what might be considered a typical visit to a MX beamline. Data can be collected and structures determined from samples that may not even be visible in a microscope. Diffraction images may not be indicative of the success of the experiment. Decision-making is sometimes the main bottleneck for data collection. Synchrotron users are not even required at synchrotrons anymore. All steps in the chain from crystal to structure can rely on robust instrumentation and automated control. In short, even complex MX experiments can become automated measurements. Synchrotron facilities, rather than just providing photons and raw data, are nowadays offering a full "user experience" and ready-to-interpret data at the end of every beamtime. Yet, human control is always possible in all steps, and in challenging cases is still required for the success of the experiment.

Beyond the most "standard" projects, the MX field is also rapidly evolving and beamlines dedicated to specific methods, for example long wavelength, microfocus or *in situ* data collection, will lead to advances in several divergent areas. New data collection strategies are also appearing. A real, future breakthrough is to come with the advent of 4th generation sources, which will push for new instrumentation, sample handling and phasing methods. It is not only structural biology driving the advances in MX beamline instrumentation and control, but also these new developments offering opportunities that science will find the way to exploit. Conceivably, the synergy between instrumentation and science will lead to new insights into the structure and function of macromolecules in the years to come.

One of the authors (JJ) is indebted to Prof. Joan Bordas and Salvador Ferrer for previous fruitful discussions on optics, experiments and beamline management.

REFERENCES

1. J. Aishima, R.L. Owen, D. Axford, E. Shepherd, G. Winter, K. Levik, P. Gibbons, A. Ashton, G. Evans, Acta Cryst. D66 (2010) 1032e1035.
2. R.W. Alkire, G. Rosenbaum, G. Evans, J. Synchrotron Radiat. 7 (2000) 61e68.
3. J. Als-Nielsen, D. McMorrow, Elements of Modern X-ray Physics, John Wiley & Sons, 2011.
4. W.W. Arndt, A.J. Wonacott (Eds.), The Rotation Method in Crystallography: Data Collection from Macromolecular Crystals, North-Holland Publishing Co, 1977.
5. D. Axford, J. Foadi, N.-J. Hu, H.G. Choudhury, S. Iwata, K. Beis, G. Evans, Y. Alguel, Acta Crystallogr. Sect. D. 71 (2015) 1228e1237.
6. D. Axford, X. Ji, D.I. Stuart, G. Sutton, Acta Crystallogr. Sect. D. 70 (2014) 1435e1441.
7. D. Axford, R. Owen, J. Aishima, J. Foadi, M. AW, J. Robinson, J. Nettleship, R. Owens, I. Moraes, E. Fry, K. Harlos, A. Kotecha, J.M. Grimes, J. Ren, G. Sutton, T.S. Walter, D.I. Stuart, G. Evans, Acta Cryst. D68 (2012) 592e600.
8. W.J. Bartels, J. Vac. Sci. Technol. B 1 (1983) 338e345.
9. E.L. Baxter, L. Aguila, R. Alonso-Mori, C.O. Barnes, C.A. Bonagura, W. Brehmer, A.T. Brunger, G. Calero, T.T. Caradoc-Davies, R. Chatterjee, W.F. Degrado, J.S. Fraser, M. Ibrahim, J. Kern, B.K. Kobilka, A.C. Kruse, K.M. Larsson, H.T. Lemke, A.Y. Lyubimov, A. Manglik,

S.E. McPhillips, E. Norgren, S.S. Pang, S.M. Soltis, J. Song, J. Thomaston, Y. Tsai, W.I. Weis, R.A. Woldeyes, V. Yachandra, J. Yano, A. Zouni, A.E. Cohen, Acta Crystallogr. Sect. D. 72 (2016) 2e11.

10. M. Bei, M. Borland, Y. Cai, P. Elleaume, R. Gerig, K. Harkay, L. Emery, A. Hutton, R. Hettel, R. Nagaoka, D. Robin, C. Steier, Nucl. Instrum. Methods Phys. Res. Sect. A Accel. Spectrom. Detect. Assoc. Equip. 622 (2010) 518e535.
11. H.D. Bellamy, R.P. Phizackerley, S.M. Soltis, H. Hope, J. Appl. Crystallogr. 27 (1994) 967e970.
12. D.H. Bilderback, A.K. Freund, G.S. Knapp, D.M. Mills, J. Synchrotron Radiat. 7 (2000) 53e60.
13. R. Bingel-Erlenmeyer, V. Olieric, J.P.A. Grimshaw, J. Gabadinho, X. Wang, S.G. Ebner, A. Isenegger, R. Schneider, J. Schneider, W. Glettig, C. Pradervand, E.H. Panepucci, T. Tomizaki, M. Wang, C. Schulze-Briese, Cryst. Growth Des. 11 (2011) 916e923.
14. C. Bloomer, J. Brandao-Neto, G. Rehm, C. Thomas, The Proceedings of the 9th European Workshop on Beam Diagnostics and Instrumentation for Particle Accelerators (DIPAC09), May, 2009, pp. 25e27.
15. C. Bloomer, A. Dent, S. Diaz-Moreno, I. Dolbnya, U. Pedersen, G. Rehm, C. Tang, C. Thomas, J. Phys. Conf. Ser. 425 (2013) 042010.
16. S. Botha, K. Nass, T.R.M. Barends, W. Kabsch, B. Latz, F. Dworkowski, L. Foucar, E. Panepucci, M. Wang, R.L. Shoeman, I. Schlichting, R.B. Doak, Acta Crystallogr. Sect. D. 71 (2015) 387e397.
17. M.W. Bowler, M. Guijarro, S. Petitdemange, I. Baker, O. Svensson, M. Burghammer, C. Mueller-Dieckmann, E.J. Gordon, D. Flot, S.M. McSweeney, G.A. Leonard, Acta Cryst. D66 (2010) 855e864.
18. M.W. Bowler, U. Mueller, M.S. Weiss, J. Sanchez-Weatherby, T.L.M. Sorensen, M.M.G.M. Thunnissen, T. Ursby, A. Gobbo, S. Russi, M.G. Bowler, S. Brockhauser, O. Svensson, F. Cipriani, Cryst. Growth & Des. 15 (2015) 1043e1054.
19. S. Brockhauser, M. Di Michiel, J.E. McGeehan, A.A. McCarthy, R.B.G. Ravelli, J. Appl. Crystallogr. 41 (2008) 1057e1066.
20. S. Brockhauser, R.B.G. Ravelli, A.A. McCarthy, Acta Crystallogr. Sect. D. 69 (2013) 1241e1251.
21. J.C. Brooks-Bartlett, E.F. Garman, Interdiscip. Sci. Rev. 40 (2015) 244e264.
22. P. Carpentier, A. Royant, J. Ohana, D. Bourgeois, J. Appl. Cryst. 40 (2007) 1113e1122.
23. L.M.G. Chavas, L. Gumprecht, H.N. Chapman, Struct. Dyn. 2 (2015) 041709.
24. L.M.G. Chavas, Y. Yamada, M. Hiraki, N. Igarashi, N. Matsugaki, S. Wakatsuki, J. Synchrotron Radiat. 18 (2011) 11e15.
25. V. Cherezov, M.A. Hanson, M.T. Griffith, M.C. Hilgart, R. Sanishvili, V. Nagarajan, S. Stepanov, R.F. Fischetti, P. Kuhn, R.C. Stevens, J. R. Soc. Interface 6 (2009) S587eS597.
26. J. Chrin, T. Schmidt, A. Streun, D. Zimoch, Nucl. Instrum. Methods Phys. Res. Sect. A Accel. Spectrom. Detect. Assoc. Equip. 592 (2008) 141e153.
27. F. Cipriani, F. Felisaz, L. Launer, J.-S. Aksoy, H. Caserotto, S. Cusack, M. Dallery, F. di-Chiaro, M. Guijarro, J. Huet, S. Larsen, M. Lentini, J. McCarthy, S. McSweeney, R. Ravelli, M. Renier, C. Taffut, A. Thompson, G.A. Leonard, M.A. Walsh, Acta Crystallogr. Sect. D. 62 (2006) 1251e1259.
28. F. Cipriani, M. Rower, C. Landret, U. Zander, F. Felisaz, J.A. Marquez, Acta Crystallogr. Sect. D. 68 (2012) 1393e1399.
29. R.G. Closser, E.J. Gualtieri, J.A. Newman, G.J. Simpson, J. Appl. Crystallogr. 46 (2013) 1903e1906.
30. A.E. Cohen, T. Doukov, S. Michael Soltis, Protein Pept. Lett. 23 (3) (2016) 283e290.
31. A.E. Cohen, P.J. Ellis, M.D. Miller, A.M. Deacon, R.P. Phizackerley, J. Appl. Crystallogr. 35 (2002) 720e726.
32. A.E. Cohen, S.M. Soltis, A. Gonzalez, L. Aguila, R. Alonso-Mori, C.O. Barnes, E.L. Baxter, W. Brehmer, A.S. Brewster, A.T. Brunger, G. Calero, J.F. Chang, M. Chollet, P. Ehrensberger, T.L. Eriksson, Y.P. Feng, J. Hattne, B. Hedman, M. Hollenbeck, J.M. Holton, S. Keable, B.K. Kobilka, E.G. Kovaleva, A.C. Kruse, H.T. Lemke, G.W. Lin, A.Y. Lyubimov, A. Manglik, Mathews II, S.E. McPhillips, S. Nelson, J.W. Peters, N.K. Sauter, C.A. Smith, J.H. Song, H.P. Stevenson, Y.S. Tsai, M. Uervirojnangkoorn, V. Vinetsky, S. Wakatsuki, W.I. Weis, O.A. Zadvornyy, O.B. Zeldin, D.L. Zhu, K.O. Hodgson, Proc. Natl. Acad. Sci. U. S. A. 111 (2014) 17122e17127.
33. N. Coquelle, A.S. Brewster, U. Kapp, A. Shilova, B. Weinhausen, M. Burghammer, J.-P. Colletier, Acta Crystallogr. Sect. D. 71 (2015) 1184e1196.
34. J.A. Cowan, C. Nave, J. Synchrotron Radiat. 15 (2008) 458e462.
35. F. Crick, B. Magdoff, Acta Cryst. 9 (1956) 901e908.
36. L.R. Dalesio, J.O. Hill, M. Kraimer, S. Lewis, D. Murray, S. Hunt, W. Watson, M. Clausen, J. Dalesio, Nucl. Instrum. Methods A 352 (1994) 179e184.
37. Z. Dauter, Acta Crystallogr. Sect. D. 66 (2010) 389e392.
38. Z. Dauter, M. Jaskolski, A. Wlodawer, J. Synchrotron Radiat. 17 (2010) 433e444.
39. R.J. Davies, M. Burghammer, C. Riekel, J. Synchrotron. Rad. 16 (2009) 22e29.
40. S. Delageniere, P. Brenchereau, L. Launer, A.W. Ashton, R. Leal, S. Veyrier, J. Gabadinho, E.J. Gordon, S.D. Jones, K.E. Levik, S.M. McSweeney, S. Monaco, M. Nanao, D. Spruce, O. Svensson, M.A. Walsh, G.A. Leonard, Bioinformatics 27 (2011) 3186e3192.
41. D.P. DePonte, R.B. Doak, M. Hunter, Z. Liu, U. Weierstall, J.C.H. Spence, Micron 40 (2009) 507e509.
42. K. Desjardins, M. Pomorski, J. Morse, J. Synchrotron Radiat. 21 (2014) 1217e1223.
43. E.L. DeWalt, V.J. Begue, J.A. Ronau, S.Z. Sullivan, C. Das, G.J. Simpson, Acta Crystallogr. Sect. D. 69 (2013) 74e81.

44. K. Dhouib, C. Khan Malek, W. Pfleging, B. Gauthier-Manuel, R. Duffait, G. Thuillier, R. Ferrigno, L. Jacquamet, J. Ohana, J.-L. Ferrer, A. TheobaldDietrich, R. Giege, B. Lorber, C. Sauter, Lab Chip 9 (2009) 1412e1421.

45. H. Dong, Q. Xiang, Y. Gu, Z. Wang, N.G. Paterson, P.J. Stansfeld, C. He, Y. Zhang, W. Wang, C. Dong, Nature 511 (2014) 52e56.

46. E.M.H. Duke, L.N. Johnson, Proc. R. Soc. Lond. A Math. Phys. Eng. Sci. 466 (2010) 3421e3452.

47. E.F. Eikenberry, C. Bronnimann, G. Hülsen, H. Toyokawa, R. Horisberger, € B. Schmitt, C. Schulze-Briese, T. Tomizaki, Nucl. Instrum. Methods A 501 (2003) 260e266.

48. D. Einfeld, Synchrotron Radiat. News 27 (2014) 4e7.

49. D. Einfeld, M. Plesko, Nucl. Instrum. Methods Phys. Res. Sect. A Accel. Spectrom. Detect. Assoc. Equip. 335 (1993) 402e416.

50. M.J. Ellis, S.G. Buffey, M.A. Hough, S.S. Hasnain, J. Synchrotron. Rad. 15 (2008) 433e439.

51. G. Evans, D. Axford, R.L. Owen, Acta Cryst. D67 (2011) 261e270.

52. G. Evans, R.F. Pettifer, J. Appl. Crystallogr. 34 (2001) 82e86.

53. R.F. Fischetti, S. Xu, D.W. Yoder, M. Becker, V. Nagarajan, R. Sanishvili, M.C. Hilgart, S. Stepanov, O. Makarov, J.L. Smith, J. Synchrotron Radiat. 16 (2009) 217e225.

54. J. Foadi, P. Aller, Y. Alguel, A. Cameron, D. Axford, R.L. Owen, W. Armour, D.G. Waterman, S. Iwata, G. Evans, Acta Cryst. D69 (2013) 1617e1632.

55. J.S. Fraser, H. van den Bedem, A.J. Samelson, P.T. Lang, J.M. Holton, N. Echols, T. Alber, Proc. Natl. Acad. Sci. U. S. A. 108 (2011) 16247e16252.

56. M.R. Fuchs, C. Pradervand, V. Thominet, R. Schneider, E. Panepucci, M. Grunder, J. Gabadinho, F.S.N. Dworkowski, T. Tomizaki, J. Schneider, A. Mayer, A. Curtin, V. Olieric, U. Frommherz, G. Kotrle, J. Welte, X. Wang, S. Maag, C. Schulze-Briese, M. Wang, J. Synchrotron Radiat. 21 (2014a) 340e351.

57. M.R. Fuchs, R.M. Sweet, L.E. Berman, W.A. Hendrickson, O. Chubar, N. Canestrari, M. Idir, L. Yang, D.K. Schneider, J. Phys. Conf. Ser. 493 (2014b) 012021.

58. J.D. Fuerst, C. Doose, Q. Hasse, Y. Ivanyushenkov, M. Kasa, Y. Shiroyanagi, AIP Conf. Proc. 1573 (2014) 1527e1534.

59. J. Gabadinho, A. Beteva, M. Guijarro, V. Rey-Bakaikoa, D. Spruce, M.W. Bowler, S. Brockhauser, D. Flot, E.J. Gordon, D.R. Hall, B. Lavault, A.A. McCarthy, J. McCarthy, E. Mitchell, S. Monaco, C. Mueller-Dieckmann, D. Nurizzo, R.B.G. Ravelli, X. Thibault, M.A. Walsh, G.A. Leonard, S.M. McSweeney, J. Synchrotron Radiat. 17 (2010) 700e707.

60. E. Garman, Acta Cryst. D66 (2010) 339e351.

61. E. Garman, R.L. Owen, Methods in Molecular Biology, in: Macromolecular Crystallography Protocols, vol. 364, Humana Press, 2007, pp. 1e18.

62. C. Gati, G. Bourenkov, M. Klinge, D. Rehders, F. Stellato, D. Oberthur, O. Yefanov, B.P. Sommer, S. Mogk, M. Duszenko, C. Betzel, T.R. Schneider, H.N. Chapman, L. Redecke, IUCrJ 1 (2014) 87e94.

63. K.J. Gofron, N.E.C. Duke, Nucl. Instrum. Methods Phys. Res. Sect. A Accel. Spectrom. Detect. Assoc. Equip. 649 (2011) 216e218.

64. A. Gotz, E. Taurel, J. Pons, P. Verdier, J. Chaize, J. Meyer, F. Poncet, G. Heunen, € E. Gotz, A. Buteau, in: ICALEPCS2003, Gyeongju, October, 2003 € .

65. V. Grama, D. Axford, G. Duller, M. Burt, R.L. Owen, MEDSI 2014,2014.

66. T. Hara, T. Tanaka, H. Kitamura, T. Bizen, X. Marechal, T. Seike, T. Kohda, Y. Matsuura, Phys. Rev. Special Top. e Accel. Beams 7 (2004) 050702.

67. A. Harmsen, R. Leberman, G.E. Schulz, J. Mol. Biol. 104 (1976) 311e314.

68. J. Hausmann, E. Christodoulou, M. Kasiem, V. De Marco, L.A. van Meeteren, W.H. Moolenaar, D. Axford, R.L. Owen, G. Evans, A. Perrakis, Acta Crystallogr. Sect. F. 66 (2010) 1130e1135.

69. R. Hettel, J. Synchrotron Radiat. 21 (2014) 843e855.

70. M. Heymann, A. Opthalage, J.L. Wierman, S. Akella, D.M.E. Szebenyi, S.M. Gruner, S. Fraden, IUCrJ 1 (2014) 349e360.

71. K. Hirata, Y. Kawano, G. Ueno, K. Hashimoto, H. Murakami, K. Hasegawa, T. Hikima, T. Kumasaka, M. Yamamoto, J. Phys. Conf. Ser. 425 (2013) 012002.

72. K. Hirata, K. Shinzawa-Itoh, N. Yano, S. Takemura, K. Kato, M. Hatanaka, K. Muramoto, T. Kawahara, T. Tsukihara, E. Yamashita, K. Tono, G. Ueno, T. Hikima, H. Murakami, Y. Inubushi, M. Yabashi, T. Ishikawa, M. Yamamoto, T. Ogura, H. Sugimoto, J.-R. Shen, S. Yoshikawa, H. Ago, Nat. Methods 11 (2014) 734e736.

73. K. Holldack, D. Ponwitz, W.B. Peatman, Nucl. Instrum. Methods Phys. Res. Sect. A Accel. Spectrom. Detect. Assoc. Equip. 467e468 (2001) 213e220. Part 1.

74. J. Holton, T. Alber, Proc. Natl. Acad. Sci. U. S. A. 101 (2004) 1537e1542.

75. C.-Y. Huang, V. Olieric, P. Ma, E. Panepucci, K. Diederichs, M. Wang, M. Caffrey, Acta Crystallogr. Sect. D. 71 (2015) 1238e1256.

76. P. Ilinski, J. Phys. Conf. Ser. 425 (2013) 042006.

77. L. Jacquamet, J. Ohana, J. Joly, F. Borel, M. Pirocchi, P. Charrault, A. Bertoni, P. Israel-Gouy, P. Carpentier, F. Kozielski, D. Blot, J.L. Ferrer, Structure 12 (2004a) 1219e1225.

78. L. Jacquamet, J. Ohana, J. Joly, P. Legrand, R. Kahn, F. Borel, M. Pirocchi, P. Charrault, P. Carpentier, J.L. Ferrer, Acta Cryst. D60 (2004b) 888e894.

79. D. Juers, B. Matthews, Acta Cryst. D60 (2004) 412e421.

80. J. Juanhuix, F. Gil-Ortiz, G. Cuni, C. Colldelram, J. Nicolas, J. Lid on, E. Boter, C. Ruget, S. Ferrer, J. Benach, J. Synchrotron. Rad. 21 (2014) 679e689.

81. A. Kachatkou, N. Kyele, P. Scott, R. van Silfhout, J. Synchrotron Radiat. 20 (2013) 596e602.

82. W.I. Karain, G.P. Bourenkov, H. Blume, H.D. Bartunik, Acta Cryst. Crystallogr. Sect. D. 58 (2002) 1519e1522.

83. D.J. Kissick, E.J. Gualtieri, G.J. Simpson, V. Cherezov, Anal. Chem. 82 (2010) 491e497.

84. B. Knapp, E. Marsh, F. Cipriani, D. Arneson, D. Oss, M. Liebers, E. Keller, J. Phys. Conf. Ser. 425 (2013) 012012.

85. A. Krolzig, G. Materlik, M. Swars, J. Zegenhagen, Nucl. Instrum. Methods Phys. 3Res. 219 (1984) 430e434.

86. B. Lavault, R.B.G. Ravelli, F. Cipriani, Acta Crystallogr. Sect. D. 62 (2006) 1348e1357.

87. A. le Maire, M. Gelin, S. Pochet, F. Hoh, M. Pirocchi, J.-F. Guichou, J.-L. Ferrer, G. Labesse, Acta Crystallogr. Sect. D. 67 (2011) 747e755.

88. Q. Liu, T. Dahmane, Z. Zhang, Z. Assur, J. Brasch, L. Shapiro, F. Mancia, W.A. Hendrickson, Science 336 (2012) 1033e1037.

89. A.Y. Lyubimov, T.D. Murray, A. Koehl, I.E. Araci, M. Uervirojnangkoorn, O.B. Zeldin, A.E. Cohen, S.M. Soltis, E.L. Baxter, A.S. Brewster, N.K. Sauter, A.T. Brunger, J.M. Berger, Acta Crystallogr. Sect. D. 71 (2015) 928e940.

90. A.A. MacDowell, R.S. Celestre, M. Howells, W. McKinney, J. Krupnick, D. Cambie, E.E. Domning, R.M. Duarte, N. Kelez, D.W. Plate, C.W. Cork, T.N. Earnest, J. Dickert, G. Meigs, C. Ralston, J.M. Holton, T. Alber, J.M. Berger, D.A. Agard, H.A. Padmore, J. Synchrotron Radiat. 11 (2004) 447e455.

91. J.T. Madden, S.J. Toth, C.M. Dettmar, J.A. Newman, R.A. Oglesbee, H.G. Hedderich, R.M. Everly, M. Becker, J.A. Ronau, S.K. Buchanan, V. Cherezov, M.E. Morrow, S. Xu, D. Ferguson, O. Makarov, C. Das, R. Fischetti, G.J. Simpson, J. Synchrotron Radiat. 20 (2013) 531e540.

92. M. Marinelli, E. Milani, G. Prestopino, C. Verona, G. Verona-Rinati, M. Angelone, M. Pillon, V. Kachkanov, N. Tartoni, M. Benetti, D. Cannata, F. Di Pietrantonio, J. Synchrotron Radiat. 19 (2012) 1015e1020.

93. T. Matsumoto, Y. Furukawa, M. Ishii, Proceedings of ICALEPCS2013, San Francisco, USA, 2013, 944e947.

94. T. Mcphillips, S. Mcphillips, H. Chiu, A. Cohen, A. Deacon, P. Ellis, E. Garman, A. Gonzalez, N. Sauter, R. Phizackerley, S. Soltis, P. Kuhn, J. Synchrotron. Rad. 9 (2002) 401e406.

95. S. McSweeney, in: R. Read, A.G. Urzhumtsev, V.Y. Lunin (Eds.), NATO Science for Peace and Security Series a: Chemistry and Biology, Advancing Methods for Biomolecular Crystallography, Springer, Netherlands, 2013, pp. 47e57.

96. A. Meents, B. Reime, N. Stuebe, P. Fischer, M. Warmer, D. Goeries, J. Roever, J. Meyer, J. Fischer, A. Burkhardt, I. Vartiainen, P. Karvinen, C. David, 2013, pp. 88510K-88510K-88517.

97. R.H. Menk, D. Giuressi, F. Arfelli, L. Rigon, AIP Conf. Proc. (2006) 1109e1112.

98. S. Monaco, E. Gordon, M.W. Bowler, S. Delageniere, M. Guijarro, D. Spruce, O. Svensson, S.M. McSweeney, A.A. McCarthy, G. Leonard, M.H. Nanao, J. Appl. Crystallogr. 46 (2013) 804e810.

99. S. Muchmore, J. Olson, R. Jones, J. Pan, M. Blum, J. Greer, S. Merrick, P. Magdalinos, V. Nienaber, Structure 8 (2000) R243er246.

100. C. Mueller, A. Marx, S.W. Epp, Y. Zhong, A. Kuo, A.R. Balo, J. Soman, F. Schotte, H.T. Lemke, R.L. Owen, E.F. Pai, A.R. Pearson, J.S. Olson, P.A. Anfinrud, O.P. Ernst, R.J.D. Miller, Struct. Dyn. 2 (2015) 054302.

101. M. Mueller, M. Wang, C. Schulze-Briese, Acta Cryst. D68 (2011) 42e56.

102. C. Mueller-Dieckmann, M. Bowler, P. Carpentier, D. Flot, A. McCarthy, M. Nanao, D. Nurizzo, P. Pernot, A. Popov, A. Round, A. Royant, D. de Sanctis, D. von Stetten, G. Leonard, Eur. Phys. J. Plus 130 (2015) 1e11.

103. E.M. Muller, J. Smedley, J. Bohon, X. Yang, M. Gaowei, J. Skinner, G. De Geronimo, M. Sullivan, M. Allaire, J.W. Keister, L. Berman, A. Heroux, J. Synchrotron Radiat. 19 (2012) 381e387.

104. T.D. Murray, A.Y. Lyubimov, C.M. Ogata, H. Vo, M. Uervirojnangkoorn, A.T. Brunger, J.M. Berger, Acta Crystallogr. Sect. D. 71 (2015) 1987e1997.

105. V. Mykhaylyk, A. Wagner, J. Phys. Conf. Ser. 425 (2013) 012010.

106. N. Nakamura, T. Katsura, Particle Accelerator Conference, 1993., Proceedings of the 1993, 1993, vol. 2253, pp. 2257e2259.

107. C. Nave, Acta Crystallogr. Sect. D. 55 (1999) 1663e1668.

108. P. Nogly, D. James, D. Wang, T.A. White, N. Zatsepin, A. Shilova, G. Nelson, H. Liu, L. Johansson, M. Heymann, K. Jaeger, M. Metz, C. Wickstrand, W. Wu, P. Bath, P. Berntsen, D. Oberthuer, V. Panneels, V. Cherezov, H. Chapman, G. Schertler, R. Neutze, J. Spence, I. Moraes, M. Burghammer, J. Standfuss, U. Weierstall, IUCrJ 2 (2015) 168e176.

109. R.L. Owen, D. Axford, J.E. Nettleship, R.J. Owens, J.I. Robinson, A.W. Morgan, A.S. Dore, G. Lebon, C.G. Tate, E.E. Fry, J. Ren, D.I. Stuart, G. Evans, Acta Cryst. D68 (2012) 810e818.

110. R.L. Owen, N. Paterson, D. Axford, J. Aishima, C. Schulze-Briese, J. Ren, E.E. Fry, D.I. Stuart, G. Evans, Acta Crystallogr. Sect. D. 70 (2014) 1248e1256.

111. R.L. Owen, A.R. Pearson, A. Meents, P. Boehler, V. Thominet, C. Schulze-Briese, J. Synchrotron. Rad. 16 (2009) 173e182.

112. R.L. Owen, D.A. Sherrell, Acta Cryst. D72 (2016) 388e394. http://dx.doi.org/ 10.1107/ S2059798315021555.

113. A.R. Pearson, R. Pahl, E.G. Kovaleva, V.L. Davidson, C.M. Wilmot, J. Synchrotron. Rad. 14 (2007) 92e98.

114. A. Perrakis, F. Cipriani, J.-C. Castagna, L. Claustre, M. Burghammer, C. Riekel, S. Cusack, Acta Cryst. D55 (1999) 1765e1770.

115. J. Pflugrath, Acta Cryst. D55 (1999) 1718e1725.

116. J. Pflugrath, Acta Crystallogr. Sect. F. 71 (2015) 622e642.

117. J.C. Phillips, A. Wlodawer, M.M. Yevitz, K.O. Hodgson, Proc. Natl. Acad. Sci. 73 (1976) 128e132.

118. G. Pompidor, F.S.N. Dworkowski, V. Thominet, C. Schulze-Briese, M.R. Fuchs, J. Synchrotron Radiat. 20 (2013) 765e776.
119. S.B. Pothineni, T. Strutz, V.S. Lamzin, Acta Crystallogr. Sect. D. 62 (2006) 1358e1368.
120. G. Rehm, J. Phys. Conf. Ser. 425 (2013) 042001.
121. C. Riekel, M. Burghammer, G. Schertler, Curr. Opin. Struct. Biol. 15 (2005) 556e562.
122. M. Robertson, Nature 285 (1980) 358e359.
123. P. Roedig, I. Vartiainen, R. Duman, S. Panneerselvam, N. Stübe, O. Lorbeer, M. Warmer, G. Sutton, D.I. Stuart, E. Weckert, C. David, A. Wagner, A. Meents, Sci. Rep. 5 (2015).
124. G. Rosenbaum, K. Holmes, J. Witz, Nature 230 (1971) 434e437.
125. A. Royant, P. Carpentier, J. Ohana, J. McGeehan, B. Paetzold, M. NoirclercSavoye, X. Vernede, V. Adam, D. Bourgeois, J. Appl. Cryst. 40 (2007) 1105e1112.
126. K. Sakai, Y. Matsui, T. Kouyama, Y. Shiro, S.-i. Adachi, J. Appl. Cryst. 35 (2002) 270e273.
127. J. Sanchez-Weatherby, M.W. Bowler, J. Huet, A. Gobbo, F. Felisaz, B. Lavault, R. Moya, J. Kadlec, R.B.G. Ravelli, F. Cipriani, Acta Crystallogr. Sect. D. 65 (2009) 1237e1246.
128. M. Sato, N. Shibata, Y. Morimoto, Y. Takayama, K. Ozawa, H. Akutsu, Y. Higuchi, N. Yasuoka, J. Synchrotron. Rad. 11 (2004) 113e116.
129. W. Schildkamp, C. Pradervand, Rev. Sci. Instrum. 66 (1995) 1956e1959.
130. H. Sehr, C. Schulze-Briese, C. Pradervand, H. Schift, J. Gobrecht, Technical Digest of Eurosensors XVIII, 2004, pp. 212e215.
131. D. Shu, T.M. Kuzay, Y. Fang, J. Barraza, T. Cundiff, J. Synchrotron Radiat. 5 (1998) 636e638.
132. O. Singh, G. Decker, 2001 Particle Accelerator Conference, PAC 2001. Proceedings of the 2001, 2001, vol. 531, pp. 539e543.
133. J.L. Smith, R.F. Fischetti, M. Yamamoto, Curr. Opin. Struct. Biol. 22 (2012) 602e612.
134. G. Snell, C. Cork, R. Nordmeyer, E. Cornell, G. Meigs, D. Yegian, J. Jaklevic, J. Jin, R. Stevens, T. Earnest, Structure 12 (2004) 537e545.
135. J. Song, D. Mathew, S.A. Jacob, L. Corbett, P. Moorhead, S.M. Soltis, J. Synchrotron. Rad. 14 (2007) 191e195.
136. D. Stoner-Ma, J.M. Skinner, D.K. Schneider, M. Cowan, R.M. Sweet, A.M. Orville, J. Synchrotron. Rad. 18 (2010) 37e40.
137. L. Struder, C. Fiorini, E. Gatti, R. Hartmann, P. Holl, N. Krause, P. Lechner, A. Longoni, G. Lutz, J. Kemmer, N. Meidinger, M. Popp, H. Soltau, U. Weber, C. von Zanthier, J. Synchrotron Radiat. 5 (1998) 268e274.
138. P.F. Tavares, S.C. Leemann, M. Sjostrom, A. Andersson, J. Synchrotron Radiat. 21 (2014) 862e877.
139. G. Ueno, H. Kanda, T. Kumasaka, M. Yamamoto, J. Synchrotron Radiat. 12 (2005) 380e384.
140. R. van Silfhout, A. Kachatkou, E. Groppo, C. Lamberti, W. Bras, J. Synchrotron Radiat. 21 (2014) 401e408.
141. G.B.M. Vaughan, J.P. Wright, A. Bytchkov, M. Rossat, H. Gleyzolle, I. Snigireva, A. Snigirev, J. Synchrotron Radiat. 18 (2011) 125e133.
142. X. Vernede, B. Lavault, J. Ohana, D. Nurizzo, J. Joly, L. Jacquamet, F. Felisaz, F. Cipriani, D. Bourgeois, Acta Crystallogr. Sect. D. 62 (2006) 253e261.
143. C. Vonrhein, C. Flensburg, P. Keller, A. Sharff, O. Smart, W. Paciorek, T. Womack, G. Bricogne, Acta Crystallogr. Sect. D. 67 (2011) 293e302.
144. M. Walsh, I. Dementieva, G. Evans, R. Sanishvili, A. Joachimiak, Acta Cryst. D55 (1999) 1168e1173.
145. S. Waltersperger, V. Olieric, C. Pradervand, W. Glettig, M. Salathe, M.R. Fuchs, A. Curtin, X. Wang, S. Ebner, E. Panepucci, T. Weinert, C. Schulze-Briese, M. Wang, J. Synchrotron Radiat. 22 (2015) 895e900.
146. A.J. Warren, W. Armour, D. Axford, M. Basham, T. Connolley, D.R. Hall, S. Horrell, K.E. McAuley, V. Mykhaylyk, A. Wagner, G. Evans, Acta Crystallogr. Sect. D. 69 (2013) 1252e1259.
147. E. Weckert, IUCrJ 2 (2015) 230e245.
148. T. Weitkamp, B. Nohammer, A. Diaz, C. David, E. Ziegler, Appl. Phys. Lett. 86 € (2005) 054101.
149. F. Willeke, 2015, Brookhaven National Laboratory (BNL).
150. G. Winter, J. Appl. Crystallogr. 43 (2010) 186e190.
151. U. Zander, G. Bourenkov, A.N. Popov, D. de Sanctis, O. Svensson, A.A. McCarthy, E. Round, V. Gordeliy, C. Mueller-Dieckmann, G.A. Leonard, Acta Crystallogr. Sect. D. 71 (2015) 2328e2343.
152. O.B. Zeldin, M. Gerstel, E.F. Garman, J. Appl. Crystallogr. 46 (2013) 1225e1230.
153. T. Zhou, W. Ding, M. Gaowei, G. De Geronimo, J. Bohon, J. Smedley, E. Muller, J. Synchrotron Radiat. 22 (2015) 1396e1402.
154. R. Signorato, O. Hignette, J. Goulon, J. Synchrotron Rad 5 (1998) 797e800.
155. L. Peverini, I.V. Kozhevnikov, A. Rommeveaux, P.V. Vaerenbergh, L. Claustre, S. Guillet, J.-Y. Massonnat, E. Ziegler, J. Susini, Nucl. Instrum. Methods Phys. Res. A 616 (2010) 115e118.
156. K. Yamauchi, H. Mimura, K. Inagaki, Y. Mori, Rev. Sci. Instrum 73 (2002) 4028.
157. F. Siewert, J. Buchheim, T. Zeschke, M. Stormer, G. Falkenbergc, R. Sankari, On € the characterization of ultra-precise X-ray optical components: advances and challenges in ex situ metrology, J. Synchrotron Rad 21 (2014) 968e975.
158. A. Snigirev, V. Kohn, I. Snigireva, B. Lengeler, Nature 384 (07 November 1996) 49e51.
159. M.W. Bowler, D. Nurizzo, R. Barrett, A. Beteva, M. Bodin, H. Caserotto, S. Delageniere, F. Dobias, D. Flot, T. Giraud, N. Guichard, M. Guijarro, M. Lentini, G.A. Leonard, S. McSweeney, M. Oskarsson, W. Schmidt, A. Snigirev, D. von Stetten, J. Surr, O. Svensson, P. Theveneau, C. MuellerDieckmann, J. Synchrotron Rad 22 (2015) 1540e1547. http://dx.doi.org/10. 1107/S1600577515016604.

CHAPTER 16

CUSTOM REAL-TIME ULTRASONIC INSTRUMENTATION FOR SIMULTANEOUS MIXTURE AND FLOW ANALYSIS OF BINARY GASES IN THE CERN ATLAS EXPERIMENT

M. Alhroob [a] , M. Battistin [b] , S. Berry [b] , A. Bitadze [c] , P. Bonneau [b] , G. Boyd [a] , O. Crespo-Lopez [b] , C. Degeorge [d] , C. Deterre [e] , B. Di Girolamo [b] , M. Doubek [f] , G. Favre [b] , G. Hallewell [g] , A. Hasib [a] , S. Katunin [h] , D. Lombard [b] , A. Madsen [e] , S. McMahon [i] , K. Nagai [j] , A. O'Rourke [e,c] , B. Pearson [a] , D. Robinson [k] , C. Rossi [l] , A. Rozanov [g] , E. Stanecka [m], M. Strauss [a] , V. Vacek [f] , R. Vaglio [b] , J. Young [a] , L. Zwalinski [b]

[a]Department of Physics and Astronomy, University of Oklahoma, Norman, OK 73019, USA

[b]CERN, 1211 Geneva 23, Switzerland

[c]School of Physics and Astronomy, University of Manchester, Oxford Road, Manchester M13 9PL, UK

[d]Physics Department, Indiana University, Bloomington, IN 47405, USA

[e]DESY, Notkestraße 85, Hamburg 22607, Germany

[f]Czech Technical University, Technická 4, 166 07 Prague 6, Czech Republic

[g]Centre de Physique des Particules de Marseille, 163 Avenue de Luminy, 13288 Marseille Cedex 09, France

[h]B.P. Konstantinov Petersburg Nuclear Physics Institute (PNPI), 188300 St. Petersburg, Russia

[i]Rutherford Appleton Laboratory, Science & Technology Facilities Council, Chilton, Didcot OX11 OQX, UK [j]Department of Physics, Oxford University, Keble Road, Oxford OX1 3RH, UK

[k]Department of Physics and Astronomy, University of Cambridge, Cambridge, UK

[l]INFN Genova, Via Dodecaneso 33, 16146 Genova, Italy

[m]Institute of Nuclear Physics PAS, ul. Radzikowskiego 152, 31-342 Kraków, Poland

Citation: M. Alhroob, M. Battistin, S. Berry, A. Bitadze, P. Bonneau, G. Boyd, O. Crespo-Lopez, C. Degeorge, C. Deterre, B. Di Girolamo, M. Doubek, G. Favre, G. Hallewell, A. Hasib, S. Katunin, D. Lombard, A. Madsen, S. McMahon, K. Nagai, A. O'Rourke, B. Pearson, D. Robinson, C. Rossi, A. Rozanov, E. Stanecka, M. Strauss, V. Vacek, R. Vaglio, J. Young, L. Zwalinski, Custom real-time ultrasonic instrumentation for simultaneous mixture and flow analysis of binary gases in the CERN ATLAS experiment, http://dx.doi.org/10.1016/j.nima.2016.04.104.

ABSTRACT

Custom ultrasonic instruments have been developed for simultaneous monitoring of binary gas mixture and flow in the ATLAS Inner Detector. Sound transit times are measured in opposite directions in flowing gas. Flow rate and sound velocity are respectively calculated from their difference and average. Gas composition is evaluated in real-time by comparison with a sound velocity/composition database, based on the direct dependence of sound velocity on component concentrations in a mixture at known temperature and pressure. Five devices are integrated into the ATLAS Detector Control System. Three instruments monitor coolant leaks into N_2 envelopes of the silicon microstrip and Pixel detectors. Resolutions better than $\pm 2\times10^{-5}$ and $\pm 2\times10^{-4}$ are seen for C_3F_8 and CO_2 leak concentrations in N_2 respectively. A fourth instrument detects sub-percent levels of air ingress into the C_3F_8 condenser of the new thermosiphon coolant recirculator. Following extensive studies a fifth instrument was built as an angled sound path flowmeter to measure the high returning C_3F_8 vapour flux (~1.2 kgs^{-1}). A precision of <2.3% FS for flows up to 10 ms^{-1}was demonstrated. These instruments have many potential applications where continuous binary gas composition measurement is required, including hydrocarbon and anaesthetic gas mixtures.

INTRODUCTION

Custom ultrasonic ("sonar") instruments have been developed for the measurement of binary gas mixtures. The instruments are used for continuous real-time flowmetry and leak monitoring for the cooling circuits of the Inner Detector of the ATLAS experiment at the CERN Large Hadron Collider.

ATLAS is a particle physics detector, consisting of a series of sub-detectors arranged around the proton beam collision point. The ATLAS Inner Detector is used for tracking charged particles, and is situated inside a 2 Tesla solenoidal magnetic field. It comprises silicon pixel and microstrip (SCT) sensors, and a Transition Radiation Tracker. The silicon sensors are maintained at a low temperature to mitigate the effects of radiation damage[1], [2], [3] and [4]. The heat from the silicon substrates and their local electronics is evacuated using an evaporative cooling system based on octafluoropropane (C_3F_8) [5]. A new pixel layer, the Insertable B-Layer (IBL), was recently added closest to the beam pipe. The IBL is cooled using a CO_2 evaporative cooling system. Three ultrasonic instruments [6], [7] and [8] monitor the nitrogen-purged anti-humidity envelopes of the SCT, Pixel and IBL subdetectors for leaks from their cooling circuits.

A new C_3F_8 thermosiphon circulation system is being commissioned in 2016 [9] and [10]to replace the present compressor-driven system. This utilises the 90 m depth of the ATLAS experimental cavern to generate the hydrostatic pressure needed to circulate the liquid coolant through the on-detector cooling channels, where it evaporates and returns as vapour to an above-ground condenser. Two further ultrasonic instruments directly monitor the thermosiphon; one measures the flow in the ascendent vapour return tube, while another checks for air ingress into the condenser. The positions of the instruments in the ATLAS installation are shown in Fig. 1.

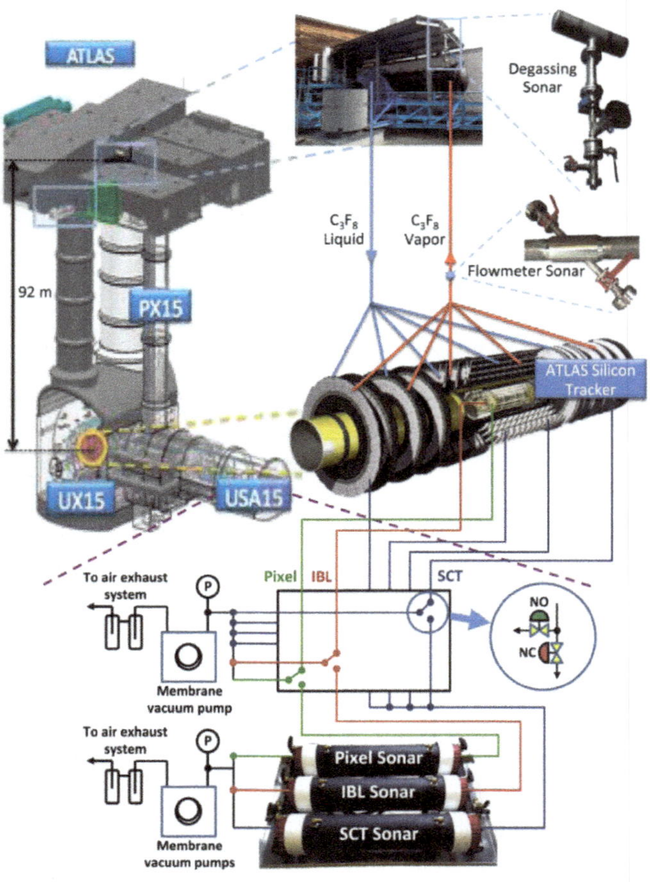

Figure 1: The locations of the ultrasonic instruments and their connecting structures within the ATLAS inner detector and thermosiphon.

In this paper, the principle and design of the ultrasonic instruments are described, along with their various applications in the ATLAS cooling system. Their integration into the Detector Control System (DCS) is also discussed.

THE INSTRUMENTS

The sonar instruments use transit time and sound velocity measurements to calculate binary gas compositions and flow rate. Composition measurement in a binary gas mixture exploits the relationship between sound velocity and molar composition at known temperature and pressure. This relationship is illustrated for mixtures of C_3F_8 and N_2 at several temperatures in Fig. 2 (see also Section 2.1).

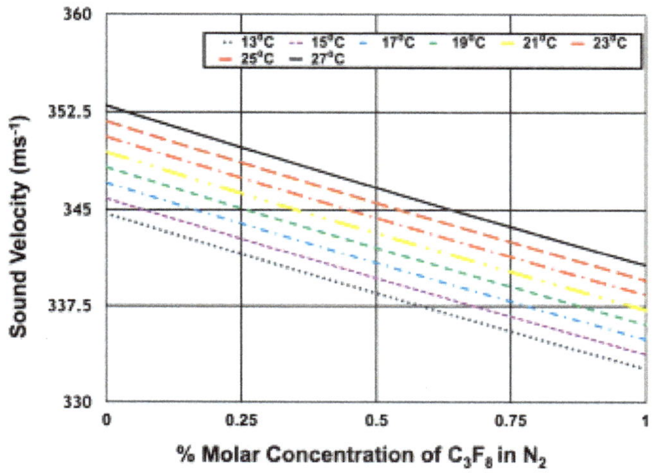

Figure 2: Examples of the variation of sound velocity with molar concentration of C_3F_8 in N_2 (from the online database) at several temperatures, at a pressure of 1000 $mbar_{abs}$.

The five installed instruments are based on custom electronics and hardware operating with pairs of 50 kHz capacitive ultrasonic transducers (SensComp model 600) placed within gas enclosures. The transducers operate as transceivers with a DC bias voltage of around +300V. In transmit mode, they are excited by down-going (300 V–0 V) square-wave pulses generated from a gated LV pulse output of a DsPIC33F microcontroller. The receiving chain has two amplification stages followed by a comparator, also implemented in the microcontroller. Transmission synchronously starts a 40 MHz transit time clock in the microcontroller, which is stopped by the first pulse crossing the user-definable comparator threshold. A mode of operation based on time-of-flight, rather than resonant frequency, is chosen to allow flowmetry in the same electronics.

Synchronous analogue temperature and pressure measurements use a multiplexed ADC implemented in an Analog Devices ADμC 847 microcontroller. Transit times, temperatures and pressures are averaged, time-stamped and pipelined by a FIFO memory implemented in a dsPIC33F microcontroller for transmission to a SCADA (Supervisory Control and Data Acquisition) computer. Communication uses ModbusTCP/IP over ethernet, implemented using the MBED LPC1768 platform.

Mixture and flow calculations are made on a Dell PowerEdge R610 SCADA computer using custom software implemented in Siemens SIMATIC WinCC[11] running on Linux. This software allows control of the instruments and archiving of data.

The installed instruments use two different configurations of acoustic path relative to the gas flow direction: co-axial and angled-crossing. Four instruments in co-axial geometry are optimised for binary concentration measurements in static gas or at low flow. The fifth instrument, shown in Fig. 3, has an acoustic path crossing the gas flow at 45°. It is optimised for flowmetry at high C_3F_8 flow speeds of around 22 ms^{-1}, corresponding to mass flow rates of around 1.2 kgs^{-1}.

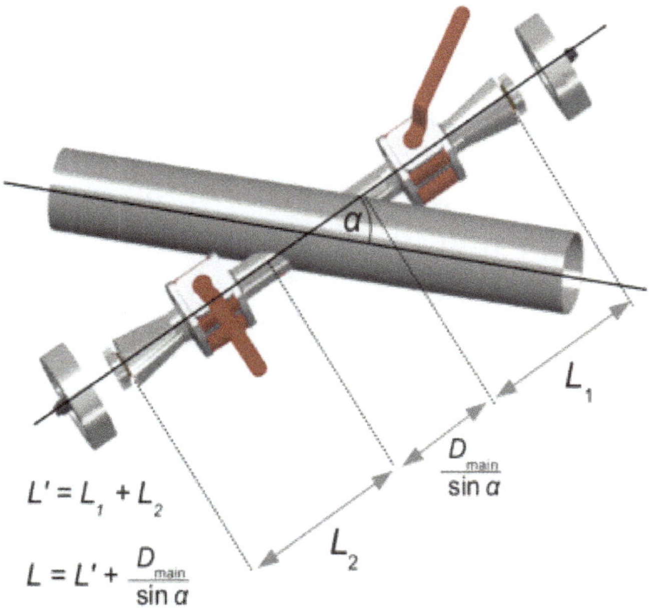

Figure 3: The ultrasonic flowmeter installed in the C_3F_8 vapour return tube to the thermosiphon condenser. The acoustic path crosses the gas flow at 45°.

Gas Composition Analysis

Binary gas composition is determined by comparison of the measured sound velocity with a stored database of sound velocity as a function of composition for the gas pairs over a range of temperature and pressure covering the process conditions [8].

In an instrument containing either static gas, or where the moving gas fills the entire acoustic path length, L, the sound velocity, vs, can be determined from the bidirectional transit times t_{up} and t_{down} of the sound pulses as

$$v_s = \frac{2L(t_{up} + t_{down})}{4t_{up}\, t_{down}}. \qquad (1)$$

The three aspirating sonars and the degassing sonar of the thermosiphon condenser (Fig. 1) use a co-axial geometry.

The measured values of temperature, pressure and sound velocity are then compared to a database of measured values created with NIST REFPROP [12] to determine the gas composition. In the case of C_3F_8 in N_2, the database is created in temperature and pressure intervals of 0.5°c and 20 mbar. An on-line algorithm implemented in WinCC interpolates between the values in the database that are closest to the measured temperature and pressure to calculate the molar composition of the binary mixture. This requires an interpolation in two dimensions using four values retrieved from the database

The precision with which the gas composition can be measured depends on the difference between the molecular weights of the two gas components and on the precision of the sound velocity measurement. The latter depends on the precisions of the known transducer spacing (± 0.1 mm), the temperature and pressure in the sonar instrument (±0.1 °C and ±1 mbar, respectively) and the precision of the transit time measurement (± 25 ns). Their combination results in an uncertainty on the measured sound velocity, δv_s, of $\pm 0.025\ ms^{-1}$. The uncertainty on the gas composition measurement can be defined as $\delta(conc)=\delta v_s/m$, where m is the local gradient of the curve of sound velocity against concentration. The average gradient for C_3F_8 in N_2 in the interval 0–0.1% concentration is − 12.55 ms^{-1} per percent of C_3F_8 (Fig. 2). From this the precision of the concentration measurement is determined to be $\pm 0.002\%$.

Gas Flow Calculation

The difference between the transit times in opposite directions is used to calculate the gas flow rate. For the angled ultrasonic flowmeter installed in the return vapour path to the thermosiphon condenser (Fig. 3), the moving gas acoustic path segment is at an angle $\alpha = 45°$ to the gas flow. The length of this acoustic path is defined as $D_{main}/\sin_\alpha$, where D_{main} is the diameter of the return gas tube, which is around 135 mm. The remaining length of the sound path, L', has components L_1 and L_2 on either side of the gas flow tube.

The gas flow velocity, vg (ms^{-1}), can be calculated using the sound velocity measured within the gas, vs, as

$$v_g = \frac{v_s\left(\frac{D_{main}}{\sin\alpha} + L' - v_s t_{down}\right)}{\cos\alpha\,(v_s t_{down} - L')}.$$

The sound velocity can be calculated as the physical root of

$$v_s = \frac{A \pm \sqrt{A^2 - 16L' t_{up} t_{down}\left(L' + \frac{D_{main}}{\sin\alpha}\right)}}{4 t_{up} t_{down}},$$

where the definition of A is

$$A = (t_{up} + t_{down})\left(2L' + \frac{D_{main}}{\sin\alpha}\right).$$

An RMS precision of $\pm 2.3\%$ for full scale (FS) for flows up to 10 ms^{-1} was demonstrated in pre-installation tests with air [8]. The flowmeter will soon be brought into operation in the closed thermosiphon circuit with C_3F_8.

APPLICATION IN THE ATLAS DETECTOR

Implementation

The configuration of the five sonar instruments deployed in the ATLAS detector is shown in Fig. 1. Three devices aspirate N_2 environmental gas from the anti-humidity enclosures of the IBL, Pixel and SCT detectors to monitor the presence of leaks from their respective cooling systems. The SCT environmental volume is monitored at four different points – two in the endcap regions, and two in the barrel region. Gas from each region is monitored for 4 hours at a time in a 16 h continuous cycle. Two sonar instruments are implemented in the new C_3F_8 thermosiphon coolant recirculation system, which is being commissioned for the Pixel and SCT detectors.

In addition to the ultrasonic flowmeter discussed in Section 2.2, a "degassing" sonar monitors C_3F_8 vapour for possible air contamination in the condenser, reducing thermosiphon performance. Air would be expected to accumulate in the condenser; not only is it the highest part of the system, but also the lowest temperature and pressure point in the cooling circuit. The degassing sonar can trigger the venting of accumulated air when this exceeds safe limits. The air/C_3F_8 molar composition is measured in the same way as described in Section 2.1.

The precision parameters and the average slope of the sound velocity vs. composition curve in the range 0–30% air in C_3F_8 over the temperature and pressure range of interest results in a mixture resolution of $\pm 9.4\times10^{-4}$[6].

A user interface and Finite State Machine (FSM) has been created and integrated into the ATLAS DCS. This allows the systems to be monitored and controlled. An alarm system is also in place, which warns of any anomalies, such as a large increase in coolant leak, or hardware errors.

Recent Measurements

The aspirating sonar instruments monitored the switch-on of the evaporative cooling systems of the ATLAS Inner Detector silicon tracker during January 2016. The instrument monitoring the N_2-purged envelope of the IBL subdetector showed that its on-detector cooling channels were very leak-tight; no increase in CO_2 concentration was seen. The instruments monitoring the SCT and Pixel sub-detectors observed changes in the concentration of C_3F_8 coolant in their envelopes.

Fig. 4 shows the concentration of C_3F_8 in the N_2 flushing gas around the Pixel detector and its on-detector cooling tubes. A steep rise in C_3F_8 concentration is observed following the simultaneous start of all 88 Pixel detector cooling circuits, followed by a slower increase to a concentration of about 0.14%.

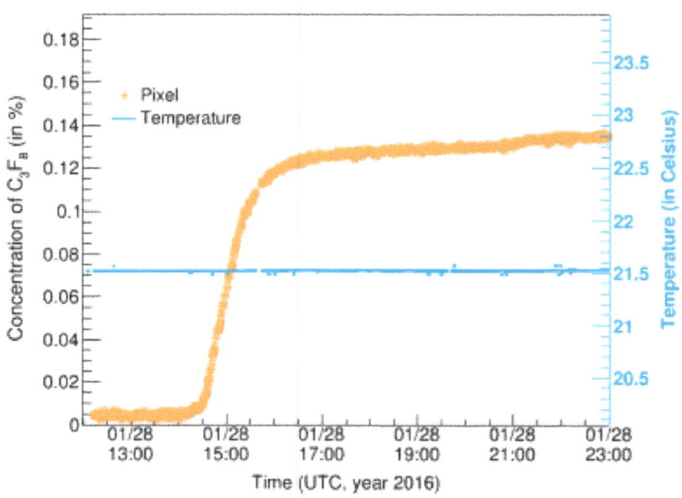

Figure 4: C_3F_8 concentration in the N_2-flushed anti-humidity enclosure of the ATLAS Pixel detector during cooling restart on January 28, 2016.

Fig. 5 shows the apparent C_3F_8 concentration in N_2 flushing gas in the four enclosures of the SCT subdetector before and after cooling restart. The SCT sonar tube was monitoring gas aspirated from the SCT barrel "level 7" during the cooling restart, revealing a sharp spike in C_3F_8 concentration as the on-detector cooling circuits were initially operated at a higher evaporation pressure before being reduced to the normal operating value of $1.8bar_{abs}$. The apparent C_3F_8 concentration in the SCT barrel increased significantly following cooling restart: from 0.005% to 0.013% in gas aspirated from the SCT barrel "level 1", and from 0.006% to 0.014% for SCT barrel "level 7". No significant increase was seen in the apparent C_3F_8 concentration in either of the SCT endcaps following cooling turn on. Endcap "A" showed a high apparent C_3F_8 concentration before and after the cooling restart, due to a known dry air ingress from the Inner Detector external envelope purge system at the time. Since binary gas analysis is a rapid hypothesis-dependent diagnostic, contamination increases in a heavy "search" gas can be mimicked by higher concentrations of a lighter contaminant. Warnings given by the instrument can indicate the need for further investigation with more expensive multi-gas sensitive instrumentation, including gas chromatography.

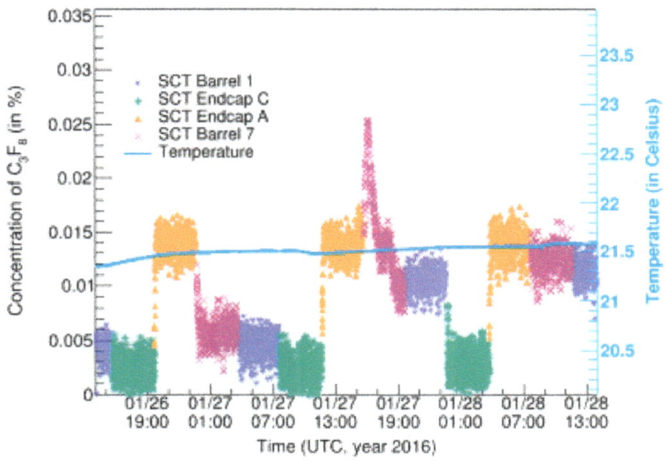

Figure 5: The concentration of C_3F_8 in N_2 for four different regions of the anti-humidity enclosure of the ATLAS SCT detector is observed during the cooling restart around 15:00 UTC on January 27, 2016.

SUMMARY

Ultrasonic instrumentation for measurement and real-time monitoring of gas flow and binary gas mixture composition has been developed. Five devices are used in applications related to the evaporative cooling system of the inner detector of the ATLAS experiment, and are integrated into the ATLAS DCS. The monitoring of the cooling of the ATLAS inner detector in January 2016 allowed further validation of the effectiveness of these instruments and the online analysis software. Precisions of $\pm 0.022\%$ of CO_2 in N_2 and $\pm 0.002\%$ of C_3F_8 in N_2 have been experimentally verified. An instrument to detect air ingress into the condenser of the new thermosiphon can operate with a molar precision of $\pm 9.4 \times 10^{-4}$. Instruments of this type have other potential applications where continuous real-time flow and binary gas composition measurements are required, for instance hydrocarbon and anaesthetic gas mixture analysis.

ACKNOWLEDGEMENTS

A. O'Rourke would like to acknowledge support from the Helmholtz Association.

REFERENCES

1. ATLAS Inner Detector, Technical Design Report, vol. 1, CERN-LHCC-97-16.
2. ATLAS Inner Detector, Technical Design Report, vol. 2, CERN-LHCC-97-17.
3. ATLAS Pixel Detector, Technical Design Report, CERN-LHCC-98-13.
4. G. Beck, G. Viehhauser, Analytic model of thermal runaway in silicon detectors, Nucl. Instrum. Methods A 618 (2010) 131–138, http://dx.doi.org/10.1016/j.nima.2010.02.264.
5. D. Attree, et al., The evaporative cooling system for the ATLAS inner detector, J. Instrum. 3 (2008) P07003, http://dx.doi.org/10.1088/1748-0221/3/07/P07003.
6. R. Bates, et al., A combined ultrasonic flow meter and binary vapour mixture analyzer for the ATLAS silicon tracker, J. Instrum. 8 (2013) P02006, http://dx. doi.org/10.1088/1748-0221/8/02/P02006 arXiv:1210.4835.
7. M. Alhroob, et al., Development of a custom on-line ultrasonic vapour analyzer and flow meter for the ATLAS inner detector, with application to Cherenkov and gaseous charged particle detectors, J. Instrum. 10 (03) (2015) C03045, http://dx.doi.org/10.1088/1748-0221/10/03/C03045.
8. M. Alhroob, et al., Implementations of custom sonar instruments for binary gas mixture and flow analysis in the ATLAS experiment at the CERN LHC, IEEE Trans. Nucl. Sci. (2016) (Submitted for publication).
9. J. Botehlo-Direito, et al., Full scale thermosiphon technical note, Technical Report, 1083852, CERN, 2010.
10. M. Battistin, The thermosiphon cooling system of the ATLAS experiment at the CERN Large Hadron Collider, Int. J. Chem. React. Eng. 13 (2015) 511–521, http: //dx.doi.org/10.1515/ijcre-2015-0022.
11. Siemens, SIMATIC WinCC, ⟨http://siemens.com/wincc⟩.
12. E.W. Lemmon, M.L. Huber, M.O. McLinden, NIST Standard Reference Database 23: Reference Fluid Thermodynamic and Transport Properties – REFPROP, Version 9.0, National Institute of Standards and Technology, Standard Reference Data Program, Gaithersburg, 2010.

CHAPTER 17

EXPERIMENTAL VALIDATION OF A COMPOUND CONTROL SCHEME FOR A TWO-AXIS INERTIALLY STABILIZED PLATFORM WITH MULTI-SENSORS IN AN UNMANNED HELICOPTER-BASED AIRBORNE POWER LINE INSPECTION SYSTEM

Xiangyang Zhou[1], Yuan Jia[1], Qiang Zhao[2] and Ruixia Yu[3]

[1]School of Instrumentation Science and Opto-electronics Engineering, Beihang University (BUAA), Beijing 100191, China

[2]Shanghai Institute of Satellite Engineering, Shanghai 200240, China

[3]School of Mechanical Engineering, University of Science and Technology Beijing, Beijing 100083, China

ABSTRACT

A compound control scheme is proposed to achieve high control performance for a two-axis inertially stabilized platform (ISP) with multi-sensors applied to an unmanned helicopter (UH)-based airborne power line inspection (APLI) system. Compared with the traditional two closed-loop control scheme that is composed of a high-bandwidth rate loop and a lower bandwidth position loop, a new current loop inside rate loop is particularly designed to suppress the influences of voltage fluctuation from power supply and motor back electromotive force (BEMF) on control precision. In this way, the stabilization accuracy of the ISP is greatly improved. The rate loop, which is the middle one, is used to improve sensor's stability precision through compensating for various disturbances. To ensure the pointing accuracy of the line of sight (LOS) of multi-sensors, the position loop is designed to be the outer one and acts as the main feedback path, by which the accurate pointing angular position is achieved. To validate the scheme, a series of experiments were carried out. The results show that the proposed compound control scheme can achieve reliable control precision and satisfy the requirements of real APLI tasks.

INTRODUCTION

The electric power distribution system is an important part of electrical power systems for delivering electricity to consumers [1]. Preventive and breakdown maintenance of the power grid are typically a costly legal public safety responsibility in most countries [2]. In order to detect defects as early as possible and to efficiently plan the required maintenance activities, distribution networks are regularly inspected [3]. Power line inspection is a vital function for electricity supply [4]. So far, the inspection of power line corridors is mainly carried out visually or by helicopters. However, visual inspection is very labor-intensive and inefficient [5]. In comparison to the foot patrol inspection, inspection by helicopter is more efficient [6]. In recent years, due to their potential applications in different fields, great interest in the utilization of unmanned aerial vehicles (UAVs) has arisen [7]. As a new method, UAV-based power line inspection utilizes modern flight control techniques, image photography and recognition techniques to carry out fast inspections

Citation: Zhou, X.; Jia, Y.; Zhao, Q.; Yu, R. Experimental Validation of a Compound Control Scheme for a Two-Axis Inertially Stabilized Platform with Multi-Sensors in an Unmanned Helicopter-Based Airborne Power Line Inspection System. Sensors 2016, 16, 366. DOI: 10.3390/s16030366

from high altitude and over far distances [8]. Therefore, unmanned helicopter (UH)-based power line inspection is becoming very attractive for it is more reliable, effective and achieves detection of more faults [2,3,9].

In an airborne power line inspection (APLI) system, an unmanned helicopter (UH) carrying imaging sensors such as a CCD camera and an infrared scanner moves along the power line. The conditions of the power line can then be estimated based on the obtained images [10]. An APLI system generally consists of an autonomous helicopter, an inertially stabilized platform (ISP), a position and orientation system (POS) and multiple sensors, such as a visible light camera, infrared camera, ultraviolet camera, laser scanner, and so on. The system can intelligently detect the power transmission line faults. All multi-sensors are installed in the ISP, including the inertial measurement unit (IMU) of POS, so the field vision of the multiple sensors can be regulated to the required direction and target with the output of POS [11].

A two-axis ISP with high control performance, which is used to support and stabilize the multiple sensors so that the sensors' line of sight (LOS) can track the APLI target accurately in real time, is indispensable for an APLI system [7]. The control task for such an ISP is to keep the LOS (optical axis) of the optical or optoelectronic payload still, even in the presence of some unwanted rotational motion of the carrier [12]. Besides, with the help of a two-axis ISP, the sensors' LOS can track the lines and remain steady all along by rejecting various disturbing torques, such as vibration, air turbulence, wind gusts and so on. Therefore, there is much interest among researchers to investigate control methods with higher accuracy and stability by rejecting various disturbances [13]. In [12], a feedforward scheme for a two-axis ISP to reject a periodic disturbing torque acting on the payload due to static mass unbalance is reported. In [14], an adaptive backstepping control method based on the LuGre model is put forward to decrease the influences of friction on control precision of a three-axis ISP. In [15], the extended Kalman Filter (EKF) is used to compensate friction disturbance for a second order system with low-pass PD controller. In [16], a robust adaptive output feedback controller is applied to a cart-crane system. In [17], an acceleration–based feedforward approach is proposed to realize imbalance torque compensation of a three-axis ISP. In [18], a model-based feedforward compensation approach is applied to a rotary table system. In [19], an anti-disturbance compensation control algorithm based on adaptive robust control idea is applied for a three-axis swing turntable system. In addition, due to the existing nonlinearity and time-varying uncertainty in the practical engineering application, improved PID controller is usually adopted to compose a complex advanced PID control method [20].

In order to achieve the necessary control performance required by an APLI system, a two-axis ISP with both of high control precision and stability is crucial in an inspection task, meaning a comprehensive control scheme of ISP should be successfully designed. Typically, the control system of an ISP is configured as a high-bandwidth rate loop inside a lower bandwidth pointing or tracking position loop [21]. The ISP might be viewed as a means for removing high-frequency disturbances and controlling the LOS, whereas the pointing and tracking loops have the task of removing the lower frequency parallactic motion and perhaps any bias or drift in the ISP rate loop [22]. Therefore, a dual closed-loop PID control scheme is widely used as a main method for the control system design of multi-axis ISP that was surveyed in [21,22]. However, in the dual closed-loop scheme, the disturbances caused by voltage fluctuation from power supply and motor back electromotive force (BEMF) are not compensated by an effective closed-loop, which can influence the stabilization accuracy of ISP. Since the motor armature current will keep following the change of the voltage fluctuation, thus, the control precision and stability of ISP will decrease. Therefore, to improve the dynamic performance of the system, the motor armature current should be kept stable in the design of control system to ensure the linearity of control torque. It has been proved that suppressing the torque ripple from the motor drive of a servo system can significantly improve system performance by reducing speed fluctuations. Therefore, in the design of control scheme of two-axis ISP, the disturbances of motor unstable armature current caused by voltage fluctuation and BEMF should be compensated on the basis of dual closed-loop scheme.

In this paper, to reduce the influences of voltage fluctuation of power supply and motor BEMF on the stabilization accuracy of a two-axis ISP, on the basis of dual closed-loop PID control scheme, a specialized current feedback closed-loop is introduced to improve the dynamic and static performance of rate loop. Together with original dual closed-loops, a compound control scheme is configured. Through this scheme, the linearity of control torque is improved. To verify the methods, serial experiments are carried out to a two-axis ISP for APLI applications.

BACKGROUND ANALYSIS

APLI System

Figure 1 shows an APLI system and a two-axis ISP with multiple sensors. An unmanned helicopter (UH) is usually chosen as a mobile aviation aircraft which can take off and land freely in a complex geographic environment. Imaging sensors such as a CCD camera, infrared scanner and ultraviolet scanner are carried by the ISP, which is mounted between the sensors and the aviation platform. The sensors and ISP are assembled as a pod system hung below the helicopter. The pod system can communicate with a supervisory computer. The control system of ISP can get information from the supervisory computer and other inertial sensors to realize stabilizing function for the LOS of camera [6].

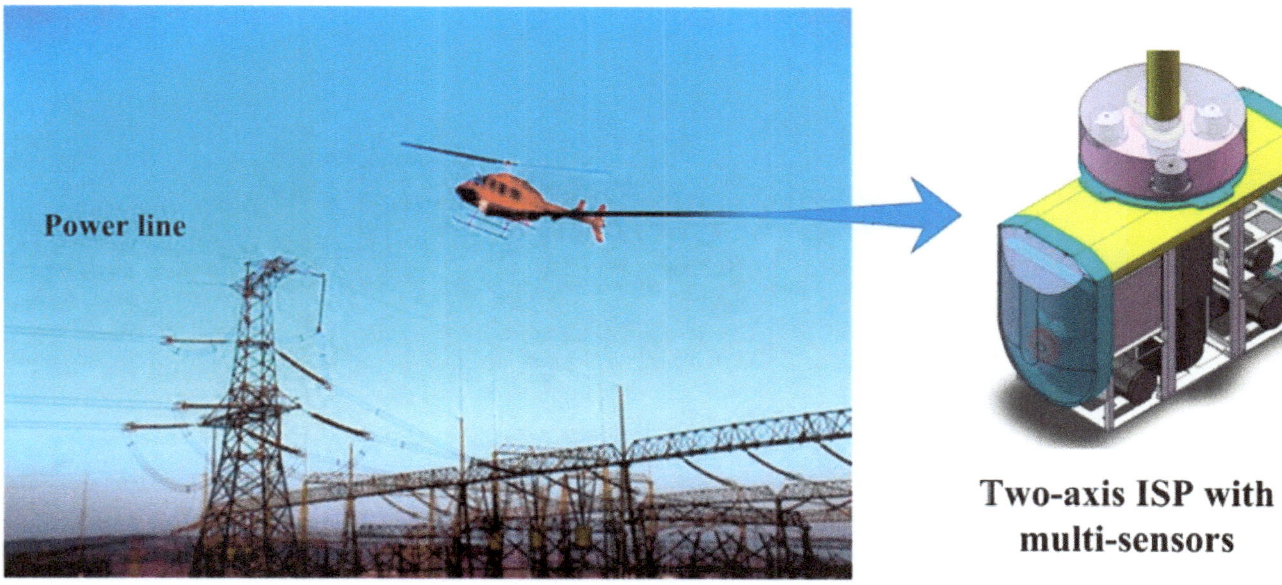

Figure 1: Schematic diagram of an APLI system and the two-axis ISP with multi-sensors.

Structure of Two-Axis ISP System

As shown in Figure 2, for a two-axis ISP system, the inner gimbal allows elevation of the payload, and the outer gimbal allows a rotation in azimuth angle. The payload consists of two CCD cameras, an infrared scanner, an ultraviolet scanner, and a laser scanner. Torque motors and gear linkages are used to drive two gimbals' rotation. Two fiber-optic gyroscopes, one POS and two encoders are attached to the payload to measure its rotation angles, rates and accelerations [10].

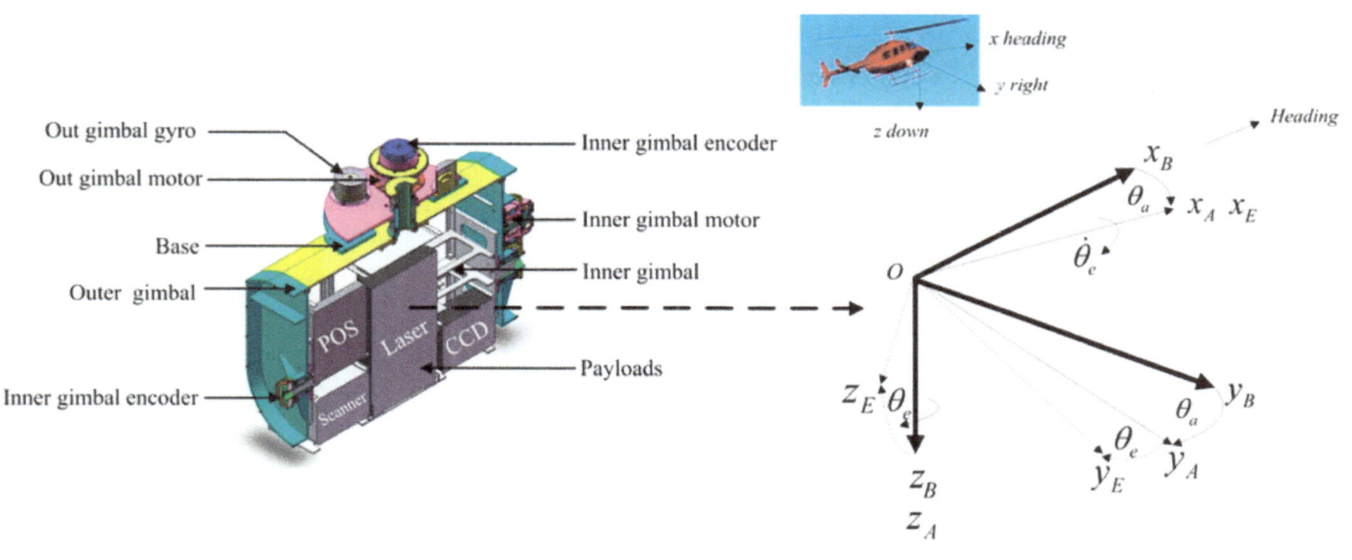

Figure 2: The structural diagram of a two-axis ISP with multi-sensors in an APLI system.

The POS, which is mainly composed of three main components, i.e., inertial measurement unit (IMU), GPS receiving antenna and data processing system, is used to provide an accurate reference of position and attitude in inertial space for control system of ISP and imaging sensors through measuring imaging sensor's angular movement.

Working Principle of Two-Axis ISP System

Figure 3 shows the structure diagram of a two-axis ISP control system in the APLI system. We can see that the ISP consists of two gimbals, which are an azimuth gimbal (A-gimbal) and a pitch gimbal (P-gimbal) [7,13]. The P-gimbal is assembled on the A-gimbal and can rotate around the Y_p axis. The A-gimbal is assembled on the base of the aviation platform and can rotate around the Z_a axis. G_p and G_a stand for the rate gyros that measure the inertial angular rate of the P-gimbal and A-gimbal, respectively. E_p and E_a respectively stand for the photoelectric encoders installed on the P-gimbal and A-gimbal, which are used for measuring the relative angles between gimbals. M_z and M_y respectively stand for the gimbal servo motors which drive the A-gimbal and P-gimbal to keep them steady in inertial space. Ay represents the accelerometer installed on the P-gimbal that is used to measure the gimbal's rotary angular acceleration.

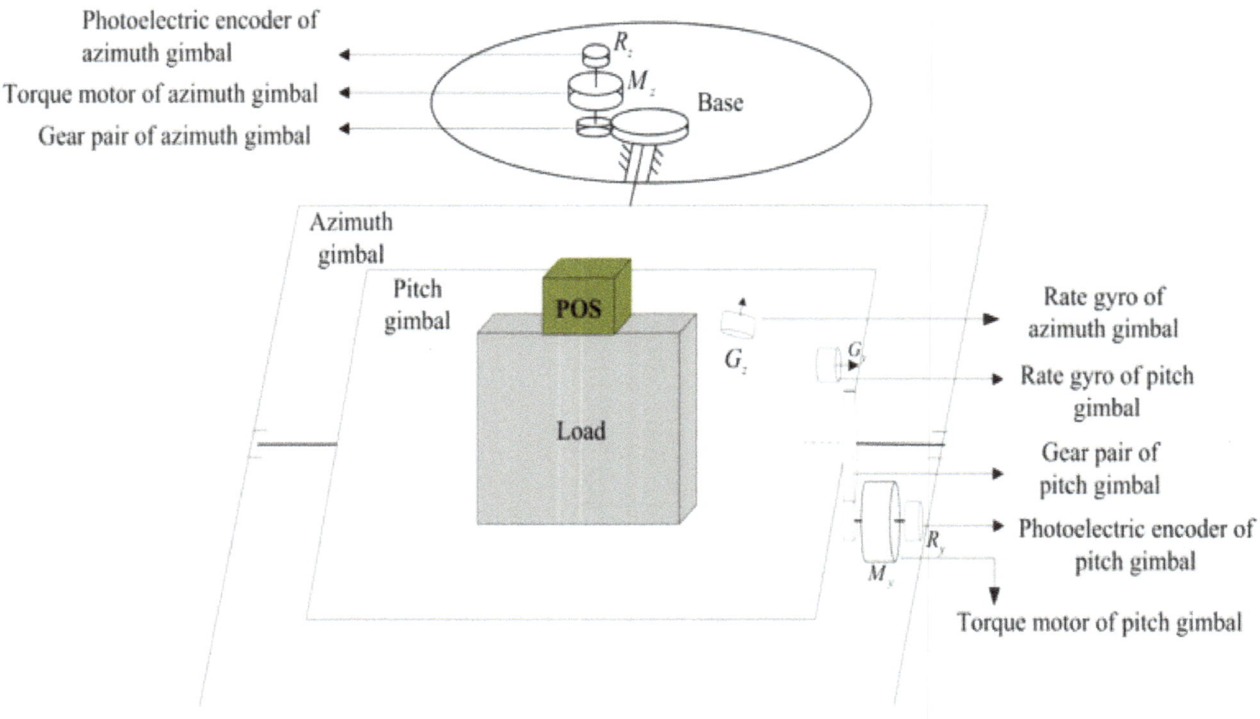

Figure 3: The structure diagram of two-axis ISP control system in APLI system.

CONTROL SCHEME OF TWO-AXIS ISP

The Traditional Dual Closed-Loop Control Structure

Figure 4 shows the schematic diagram of a traditional dual closed-loop control structure. We can see that this structure is composed of the inner rate loop and outer position loop. The inner rate loop uses a rate gyro to measure the angular rate of the gimbal in the inertial space, which is used to compensate the difference between the rate command input and the angular rate of the gimbal, eventually improving the steady-state precision. As the main feedback path, the outer position loop takes the attitude angle measured by POS as accurate reference to ensure the accurate pointing of the LOS. However, since there is no current loop inside, the influences of voltage fluctuation and the motor BEMF cannot be suppressed, eventually leading to decreased control precision and destabilization of the ISP.

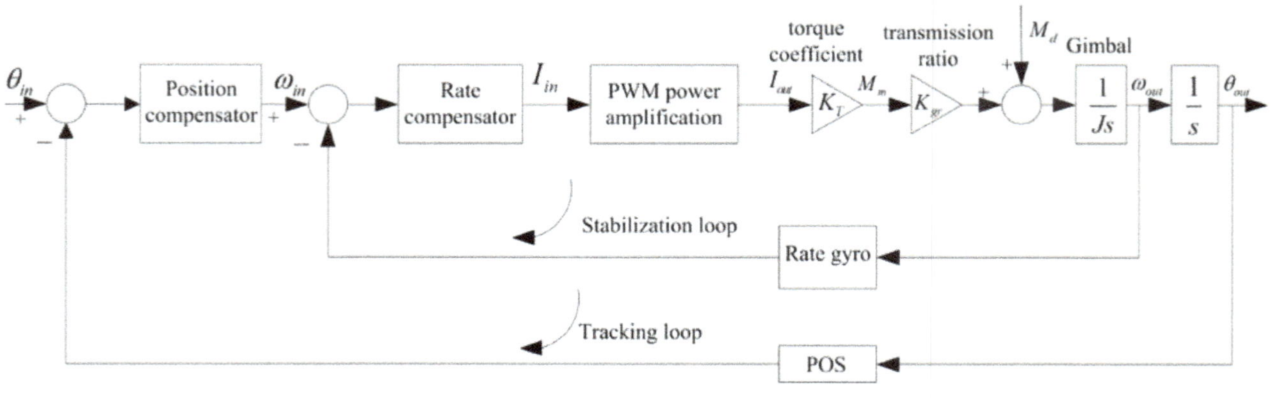

Figure 4: Schematic diagram of dual closed-loop PID control structure.

The Three Closed-Loop Compound Control Structure

Figure 5 shows the control structure of the proposed three closed-loop compound control scheme. On the basis of dual closed-loop PID control scheme, a specialized current feedback closed-loop is designed to improve the dynamic and static performance of rate loop, which is highlighted in a red dotted box in Figure 5. Thus, the control system structure is composed from the inside to the outside of a current loop, rate loop and position loop, respectively. By adding the current loop, the disadvantages of a dual closed-loop PID control structure are overcome. As a result, the control precision and stabilization of ISP are improved since the influences of voltage fluctuation of power supply and the motor EMF are suppressed by the added current loop inside.

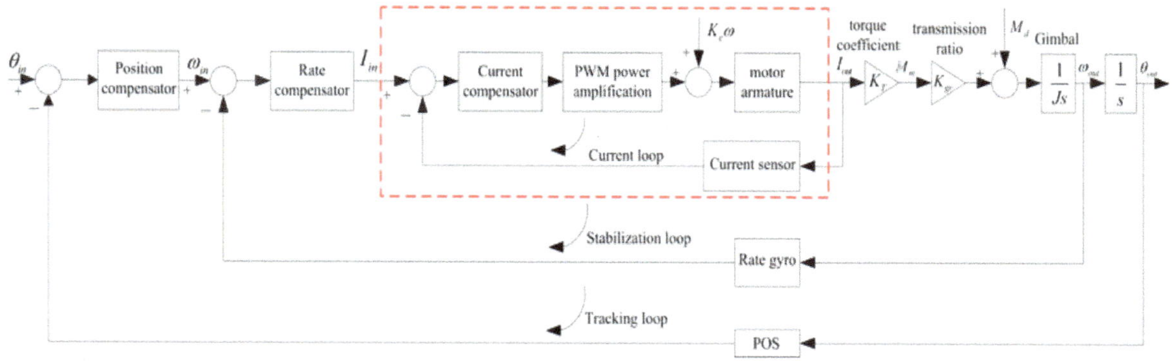

Figure 5: Schematic diagram of three closed-loop compound control structure.

Figure 6 shows the block diagram of the proposed three closed-loop compound control structure. The blocks of G-pos, G-spe and G-cur separately represent the controllers in the position loop, rate loop and current loop; the PWM block represents the power amplification used for amplifying the current to drive the torque motor; the symbol Lrepresents the inductance of a torque motor and R represents the resistance; the symbol K_t represents the torque coefficient of the motor and N is the transition ratio from the torque motor to the gimbals; J_m represents the moment of inertia of the motor and J_l represents the moment of inertia of the gimbals along the rotation axis.

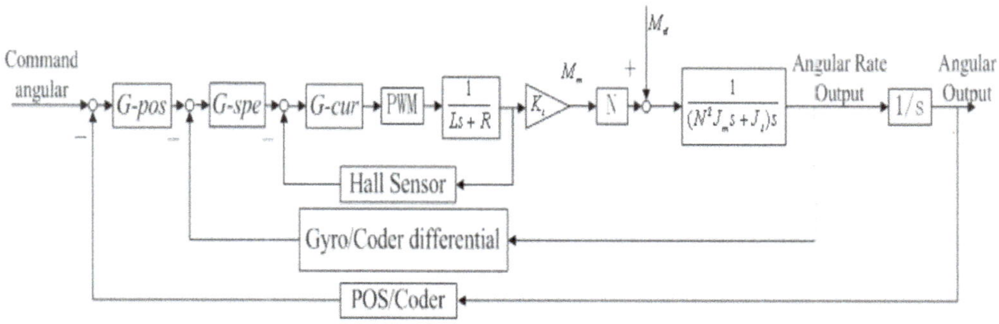

Figure 6: Block diagram of three-loop compound PID control structure of two-axis ISP control system in APLI system.

For the structure of the three closed-loop compound control, the system bandwidths of each loop increase from the outer loop to the inner loop, meaning the current loop has the highest bandwidth among the three closed-loops, leading to the rapidest response speed. As for each gimbal control system, the function of rate loop, i.e., stabilization loop, is to minimize the effects of external disturbances, in which a rate gyro is used as feedback sensor to directly measure inertial LOS rate. Therefore, the compensated torques for disturbance rejection are determined by the stabilization controller in response to the combined command input and LOS inertial rate feedback [21]. The function of position loop is to keep tracking the position objective in real time, in which a POS is used as the position sensor.

Design of the Current Loop Controller

Figure 7 shows the structural diagram of the current loop for a pitch system. A PID controller is adopted in the design. We can see that the drive model of the motor power is composed of the motor model, the current detection model, the current controller model and PWM power drive model. In Figure 7, I_{in} and I_{out} stand for the input current and output current, respectively. $U/T_{pwn}s+1$ is the transfer function of the PWM power drive unit, in which U is the power supply voltage and T_{pwm} is the PWM switching period. $K_e\omega$ stands for the motor counter electromotive force, in which ω and K_e are the motor speed and BEMF coefficient, respectively. K_m/T_es+1 is the transfer function of the motor armature, in which T_e and K_m stand for the time constant of the armature magnetic and armature current coefficient, respectively.

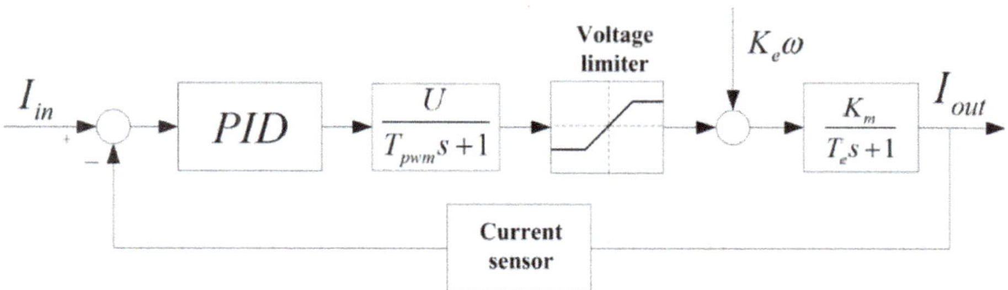

Figure 7: The structural diagram of current loop.

In order to get a better response speed and stability precision, the current loop controller is designed with a PI serial correction. Assume that the current controller is $K_I(\tau_I s+1/s)$, where τ_I is set to equal the time constant of the magnetic armature T_e in order to offset the electromagnetic inertia of motor at a greater extent. Thus, the transfer function of the open current loop can be obtained as:

$$G_{opc} = \frac{K_I \cdot K_m \cdot U}{(T_{pwm}s+1)s} = \frac{K_I \cdot K_m \cdot U}{T_{pwm}} \cdot \frac{1}{s^2 + \frac{1}{T_{pwm}}s}$$

$$\begin{cases} 2\xi\omega_n = \dfrac{1}{T_{pwm}} \\ \omega_n^2 = \dfrac{K_I \cdot K_m \cdot U}{T_{pwm}} \end{cases} \qquad (1)$$

We can see that the current loop system is a second-order system. When the damping ratio is $\xi = 0.707$, system will be an optimized second-order system, so if ξ is substituted into Equation (1), the ω_n and PID parameter K_I are then obtained. Figure 8 shows the Bode diagram of the pitch gimbal's open current loop. We can see that the phase margin and bandwidth are 65.5 degrees and 7580 rad/s, respectively.

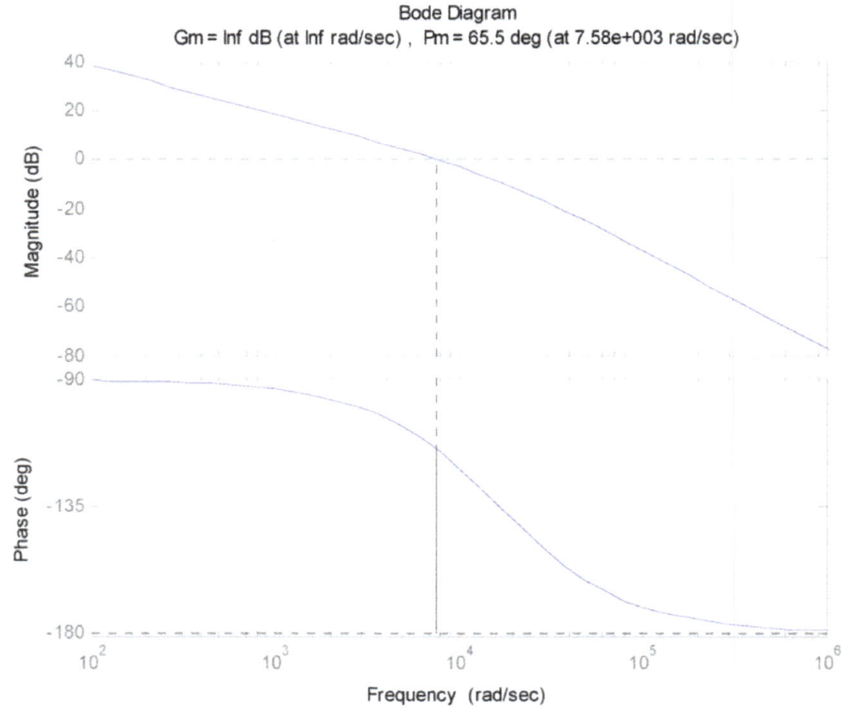

Figure 8: Simulation correction results of open current loop for pitch system.

Design of the Rate Loop Controller

Figure 9 shows the structural diagram of rate loop. Since the bandwidth of the current loop is much greater than that of the rate loop, the current loop can be approximately taken as a proportional element whose coefficient is equal to 1. Thus, the structure of the rate loop is simplified, which is helpful for the analysis and correction of the rate loop. The PI series controller is adopted in the design, which is represented as $K_\omega(\tau_\omega s+1/s)$. Besides, K_T and K_{gr} stand for the coefficient of motor torque and transmission ratio of gear linkage, respectively.The transfer function of rate gyro is $1/T_G s+1$, M_d stands for the disturbance torques, J stands for the rotational inertia of the gimbal together with the payload. The time constant T_G is small so it can be neglected.

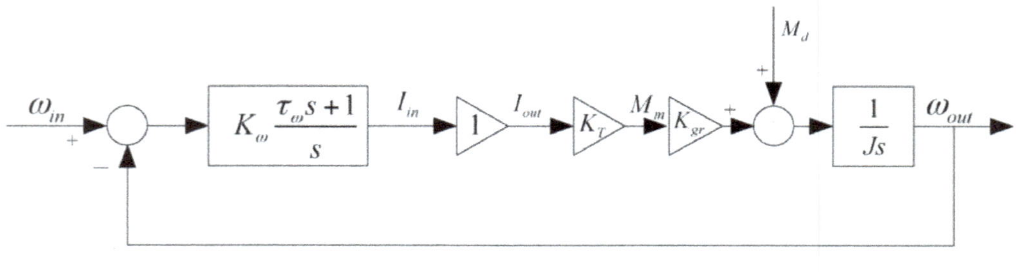

Figure 9: The structural diagram of the rate loop.

The open loop transfer function of the rate loop is as follows:

$$G_{op\omega} = K_\omega \cdot K_T \cdot K_{gr} \cdot \frac{\tau_\omega s + 1}{Js^2} \tag{2}$$

The frequency characteristic is:

$$G_{op\omega}(j\omega) = K_\omega \cdot K_T \cdot K_{gr} \cdot \frac{\tau_\omega \cdot j\omega + 1}{J \cdot j\omega \cdot j\omega} \tag{3}$$

Since the mechanical resonance frequency of the ISP, ω_R, can be obtained by modal analysis and testing, the open loop cut-off frequency of the rate loop, $\omega_{c\cdot\omega}$, can be set to be 1/5 of ω_R and then obtained as : $\omega_{c\cdot\omega}$ = 6.28 rad/s. In order to make system have better performance, such as stability, short regulation time, small overshoot, and so on, the phase margin of the open rate loop is designed to be 70 degrees. It means that when $\omega_{c\cdot\omega}$ = 6.28 rad/s, there is $\angle j \cdot 9\tau_\omega + 1 = 70°$. Thus, we get: $\tau_\omega = 1 \cdot \mathrm{tg}70°(1/6.28) = 0.4375$.

When $\omega_{c\cdot\omega} = 6.28\ \mathrm{rad/s},\ G_{op\omega}|_{(j\omega_{c\cdot\omega})} = 1$. From Equation (3), we get:

$$K_\omega \cdot K_T \cdot K_{gr} \cdot \frac{\sqrt{(\tau_\omega \omega_{c\cdot\omega})^2 + 1}}{J \cdot \omega_{c\cdot\omega} \cdot \omega_{c\cdot\omega}} = 1 \tag{4}$$

If $\omega_{c\cdot\omega}$ = 6.28 rad/s and τ_ω = 0.4375 are substituted into Equation (4), then K_ω is obtained.

Design of the Position Loop Controller

Figure 10 shows the structural diagram of position loop. The PI series controller is adopted in the design, which is represented as $K_\theta(T_{2s} + 1/T_1s + 1)$. The open loop cut-off frequency of the position loop is set to be 1/4 of $\omega_{c\cdot\omega}$, i.e., $\omega_{c\cdot\theta}$= 1.57 rad/s. As mentioned above, the open loop cut-off frequency of rate loop, $\omega_{c\cdot\omega}$, is equal to 6.28 rad/s. Since the rate loop is as an inner loop inside the tracking loop, it can be taken as a first order inertial element whose time coefficient is $T_\omega = 1/\omega_{c\cdot\omega}$ = 0.1592 s.

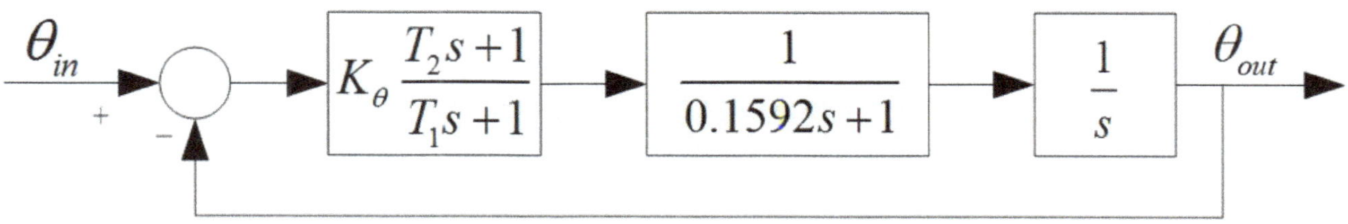

Figure 10: The structural diagram of position loop.

So the approximately equivalent closed-loop transfer function of rate loop is as follows:

$$\Phi_\omega = \frac{1}{T_\omega s + 1} \tag{5}$$

Then the transfer function of open position loop is approximately expressed as follows:

$$G_{op\theta} = K_\theta \frac{T_2 s + 1}{T_1 s + 1} \cdot \frac{1}{T_\omega s + 1} \cdot \frac{1}{s} \tag{6}$$

We set $1/T_2 = 0.1\omega_{c\cdot\theta}$ and $T_1 = 3T_2$. Since the cut-off frequency of open position loop is designed as $\omega_{c\cdot\theta}$ = 1.57 rad/s,T_1 and T_2 are obtained. Thus, two known constants T_1 and T_2 are substituted into $|G_{op\theta}|_{(j\omega_{c\cdot\theta})} = 1$, then the PID parameter K_θ is obtained.

After calculation, the bandwidth and phase margin of the pitch gimbal's open position loop are 1.57 rad/s and 72.2 degree, respectively. The bandwidth of the pitch gimbal's closed loop is 2.18 rad/s, which meets the requirement of tracking POS.

EXPERIMENTS AND RESULTS

Testing under Static Base Conditions

A series of experiments are carried out to a two-axis ISP developed for an UH-based APLI system to verify the proposed scheme. Figure 11 shows the experimental UH-based APLI system. The flight time of each inspection is about 0.5–1 h.

Figure 11: The experimental UH-based APLI system.

Figure 12 shows the picture of the two-axis ISP with multiple sensors under the static base testing conditions. In the static testing, the two-axis ISP is fixed under the bottom of the UH and tests are conducted in a stable environment. The main physical parameters of the ISP are as follows: maximum load capacity is 36 kg and the weight of ISP is 20 kg, respectively. The range of azimuth rotation angle is 0°–360°, the maximum pitch rotation angle range is ±50°.

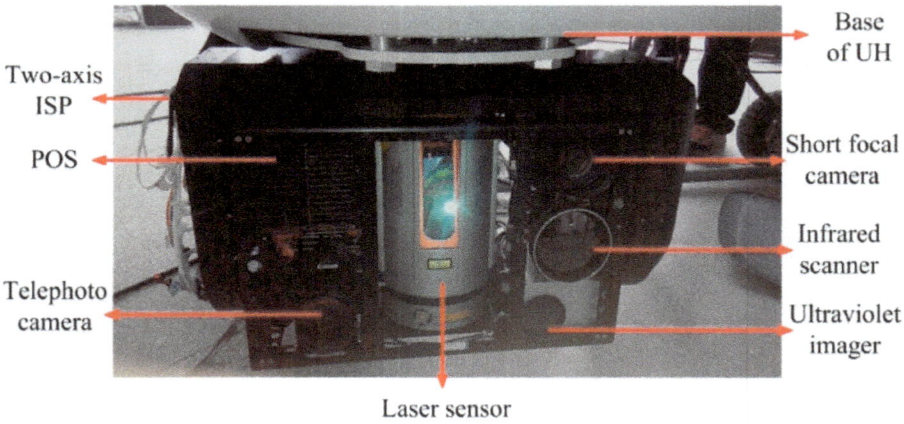

Figure 12: A picture of the two-axis ISP with multiple sensors under the static base testing conditions.

Tracking Performances

Figure 13 shows the tracking results of ISP under static base testing conditions. It can be seen that the ISP can steadily track the command instructions with high accuracy. For the pitch system, the command angles are from −20° to −80° and then reverse to 0° with a short stabilization time per 20° interval.

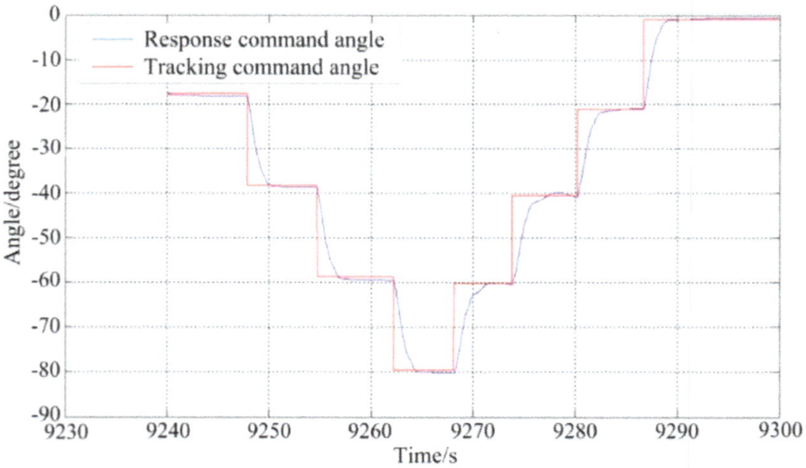

(a)

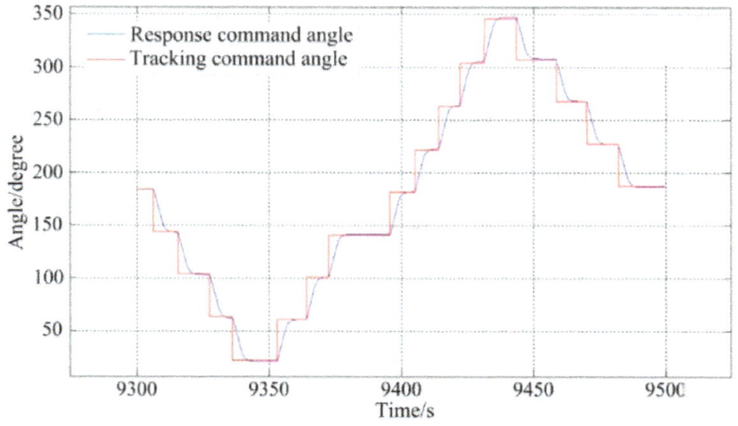

(b)

Figure 13: The tracking results of angle position under static testing: (**a**) the pitch system; (**b**) the azimuth system.

For the azimuth system, the stabilization performances is tested per 40° interval from 180° to 20° and then reverse to 320°. The maximum overshoot is less than 1.5° for both gimbal systems.

Stabilization Performance

(1) Pitch System

Figure 14 shows the stabilization results of the pitch system under static base testing conditions, which include two cases: (a) represents a command angle from 0° to −80° and (b) represents a command angle from 0° to −45°. It can be seen that the angular position errors of the pitch system under tracking −80° and −45° are less than 0.0165° (RMS) and 0.0155° (RMS), respectively, meaning that the compound control scheme can achieve a high static stabilization precision.

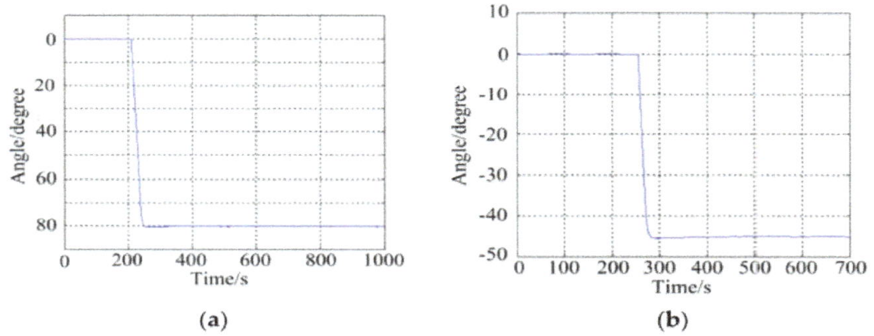

Figure 14: Stabilization results of the pitch system under static base conditions: (**a**) instruction command angle from 0° to −80°; (**b**) instruction command angle from 0° to −45°.

(2) Azimuth System

Figure 15 shows the stabilization results of the azimuth system under static base testing conditions that also include two cases: (a) represents a command angle from 260° to 180° and (b) represents a command angle from 180° to 260°. It can be seen that the angular position errors of azimuth system when tracking 180° and 260° are less than 0.0102° (RMS) and 0.0125° (RMS), respectively. Likewise, this means the compound control scheme of the azimuth system can achieve a reliable static stabilization performance.

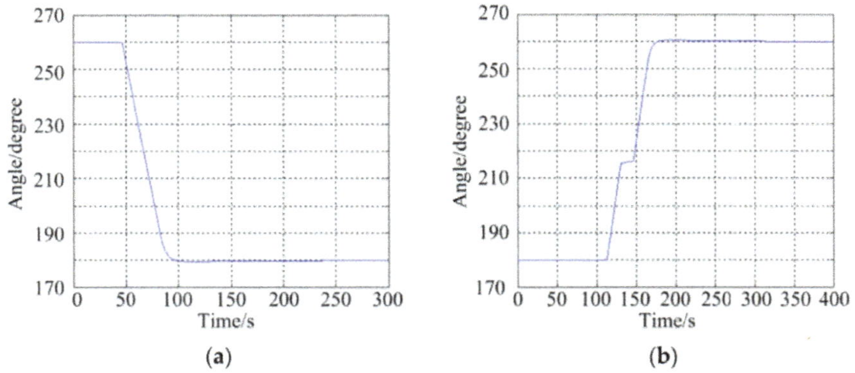

Figure 15: Stabilization results of the azimuth system under static base conditions: (**a**) command angle from 260° to 180°; (**b**) command angle from 180° to 260°.

Testing under Dynamic Base Conditions

As shown in Figure 16, to further validate the dynamic performance of the proposed scheme, dynamic experiments were carried out on a movable vehicle. In experiments, the ISP is fixed on the top of a box through a transitional support frame.

(a) (b)

Figure 16: Pictures of the experimental equipment in dynamic experiments: (**a**) moving vehicle; (**b**) two-axis ISP inside the vehicle.

Figure 17 shows the testing results in the real movable vehicle experiments. It can be seen that the stabilization errors of the pitch system and azimuth system are less than 0.5° and 0.2° in case of tracking the angular positions of −30° and 150°, respectively, meaning a reliable dynamic performance that can reject effectively random disturbances in a moving base scenario.

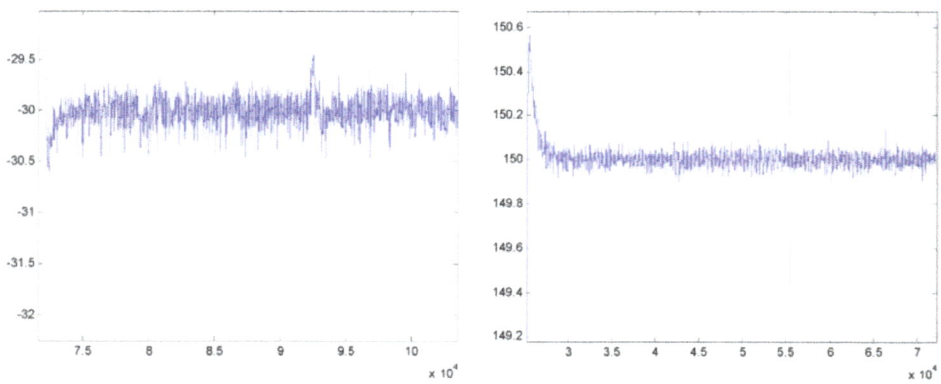

(a) (b)

Figure 17: Stabilization results in dynamic experiments: (**a**) pitch system; (**b**) azimuth system.

Testing the UH-Based APLI System in the Air

In order to verify the effectiveness of the control system, a two-axis ISP with the proposed compound control scheme was applied in a real UH-based APLI system to detect the defects of a high- voltage power line in Foshan, Guangdong, China, as shown in Figure 18.

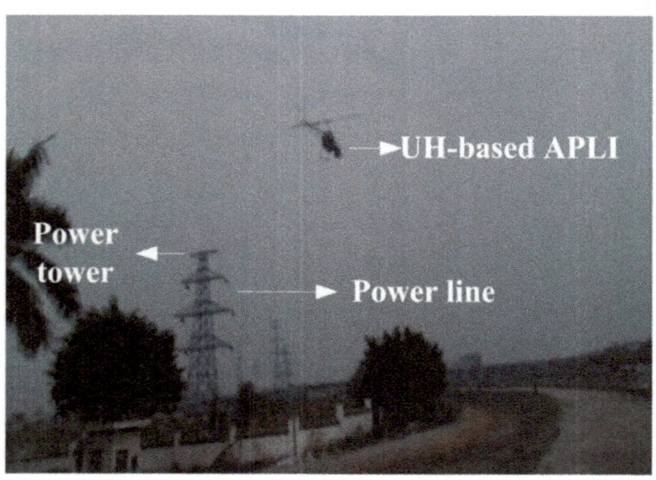

Figure 18: The picture of the experiments for a real UH -based APLI system.

Figure 19 shows the partial imaging data acquired by the APLI system. It can be seen the imaging data are clear regardless on which of the multiple sensors is used, meaning that the control performance of the two-axis ISP is reliable, and can satisfy the imaging requirements of real APLI system.

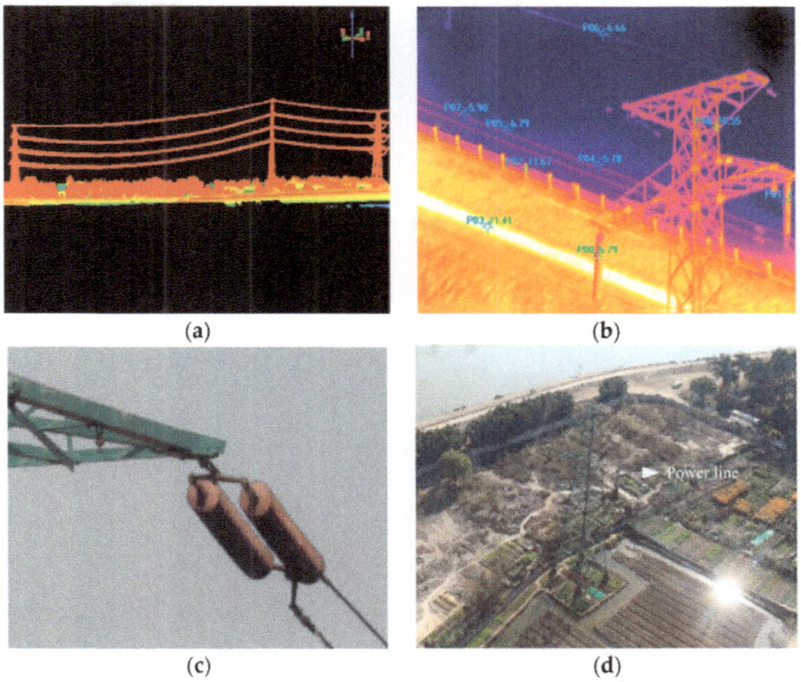

Figure 19: The partial real imaging data acquired by the APLI system using: (**a**) laser scanner; (**b**) infrared scanner; (**c**) CCD short focal distance camera; (**d**) telephoto camera.

Figure 20 shows the experimental results of the two-axis ISP of the real UH-based APLI system in the air. It can be seen that under real flight conditions, the ISP still has high tracking accuracy with the command instructions, similar to the results in the movable vehicle experiments. The steady-state tracking errors of the pitch system and azimuth are 0.3456° (RMS) and 0.3481° (RMS), respectively. The RMS values of different trails in Figure 20a,b are summarized in Table 1 and Table 2, respectively.

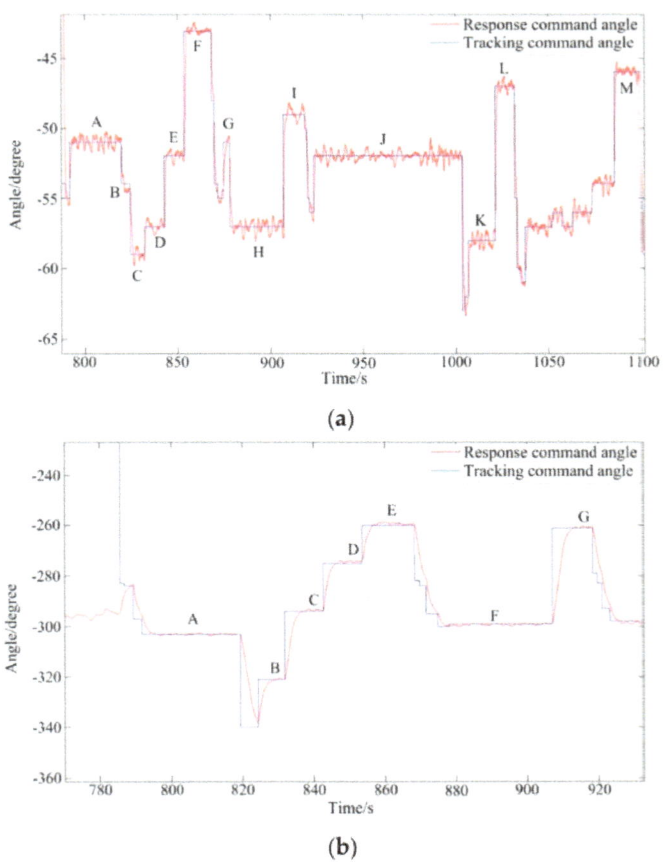

Figure 20: The tracking results of angle position of the real UH-based APLI system in air: (**a**) the pitch system; (**b**) the azimuth system.

Table 1: Steady state RMS errors of each tracking angle for pitch gimbal

Number	Tracking Angles (Degree)	RMS	Number	Tracking Angles (Degree)	RMS
A	−51	0.3982	H	−57	0.4200
B	−54	0.4642	I	−49	0.4330
C	−59	0.3773	J	−52	0.3413
D	−57	0.3563	K	−58	0.4379
E	−52	0.3088	L	−47	0.2987
F	−43	0.1925	M	−46	0.2843
G	−51	0.2819			

Table 2: Steady state RMS errors of each tracking angle for azimuth gimbal

Number	Tracking Angles (Degree)	RMS	Number	Tracking Angles (Degree)	RMS
A	−303	0.3164	E	−260	0.4105
B	−321	0.2937	F	−299	0.3997
C	−294	0.3743	G	−261	0.3162
D	−275	0.3832			

CONCLUSIONS

In this paper, to improve the stabilization accuracy of a two-axis ISP used in an UH-based APLI system, a compound control scheme is proposed. To reduce the influences of power supply voltage fluctuations and motor BEMF, a specialized current feedback closed-loop is introduced to improve the dynamic and static performance of the rate loop on the basis of a dual closed-loop PID control scheme. The control performances of the scheme are analyzed and simulated. Particularly, a series of experiments are carried out to validate the scheme. The experimental results under real flight conditions show that the ISP has high tracking accuracy with the command instructions, whose steady-state tracking errors of the pitch system and azimuth are 0.3456° (RMS) and 0.3481° (RMS), respectively. By using a two-axis ISP operated with the method, reliable imaging data of multiple sensors are acquired for a real APLI system. The results validate that the proposed compound control scheme effectively improves the stabilization accuracy of a two-axis ISP.

ACKNOWLEDGMENTS

This work was supported by the National Natural Science Foundation of China (No. 51375036, No. 51205019) and China Scholarship Council (CSC).

REFERENCES

1. Murthy, V.S.; Gupta, S.; Mohanta, D.K. Digital image processing approach using combined wavelet hidden Markov model for well-being analysis of insulators. IET Image Process. 2011, 5, 171–183. . .
2. Sharma, H.; Bhujade, R.; Adithya, V.; Balamuralidhar, P. Vision-based detection of power distribution lines in complex remote surroundings. In Proceedings of the 20th National Conference on Communications (NCC), Kanpur, India, 28 February–2 March 2014.
3. Katrasnik, J.; Pernus, F.; Likar, B. A Survey of mobile robots for distribution power line inspection. IEEE Trans. Power Deliv. 2010, 25, 485–493. . .
4. Jones, D.; Golightly, I.; Roberts, J.; Usher, K. Modeling and control of a robotic power line inspection vehicle. Computer Aided Control System Design. In Proceedings of IEEE International Conference on Control Applications, Munich, Germany, 4–6 October 2006; pp. 632–637.
5. Wang, B.H.; Han, L.; Zhang, H.L.; Wang, Q.; Li, B.Q. A flying robotic system for power line corridor inspection. In Proceedings of the IEEE International Conference on Robotics and Biomimetics (ROBIO), Guilin, China, 19–23 December 2009; pp. 2468–2473.
6. Jones, D.I.; Earp, G.K. Camera sightline pointing requirements for aerial inspection of overhead power lines.Electr. Pow. Syst. Res. 2001, 57, 73–82. . .
7. Zhou, X.Y.; Zhao, B.L.; Gong, G.H. Control parameters optimization based on co-simulation of mechatronic system for an UA-based two-axis inertially stabilized platform. Sensors 2015, 8, 20169–20192. . . .
8. Ming, L.; Sheng, G.H.; Liu, Y.D.; Jiang, X.C.; Nie, S.G.; Qu, G.Y. Research on auto-tracking algorithm for power line inspection based on unmanned aerial vehicle. In Proceedings of the 2012 Asia-Pacific, Power and Energy Engineering Conference (APPEEC), Shanghai, China, 27–29 March 2012; pp. 1–5.
9. Whitworth, C.C.; Duller, A.W.G.; Jones, D.I.; Earp, G.K. Aerial video inspection of overhead power lines. Power Eng. J. 2001, 15, 25–32. . .
10. Zhou, X.Y.; Zhang, H.Y.; Yu, R.X. Decoupling control for two-axis inertially stabilized platform based on an inverse system and internal model control. Mechatronics 2014, 24, 1203–1213. . .
11. Liu, Z.C.; Zhu, Z.S.; Mai, X.M.; Peng, X.Y.; Wang, K. Dynamic error compensation of POS in power

transmission line inspection. In Proceedings of 2013 Tbale2nd International Symposium on Instrumentation and Measurement, Sensor Network and Automation (IMSNA), Toronto, ON, Canada, 23–24 December 2013; pp. 408–411.

12. Rezac, M.; Hurak, Z. Vibration rejection for inertially stabilized double gimbal platform using acceleration feedforward. In Proceedings of the 2011 IEEE International Conference on Control Applications (CCA), Denver, CO, USA, 28–30 September 2011; pp. 363–368.
13. Zhou, X.Y.; Gong, G.H.; Li, J.P. Decoupling control for a three-axis inertially stabilized platform used for aerial remote sensing. Trans. Inst. Meas. Control 2015, 37, 1135–1145. . .
14. Liu, W.; Zhou, X.Y. Nonlinear friction modeling and adaptive compensation on an inertially stabilized platform system for aerial remote sensing application. Chin. J. Mech. Eng. 2013, 49, 122–129. (In Chinese). . .
15. Gomonwattanapanichl, O.; Pattanapukdee, A.; Mongkolwongrojn, M. Compensation and estimation of friction by using extended Kalman filter. In Proceedings of the SICE-ICASE, International Joint Conference, Busan, Korea, 18–21 October 2006; pp. 5032–5035.
16. Ohdaira, S.; Mizumoto, I.; Kohzawa, R.; Iwai, Z. Dual-rate robust adaptive control of a cart-crane system. In Proceedings of the SICE-ICASE, International Joint conference, Busan, Korea, 18–21 October 2006; pp. 1800–1805.
17. Mu, Q.Q.; Liu, G.; Zhong, M.Y.; Chu, Z.Y. Imbalance torque compensation for three-axis inertially stabilized platform using acceleration feedforward. In Proceedings of the 2012 8th IEEE International Symposium on Instrumentation and Control Technology (ISICT), London, 11–13 July 2012; pp. 157–160.
18. Ito, K.; Maebashi, W.; Ikeda, J.; Iwasaki, M. Fast and precise positioning of rotary table systems by feedforward disturbance compensation considering interference force. In Proceedings of the IECON 2011 37th Annual Conference on IEEE Industrial Electronics Society, Melbourne, VIC, Australia, 7–10 November 2011; pp. 3382–3387.
19. Chen, G.M.; Liang, L.H. Research on system of three-axis swing turntable based on robust compensation control. In Proceedings of the 2011 3rd International Workshop on Intelligent Systems and Applications (ISA), Wuhan, China, 28–29 May 2011; pp. 1–4.
20. Dan, J.G.; Kaan, E. Precision control of a T-type gantry using sensor/actuator averaging and activevibration damping. Precis. Eng. 2012, 36, 299–314. .
21. Masten, M.K. Inertially stabilized platforms for optical imaging systems. IEEE Contr. Syst. Mag. 2008, 28, 47–64. . .
22. Hilkert, J.M. Inertially stabilized platform technology Concepts and principles. IEEE Contr. Syst. Mag. 2008, 28, 26–46.

CHAPTER 18

UPGRADE OF THE CMS INSTRUMENTATION FOR LUMINOSITY AND MACHINE INDUCED BACKGROUND MEASUREMENTS

Anne Dabrowski

CERN, CH-1211 Geneva 23, Switzerland

ABSTRACT

To optimise performance with the higher luminosity, higher beam energy and shorter bunch spacing of 25 ns at the LHC after 2014, an upgrade program is performed for the detectors to measure the luminosity and machine induced background. A new detector is the pixel luminosity telescope consisting of 8 telescopes, equipped with silicon pixel sensors, on both ends of the interactions point. The Fast Beam Conditions Monitoring system, using diamond sensors, is upgraded to 24 sensors, 12 on each end of the IP. In addition, dedicated fast ASICs produced in 130 nm commercial CMOS technology and dead-time free backend electronics using FPGAs for fast signal processing are being developed and built. Also, the part of the forward HCAL used for the luminosity measurement is instrumented with new readout electronics, in microTCA standards. The machine induced background measurement will be supported by a new system of direction sensitive quartz Cherenkov counters, with excellent time resolution. A data acquisition architecture is being developed that is common for all subsystems and allows for synchronization across different hardware. The design of the new system will be presented, and a report will be given on the performance of each subsystem measured in several test-beam campaigns and prototype operation in the last LHC run.

INTRODUCTION

The Compact Muon Solenoid (CMS) detector [1] is one of two general purpose detectors at the Large Hadron Collider (LHC) [2] at CERN. During LHC Run I (2010-2012 running period), proton-proton beam collided at a centre-of-mass energy up to 8 TeV with 50 ns bunch spacing. The instantaneous luminosity reached more than $7 \times 10^{33}cm^{-2}s^{-1}$ and about 30 f b−1 of integrated luminosity was delivered to CMS. To ensure safe, high quality data-taking conditions for the CMS detector and to optimize and account for the luminosity delivered for physics during Run II, upgraded instrumentation is designed to monitor the LHC beam conditions for CMS. These systems are being installed during the 2013-2014 period of LHC scheduled maintenance, Long Shutdown 1 (LS1). The systems have been optimized to measure the per-bunch online luminosity and the machine-induced background (MIB) for the CMS experiment, for the beam conditions expected in Run II.

Citation: Anne Dabrowski, Upgrade of the CMS Instrumentation for luminosity and machine induced background measurements, Nuclear and Particle Physics Proceedings, Volumes 273–275, April–June 2016, Pages 1147-1154, ISSN 2405-6014, http://dx.doi.org/10.1016/j.nuclphysbps.2015.09.180.

The LHC will re-start operations in Run II with proton-proton collisions at a centre-of-mass energy of 13 TeV and bunch spacing of 25 ns. A peak luminosity of 1.4×10^{34} cm^{-2} s^{-1} will be achieved during the first year and then reach $2 \times 10^{34} cm^{-2} s^{-1}$ with around 50 interactions per bunch crossing, twice the instantaneous luminosity foreseen in the initial LHC technical design report. The luminosity is a key parameter of a collider experiment as higher values allow an exploration of lower cross section processes. Its precise measurement is necessary to determine cross sections. In addition, from the integrated luminosity the dose load to the detector components is inferred. The online luminometers include the Pixel Luminosity Telescope (PLT) described in Section 2, the Fast Beam Conditions Monitor (BCM1F), described in Section 3, and the luminosity based on a dedicated readout of the Hadron Forward (HF) Calorimeter described in Section 4. These online luminometers are independent and together with the highly linear offline luminosity measurement based on pixel cluster counting, allow for a cross check and reduction of the systematic errors. The higher beam intensity implies tighter collimator settings and an increased susceptibility to electron cloud effects contributing to a higher potential for MIB for CMS. The upgraded Fast Beam Conditions Monitor (BCM1F), as described in Section 3, is designed to have sensitivity at low radius where beam gas interactions near CMS will be detected.

The Beam Halo Monitor (BHM), as described in Section 5, will have sensitivity towards distant distant beam gas and beam halo interactions with the collimators. To fully exploit the redundant online luminosity and beam background measurement system, the BRIL subsystems are synchronized using common timing signals distributed by the CMS Trigger, Control and Distribution System (TCDS). A common architecture for the data acquisition is being developed based on the CMS XDAQ online software framework to ease data exchange, storage, monitoring, database management and the maintainability of the readout software across systems and will be described in detail in Section 6. This paper will present the design of the upgrades of the beam monitoring systems by the CMS BRIL project during LS1, as summarised in Figure 1.

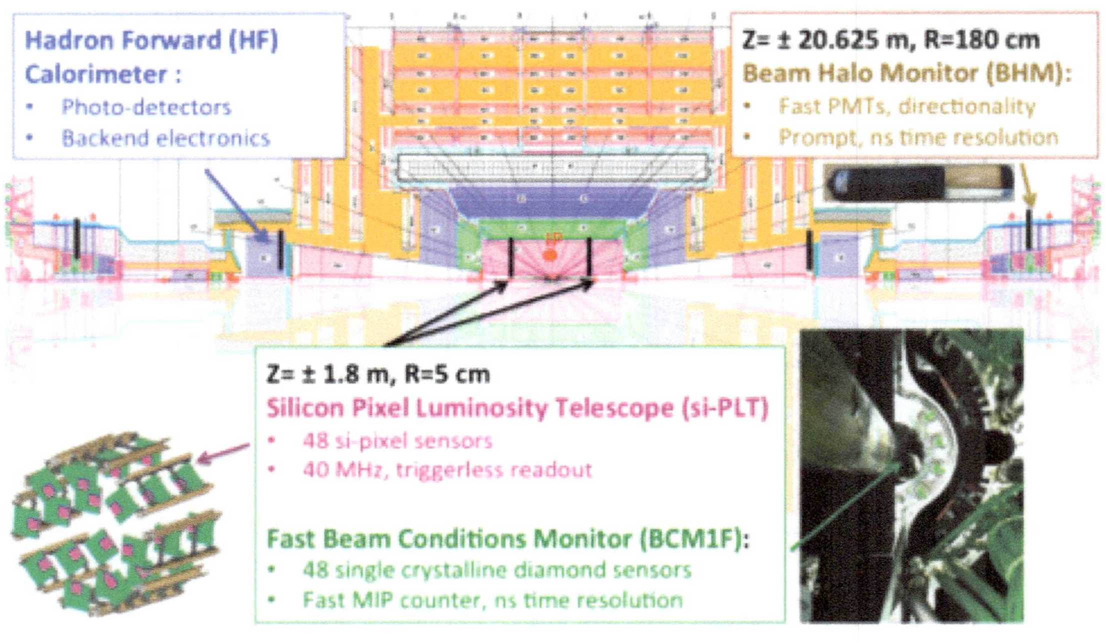

Figure 1: An overview of the upgraded CMS beam instrumentation sub-systems and their location in the CMS detector.

Pixel Luminosity

Telescope The Silicon Pixel Luminosity Telescope (PLT) is a dedicated luminometer monitor for CMS based on silicon pixel sensors [3], as used in the CMS barrel pixel detector, bump bonded to the PSI46v2 CMS pixel readout chip [4] (ROC). The functionality and readout of this detector is based on an earlier design [5] that used uncooled single crystalline CVD pixelated diamond sensors.

The PLT is comprised of two arrays of eight smallangle telescopes situated one on each end of CMS, 1.75 m from the CMS interaction point (IP). Each telescope consists of three equally spaced planes of pixel sensors with a total telescope length of 7.5 cm and are located 5 cm radially from the beam line.

The PLT is designed to provide a measurement of the bunch-by-bunch luminosity at the CMS collision point on a time scale of a few seconds and a high-precision measurement of the integrated luminosity. The primary luminosity measurement of the PLT is based on counting the number of telescopes with threefold coincidences formed from the fast-or of the column-multiplicity signal output by the PSI46v2 readout chip. The fast-or signal, clocked at the bunch crossing frequency of 40 MHz, indicates the number of double columns with pixels over threshold in each bunch crossing. In addition, the full hit information consisting of the row and column addresses and the pulse heights of all signals over threshold is readout at a lower configurable frequency of a few kHz. This full pixel readout provides tracking information and is a powerful tool for determining systematic corrections, calibrating pixel ef- ficiencies and measuring the collision point centroid as a function of time.

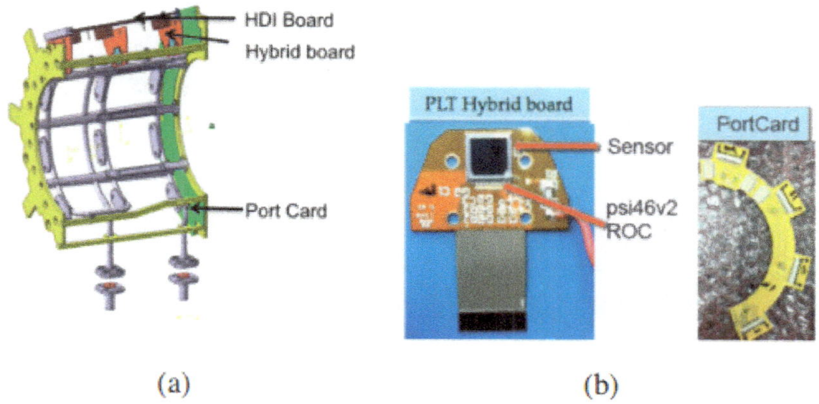

Figure 2: (a) PLT cassette providing the frame for electronics, alignment components and cooling distribution (b) An example of the silicon PLT hybrid board and port card.

The PLT frontend hybrid boards, hosting the sensor and readout chip, are mounted onto the cassette as shown in Figure 2 (a), on the IP end of the carriage, as shown in Figure 3. The mechanics for the PLT detector and the hybrid boards were designed and produced during LS1 to incorporate the new requirement for providing a cooling distribution and performant thermal contact to the silicon sensors, as shown in Figure 3.

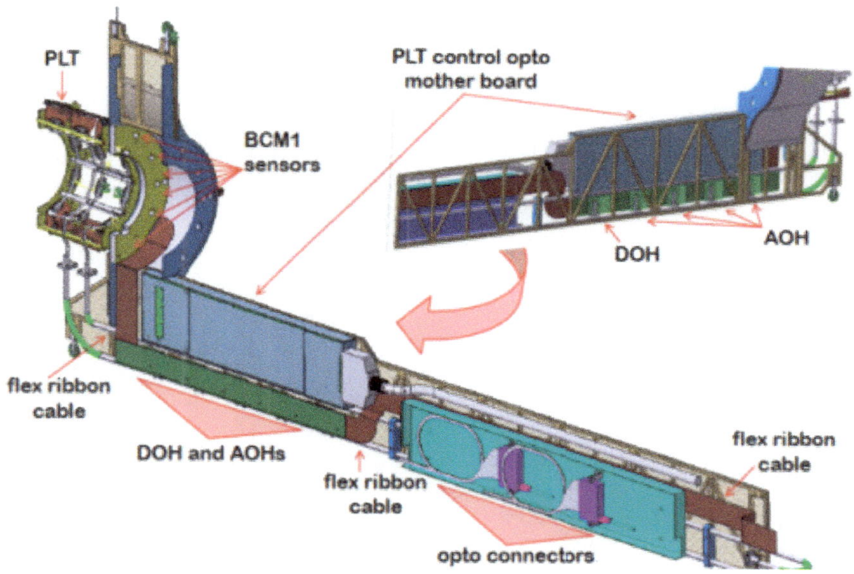

Figure 3: Drawings of the PLT and BCM1 support carriage, 1/4 of the detector system is shown, installed at a Z location of ±1.8 m from the IP, inside the pixel support tube.

The cassette was produced from a titanium alloy using the selective laser melting (SLM) technique. This structure has a hallow inner diameter of 2.8 mm allowing for the flow of C6F14 coolant qualified to a pressure of 15 bar. Three hybrid board planes form a telescope and are connected by pigtails to the High Density Interconnect (HDI), a four-layer flex circuit. The HDI houses a CMS pixel Token Bit Manager (TBM) chip that communicates with the three readout chips in the telescope and orchestrates the readout of the full sensor information. A custom PLT driver chip amplifies and outputs the three fast-or signals onto separate analog lines. The HDI also distributes low voltage power and sensor bias voltage to the hybrid boards. The HDI circuits connect to a semi-circular port card board located at the end of the support carriage with one port card for each half carriage (four telescopes), as shown in Figure 2 (b).

The port card connects to the opto board a custom circuit, located at the foot of the carriage, shown in Figure 4 (a). The opto board houses the optical hybrid circuits for the outgoing analog signals (full signal readout of the pixels and fast-ors) and for the digital control signals. From the opto board the analog signal fibers connect to a Front End Driver (FED) [6], a VME flash ADC module, located in the CMS underground service cavern. The histogramming of the 40 MHz, fast-or signals is done in the PLT FED.

As for all BRIL subsystems, the firmware in the FED has been adapted to histogram the fast-or data based on a common time intervall, so called "lumi-nibble", distributed by the CMS TCDS system. A photo of the assembled cassette mounted onto the carriage is shown in Figure 4 (b).

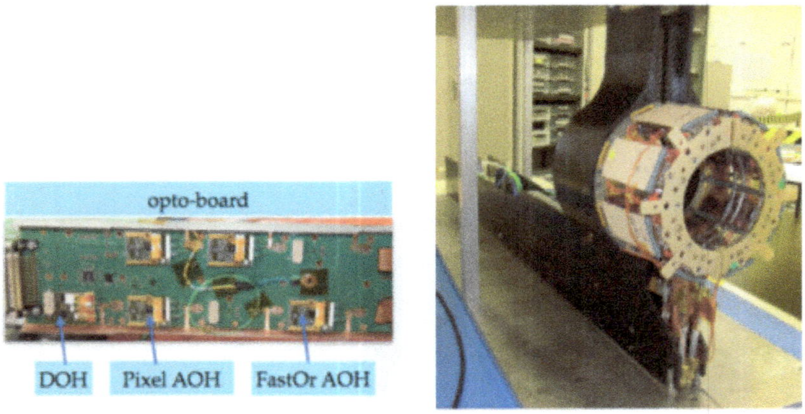

(a) (b)

Figure 4: (a) Optoboard for control and readout of full cassette (b) Two completed PLT cassette assemblies mounted to the carbon fibre support carriage.

UPGRADED FAST BEAMS CONDITIONS MONITOR

Run I Configuration

The Fast Beams Conditions Monitor (BCM1F) [7] has been in operation in CMS since 2008, successfully providing measurements of beam background [8] and a backup measurement of the online luminosity [10] for CMS. This BCM1F detector used in Run 1 consisted of 8 single crystalline CVD diamond sensors, with an active area of 5.0 × 5.0 mm2, a thickness of 500 μm, located at a distance of 1.8 m from the interaction point at a radial distance of 5 cm. The sensors were coupled to a charge-sensitive, radiation-hard amplifier and shaper ASIC of the type JK16 [11] with a peaking time of 25 ns.

Upgraded BCM1F for Run II

The upgrade of the BCM1F detector is driven by the requirement for efficient identification of collision and MIB signals at high rates and the requirement for sufficient radiation hardness. An average instantaneous charged particle flux of about $6 \times 10^7 cm^{-2}s^-1$ is predicted by FLUKA [12] calculations at the BCM1F detector location, for an instantaneous luminosity of $2 \times 10^{34} cm{-}2s^{-1}$. The corresponding integrated hadronic fluence to the detector is predicted to exceed $6 \times 10^{14} cm^{-2}$ hadrons, for 300 fb^{-1} of delivered luminosity. The upgrade of the detector comprises of two– pad sensors, dedicated frontend ASICs and the design of dead-time free backend electronics, with flexible FPGA for signal processing and sub-bunch histogramming capabilities.

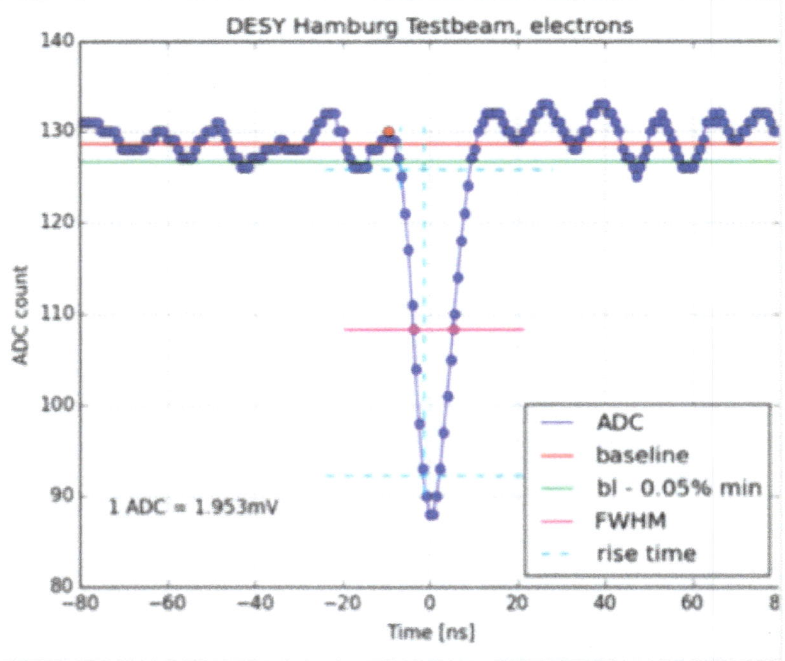

Figure 5: A MIP response from a test board containing the new fronted ASIC and a diamond sensor, measured in a test beam. The signal is digitised by an 8-bit ADCs with a 1.25 GS/s sampling rate.

To improve the time resolution of the detector, a new dedicated front-end ASIC has been developed at CERN in collaboration with the University of Science and Technology AGH Krakow. The ASIC is based on commercial IBM CMOS-8RF 130 nm technology and includes a fast trans-impedance preamplifier with an active feedback, a shaper stage and a fully differential output buffer. An charge-to-voltage conversion of 50 mV/fC is obtained which will improve the efficiency of particle detection. A FWHM of less than 10 ns is achieved for a MIP signal of 3 fC and less than 1k electrons ENC is measured. Each chip has 4 channels and a sophisticated calibration circuit to monitor the gain for 2 input levels corresponding to 1/2 and 1 MIP. The active feedback circuit allows for a fast recovery time (<30 ns), even for pulses with a large charge of up to 150 fC. The frontend was qualified in a test beam in DESY Hamburg with an electron beam, and a typical example of a MIP signal is shown in Figure 5.

The upgraded BCM1F detector acceptance is increased by a factor of three from an active area of 8 5.0 × 5.0 mm2 sensors to 24 5.0 × 5.0 mm^2 sensors, to allow for additional statistical sensitivity to monitor the MIB. Each sensor, is metalised into two pads of approximately 5.0 × 2.5 mm^2 (48 channels in total) to decrease the count rate per channel, and maintain linearity with increased pileup. The width of the gap between the two pads is 25 μm. To house and bring services to 12 channels per 1/4 of the detector, a novel single integrated PCB has been produced made from flex and rigid parts, as shown in Figure 6 (a). Four such boards are required for the full detector assembly. This integrated PCB will host all passive and active components. The PCB has a complex, multilayer layout in order to fit all lines on the PCB and a limited volume inside the pixel service tube.

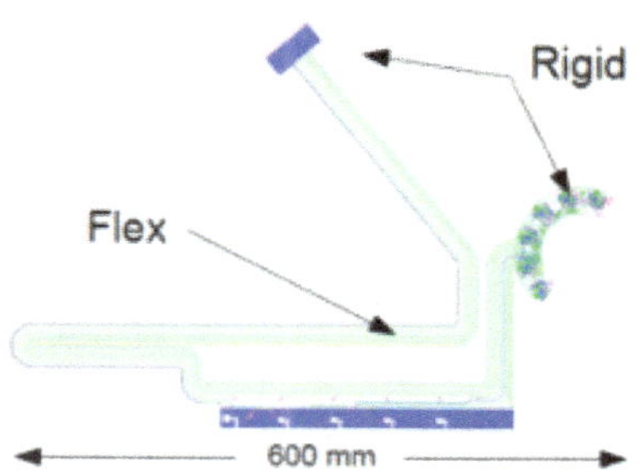

(a) BCM1 flat PCB

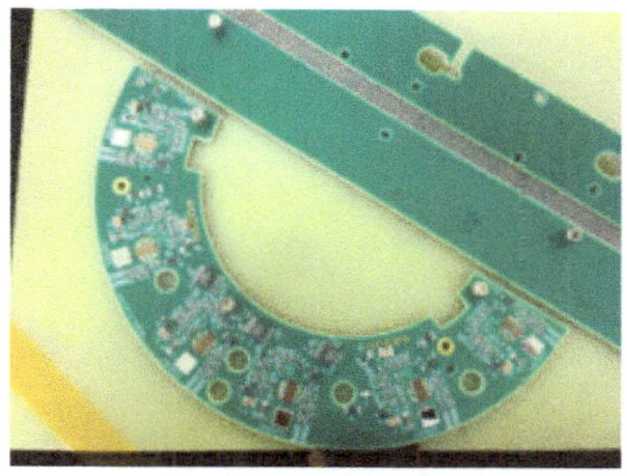

(b) BCM1 half ring

Figure 6: Single integrated PCB for the BCM1F frontend upgrade.

This allows to have a connector free system, minimising the material budget in the pixel volume. The BCM1 PCB extended and mounted onto the support carriage is shown in Figure 3.

Onto the rigid half ring part of the PCB, are mounted six sCVD diamond sensors, metalized into two independent readout pads as shown in Figure 6 (b). The two readout pads are coupled to two input channels of the BCM1F ASIC. The central rigid PCB continues to a rectangular rigid PCB on the detector arm, through a flexible PCB. Onto this rectangular part is mounted the four, 3 channel analogue optohybrids [9] for optical signal transmission and a digital optohybrid for the slow control and settings of the gain and bias of the linear laser drivers. The mounting of the analogue optohybrids on to the detector arm, which is further away from the beam line than in Run I, lowering their exposure to radiation. The HV lines to the BCM1F sensors have been qualified to 1 kV. Operating the sensors at this voltage leads to less sensitivity to future radiation damage of the sCVD sensors.

Figure 7: BCM1F VME Real-time histogram unit.

The purpose of the backend electronics is to measure the arrival time and amplitude of all signals and associate the signals to either collision products or MIB for online per bunch luminosity and beam background measurements. The BCM1F backend electronics strategy will retain the parallel path design from the Run I setup. Signals will be transmitted through the approximately 80 m of optical single mode fibers to the counting room where they are received by four 12-channel Analogue Receiver Modules (ARx12). The VME discriminator path will be used for initial running, while a MicroTCA digitizer system with fast peak-finding will be commissioned for future use.

The analog signal are fed to a VME discriminator with a double pulse resolution of 7 ns to measure the arrival time hit occupancy. The resulting digital signal are sent to a dedicated board, the Real-time histogramming unit (RHU) [13] and shown in Figure 7. The histograms are binned at four times the external bunch clock frequency, approximately 6.25 ns per bins, for 14,256 bins per orbit. The histogram integration interval is decoded from an optical timing signal received from the CMS TCDS system.

The RHU provides double buffering and dead-time free processing of signals which is an important feature of an online luminosity system. A copy of the discriminated signals are also sent to a multipurpose logic board to register hit coincidences between channels and to derive technical triggers for optimal tagging of underlying beam background events.

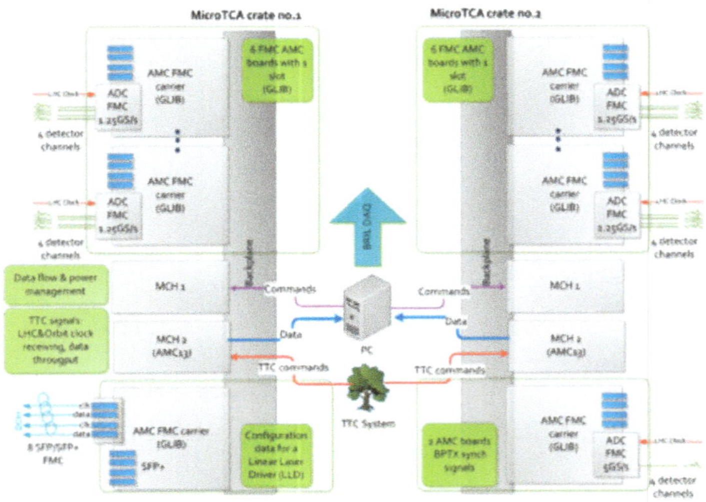

Figure 8: BCM1F MicroTCA electronics architecture and data flow.

Figure 9: BCM1F MicroTCA electronics consisting of an 8-bit ADCs mezzanine ADC with 1.25GS/s sampling rate (left) and GLIB carrier board (right).

Complementing the VME electronics, a MicroTCA backend electronics architecture [14] is deployed to monitor the performance of the system and to exploit sophisticated online fast peak-finding algorithms to resolve overlapping signals at high pileup. The trigger-less data flow and the system architecture is shown in Figure 8. The system control and communication will be operated by the specialized CMS MicroTCA Hub module AMC13 XG [15]. The module uses 10 Gb/s Ethernet based IPbus communication, suitable for transferring the large amount of the data. The analogue to digital conversion is performed by twelve 4-channel FMC mezzanines with 8-bit ADCs at a 1.25GS/s sampling rate (FMC125, 4DSP), mounted on the AMC carriers with single high-pin connector FMC connectors (GLIB), as shown in Figure 9. The external LHC clock will be provided to the mezzanines through the front panel to ensure ADC sample synchronisation. The 8-bit samples will be transmitted through FMC connectors to FPGAs mounted on the AMC carriers, that will host the firmware for the peak finding algorithms. The timing signals from the CMS TCDS will be decoded by the AMC13 modules and transmitted through the backplanes to all GLIB modules. The processing FPGA will use this information for creating the time stamps for the collected data. Amplitude and occupancy histograms will be collected during a Lumi Nibble. One orbit of a raw data per Lumi Nibble will be stored to provide an input for the offline statistic processing. During the storage time, without introducing any dead time, the data from a previous Lumi Nibble will be transmitted to the AMC13 through the backplane and subsequently to the BRIL DAQ.

HADRON FORWARD CALORIMETER UPGRADE FOR LUMINOSITY MEASUREMENT

During Run I the HF tower occupancy method was the primary method used to perform high statistics, real-time bunch-by-bunch luminosity measurements for CMS. It is based on zero counting, in which the average fraction of empty towers is used to infer the mean number of interactions per bunch crossing. A second method, called "ET sum", exploits the linear relationship between the average transverse energy per tower and the luminosity. This algorithm has not yet been used, but is under consideration for Run II. Two main activities in LS1, in the context of the HCAL Phase I upgrade program [16], affect the HF luminosity calibration and electronics: The first is the exchange of the photomultipliers and the second is an upgrade of the HF backend electronics. The upgrade of the HF backend electronics in particular implies redevelopment of the HF luminosity firmware and DAQ during LS1.

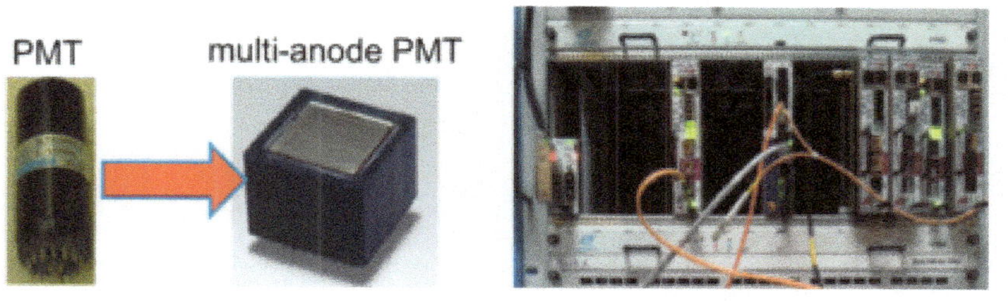

(a) HF PMT Upgrade (b) HF backend electronics crate

Figure 10: Upgrade of the HF frontend photodetectors (Hamamatsu R7600U-200-M4) and backend electronics using MicroTCA standard.

PMT Performance in Run I and Upgrade in LS1

Time-dependent variations in the energy response of the HF directly affect the luminosity measurement. The largest such effect in Run I originated from timedependent gain variations in the photomultipliers used to read out the HF. These gain changes can be at the few percent level, and translate into time dependent calibration changes of the luminosity values at the few percent level, necessitating recalibration.

The HF has undergone a complete replacement of its photodetectors during LS1 for radiation hard, Hamamatsu R7600U-200-M4 multi-anode tubes. During Run 1, a small fraction of anomalous signals associated to interactions of charged particles in the PMT windows was measured. These tubes will have a thinner optical window, reducing the anomalous signals by a factor of four. The multi-anode readout of the tubes also provides the ability to identify an anomalous signal by its timing information since these signatures are ~ 5 ns prior to the collision signal, typically in one of the readout channels only.

Backend Electronics Upgrade

The HF backend electronics hosts the luminosity firmware algorithms and performs real-time histogramming of the data used in the luminosity measurement. The HCAL back-end electronics has been upgraded in LS1 and based on commercial FPGAs and the MicroTCA standard. The main components of the backend electronics is the μHTR board and the AMC13 board. The data processing for the HF luminosity determination in the μHTR board will be split between the two FPGAs. The front FPGA will calculate the numbers of long-fiber cells over threshold and the total transverse energy observed during each bunch crossing. The front FGPA will transmit this information to the back FPGA via a unidirectional 3.6 gigabit copper link. The back FPGA then creates a set of histograms to collect the ingredients for the luminosity measurements as a function of LHC bunch number. The histogram integration period can be either internally counted or based on the Lumi Nibble timing synchronisation signals from the TCDS, decoded and distributed by the AMC13. At the end of each integration cycle, the histograms will be moved into a storage buffer, with redundant triple buffering capabilities, from which the BRIL DAQ source processor will read them via the IPBus.

The firmware and software for the μHTR has been designed to allow separate operation of the luminosity path and the trigger/DAQ path. The two data paths will be coupled only at the alignment of the data from the front-ends which will be reset only when required by clock changes or other effects. The luminosity system will have its own set of calibration lookup tables (and the software stack is being designed to allow separate xDAQ applications and RCMS controllers to be responsible for the luminosity and trigger/DAQ activities of the uHTR. This separation will provide maximum flexibility for both systems, allowing the luminosity system to have the largest possible uptime while also granting operational simplicity for the trigger/DAQ operations.

BEAM HALO MONITOR

A Beam Halo Monitoring (BHM) system [17] has been designed to provide an online bunch-by-bunch measurement of MIB arriving CMS at a radius of 1.8 m from beam axis, separately for the two beams. At this location the detectors are sensitive to muon MIB arising from interactions in collimators in the long straight sections. The system is an array of fast, directional Cherenkov detectors that are distributed azimuthally outside the rotating shielding of CMS. The longitudinal location of

20.625 m from the IP as shown in Figure 11 (left), provides a maximum time separation of MIB from the collision products. The system will flag at realtime adverse beam conditions to CMS and LHC, complementing the measurement by the diamond-sensor based BCM1F system at lower radius and closer to the IP.

BHM Frontend Design and Qualification

A BHM detector unit consists of the following components: Cherenkov radiator made from a synthetically fused quartz bar of 10 cm length and 5.1 cm diameter, black-painted on the front side; 1 mm optical coupling using RTV1345; a fast, UV sensitive photomultiplier

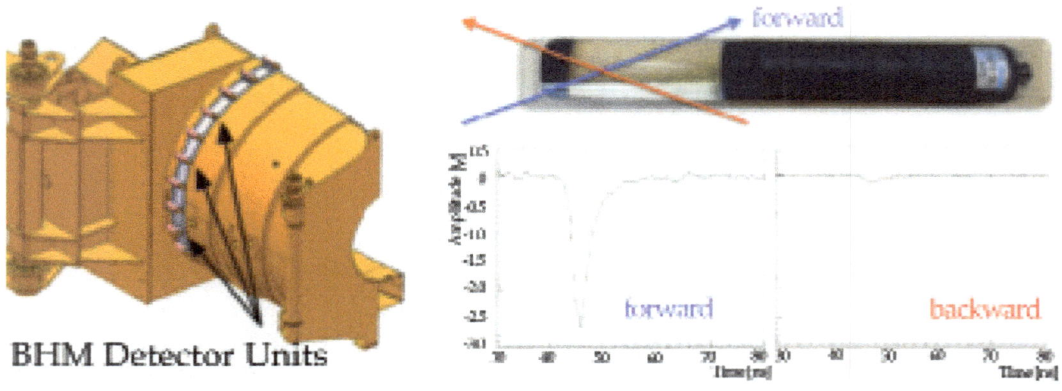

Figure 11: (left) BHM system installed around CMS rotating shielding (1/4 of installation). (right) Signal amplitude as a function of time show the different responses to forward and backward going particles.

R2059 from Hamamatsu; triple layer magnetic shielding consisting of 0.6 mm permalloy, 1 mm μ-metal and 1 cm thick steel tube. By exploiting the prompt and directional nature of Cherenkov radiation, the BHM system makes use of both timing and directionality to identify background. The digitized signal is characterized by 3.0 ns FWHM. The signal amplitude depends on the direction of the incoming particle, as shown in Figure 11 (right). Each detector unit is orientated such that the incoming beam background from the tunnel will produce a large signal relative to collision products. The magnetic shielding serves both as a shield for the PMT from the 17 mT residual magnetic field and a shielding from low energy e^-/e^+ produced by proton-proton collisions, arriving at the same direction with the MIB.

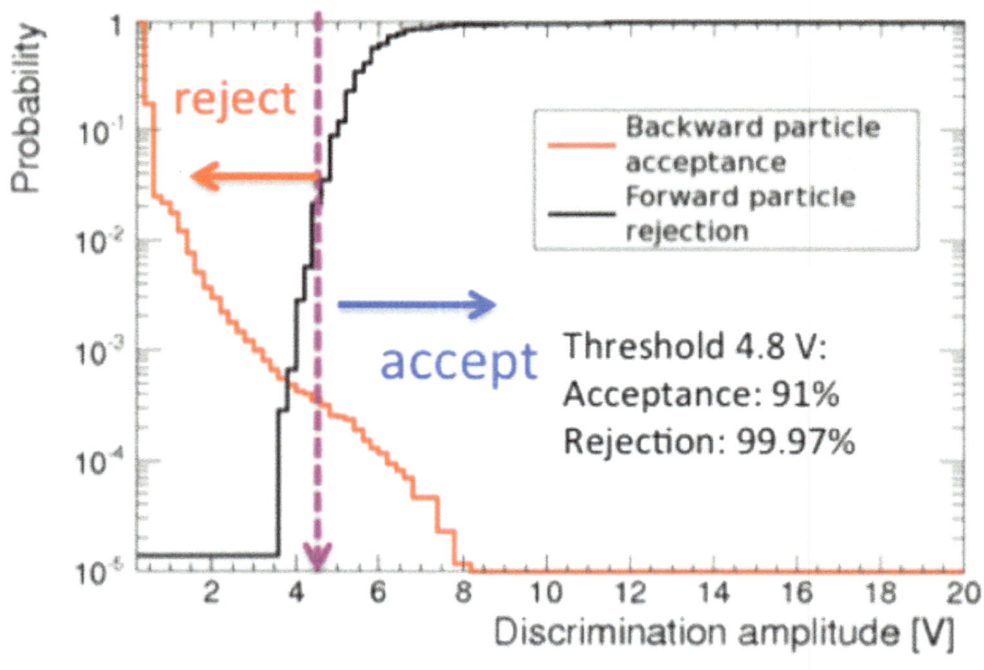

Figure 12: Normalized amplitude distributions from test beam data for forward and backward going particles.

The directional response of the BHM is used to suppress signals originating from collision products while selecting particles parallel and consistent with the direction of the incoming beam. This allows for an efficient MIB measurement. During test-beams it was validated that the suppression of the backward signal to 1 over 1000 is achievable; by setting a discrimination voltage, e.g. 4.8 V as shown in Figure 13, 91% of the forward signal is accepted, while the backward signal acceptance is suppressed to O(10-4).

The fronted units are radiation hard and are qualified for > 3000 fb^{-1} at this location.

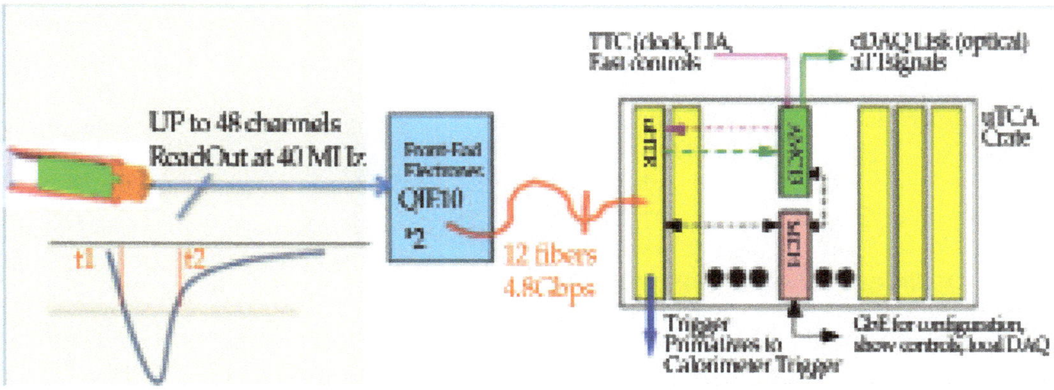

Figure 13: Overview schematic of the BHM Electronics.

BHM Electronics

The readout of the BHM detector will make use of many components developed for the CMS HCAL electronics upgrade [16], with dedicated firmware and readout adapted to the beam monitoring requirements [18]. The PMT signal will be digitized by a charge integrating ASIC (QIE10), providing both the signal rise time and the charge integrated over one bunch crossing. It ensures dead-timeless readout of signal amplitude and edge time information with 500 ps resolution. The backend electronics will record histograms with few ns binning. The beam background rates will be published to CMS and the LHC. A calibration monitoring system has been designed to generate triggered pulses of light to monitor the efficiency of the system.

DATA ACQUISITION AND SYNCHRONIZATION

The BRIL DAQ system is presently under design and builds on the XDAQ[19], a C++ framework for distributed data acquisition, and the run-control framework of CMS. Similar to CMS central DAQ, the BRIL DAQ manages distributed and heterogeneous subsystems, using a homogeneous distributed software architecture, where reliability and redundancy are important to prevent unwanted down time. The event-driven architecture enables large numbers of loosely coupled software components and services to exchange information in near real-time.

To fully exploit the redundant online luminosity and beam background monitoring system, the BRIL subsystem histogramming frontends are synchronized using common timing signals distributed by the CMS TCDS system, defining the hit count integration interval boundaries. Long command counters are also distributed for additional for inter-system synchronization. Using such a technique uniforms the accounting of delivered luminosity and the downstream data handling of the BRIL DAQ.

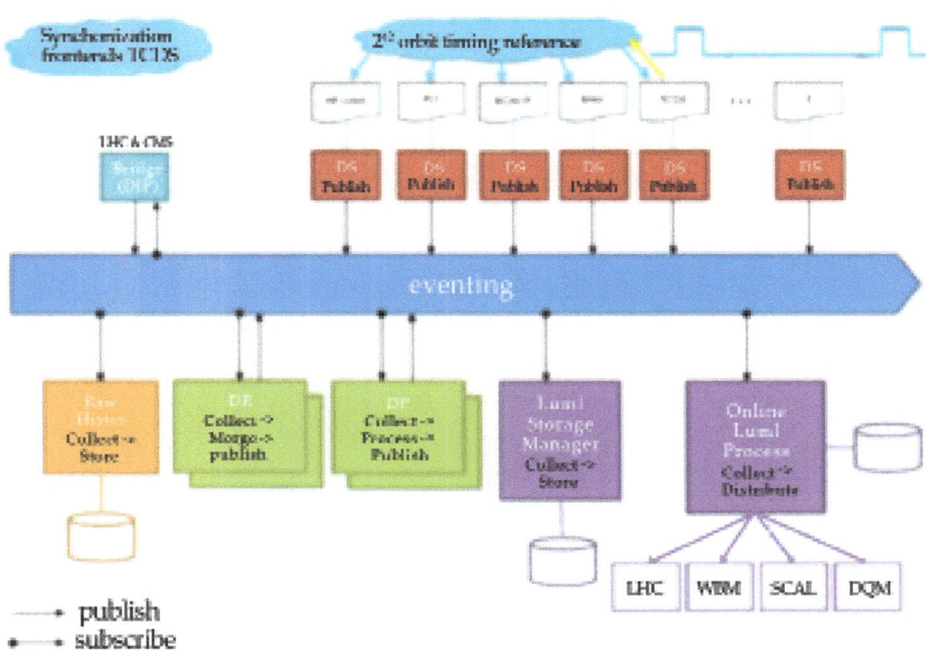

Figure 14: Schematic of the BRIL DAQ architecture based on XDAQ and a publisher / subscriber exchange of data via a common "eventing" bus.

BRIL DAQ components are logically categorized as a source, a processor or a central processor. A source is the hardware readout unit for each subsystem. A processor is responsible for local data aggregation and reduction, luminosity and beam background calculations. A processor component can collect, aggregate and further reduce data from various subsystems. Central processors serve for different purposes such as storage, luminosity selection or global online data quality monitoring. XDAQ b2in eventing in publisher/ subscriber mode is used for data transport between these software components, as shown schematically in Figure 14. There are multiple XDAQ processes running in the BRIL linux cluster. The global configuration and control of them will be achieved by a single generic function manager driving several configuration groups. The control of BRIL DAQ processes, which are stateless, and should run whenever there is beam in the LHC machine, independent of the state of the central CMS DAQ.

CONCLUSION

Upgraded beam instrumentation is being installed in LS1 to monitor the beam conditions for CMS for 2015. An online luminosity system consisting of three independent luminometers, the PLT, BCM1F and HFluminosity readout, will ensure a robust measurement and accounting of the delivered luminosity and offers an advantage for the estimation of systematic errors. The upgraded BCM1F will enhance monitor beam background particles at low radius and the novel BHM system will complement with efficient measuring of the higher radius beam background. The detector frontend readouts will be synchronized by common distributed timing signals and a data acquisition architecture is being developed to provide a com mon software framework for data processing, publishing and storage of synchronized data.

REFERENCES

1. CMS Collaboration, The CMS experiment at the CERN LHC, JINST 3 (2008) S08004, doi:10.1088/1748-0221/3/08/S08004.
2. Lyndon Evans and Philip Bryant, LHC Machine, 2008 JINST 3 S08001, doi:10.1088/1748-0221/3/08/S08001.
3. Allkofer, Y et al, Design and performance of the silicon sensors for the CMS barrel pixel detector, Nucl. Instrum. Methods Phys. Res., A 584 (2008) 25-41
4. M. Barbero et al., Design and test of the CMS pixel readout chip, Nucl. Instrum. Methods Phys. Res., A 517 (2004) 349
5. R. Hall-Wilton et al Results from a Beam Test of a Prototype PLT Diamond Pixel Telescope, Nucl. Instrum. Methods Phys. Res., A 636 (2011) S130-S136
6. M Pernicka et al The CMS Pixel FED, Topical Workshop on Electronics for Particle Physics, Prague, Czech Republic (2007), 487-491 doi 10.5170/CERN-2007-007.487
7. R. Hall-Wilton et al Fast beam conditions monitor (BCM1F) for CMS, Topical Workshop on Electronics for Particle Physics, Greece, (2008) pp. 259-263
8. A. Dabrowski et al., The Performance of the Beam Conditions and Radiation Monitoring System of CMS, proceedings IEEE Nuclear Science Symposium and Medical Imaging Conference, Valencia, Spain, 23 - 29 Oct 2011 DOI 10.1109/NSSMIC.2011.6153979
9. F. Vasey, Development of radiation-hard optical links for the CMS tracker at CERN IEEE Trans. Nucl. Sci. 45 (1998) 331- 337
10. N. Odell., Measurements of Luminosity and Normalised BeamInduced Background Using the CMS Fast Beam Condition Monitor, in proceedings of ICHEP2012, PoS(ICHEP2012)526.
11. J. Kaplon and W. Dabrowski, Fast CMOS binary front end for silicon strip detectors at LHC experiments, 2005 IEEE Transactions on Nuclear Science 52(6), pp.2713-2720.
12. A. Fasso, A. Ferrari, J. Ranft, and P.R. Sala, FLUKA: a multiparticle transport code, CERN-2005-10 (2005), INFN /TC 05 /11, SLAC-R-773
13. M. Penno, A Real-time Histogramming Unit for Luminosity and Beam Background Measurements for each Bunch Crossing at the CMS Experiment, TWEPP 2013 - Topical Workshop on Electronics for Particle Physics, Perugia, Italy
14. A Zagozdzinska et al, Architecture of the upgraded BCM1F Backend Electronics for Beam Conditions and Luminosity Measurement, TWEPP 2014 - Topical Workshop on Electronics for Particle Physics, Proceedings, to be published in JINST
15. E Hazen et al The AMC13XG: a new generation clock/ timing/DAQ module for CMS MicroTCA (2013) JINST 8 C12036 doi:10.1088/1748-0221/8/12/C12036
16. J. Mans et al, CMS Technical Design Report for the Phase 1 Upgrade of the Hadron Calorimeter, CERN-LHCC-2012-015, CMS-TDR-010
17. S. Orfanelli et al, Design of a Novel Cherenkov Detector System for Machine Induced Background Monitoring in the CMS Cavern, Proceedings, 2nd International Beam Instrumentation Conference, Oxford UK, (2013) isbn = 978-3-95450-127-4
18. N. Tosi et al, Electronics and Calibration system for the CMS Beam Halo Monitor, Proceedings, TIPP'14 conference, (2014)
19. J. Gutleber, S. Murray, L. Orsini, Towards a homogeneous architecture for high-energy physics data acquisition systems Comput. Phys. Commun. 153 (2003), 155163 DOI: 10.1016/S0010- 4655(03)00161-9

CHAPTER 19

TOOLS FOR AUTOMATING THE IMAGING OF ZEBRAFISH LARVAE

Rock Pulak

Union Biometrica, Inc., 84 October Hill Rd., Holliston, MA 01746, USA

ABSTRACT

The VAST BioImager system is a set of tools developed for zebrafish researchers who require the collection of images from a large number of 2–7 dpf zebrafish larvae. The VAST BioImager automates larval handling, positioning and orientation tasks. Color images at about 10 lm resolution are collected from the on-board camera of the system. If images of greater resolution and detail are required, this system is mounted on an upright microscope, such as a confocal or fluorescence microscope, to utilize their capabilities. The system loads a larvae, positions it in view of the camera, determines orientation using pattern recognition analysis, and then more precisely positions to user-defined orientation for optimal imaging of any desired tissue or organ system. Multiple images of the same larva can be collected. The specific part of each larva and the desired orientation and position is identified by the researcher and an experiment defining the settings and a series of steps can be saved and repeated for imaging of subsequent larvae. The system captures images, then ejects and loads another larva from either a bulk reservoir, a well of a 96 well plate using the LP Sampler, or individually targeted larvae from a Petri dish or other container using the VAST Pipettor. Alternative manual protocols for handling larvae for image collection are tedious and time consuming. The VAST BioImager automates these steps to allow for greater throughput of assays and screens requiring high-content image collection of zebrafish larvae such as might be used in drug discovery and toxicology studies.

INTRODUCTION

Zebrafish (*Danio rerio*) have become an important vertebrate model organism for biomedical research [1] and [2]. Zebrafish embryos develop along a path shared with other vertebrates. Structural, functional and genetic similarities have led various groups to construct human disease models that have been used in phenotypic screening approaches for drug discovery and toxicology testing [3] and [4]. Additional features that make zebrafish ideal for this type of research includes their small size, optical clarity, that embryonic development occurs outside the mother, and the availability of a collection of molecular tools allowing genetic changes that mimic human disease states to be made[5] and [6].

Microscopy is an important tool for zebrafish researchers. The fertilized embryo starts out at about 1.2 mm diameter and this small size necessitates visual magnification to observe the development and function of the different tissues and organ systems. Imaging – collecting pictures and videos – provides

Citation: Rock Pulak, Tools for automating the imaging of zebrafish larvae, Methods, Volume 96, 1 March 2016, Pages 118-126, http://dx.doi.org/10.1016/j.ymeth.2015.11.021.

the means of capturing the data for further analysis and archiving of the results of experiments.

Many advances have been made in microscopy over the years that allow for incredibly detailed image collection, especially fluorescence imaging and confocal imaging. However, traditional manual protocols for handling larvae for image collection are tedious and time consuming [10]. To automate image collection in a high throughput setting, methods for handling zebrafish larvae needed to be developed. The concept for one approach was proposed by Yanik [7] and [8] and involved automating a fluidics system to sequentially load young larvae to a capillary tube for orienting and imaging. This approach has been developed into a collection of tools called the VAST BioImager. It automates the handing of 2–7 dpf larvae for high throughput and high content image collection. The larvae to be imaged may be live, anesthetized, or fixed. This report describes the VAST BioImager system and how it works to capture images from many young larvae in an automated and high throughput process (see Fig. 1, Fig. 2, Fig. 3, Fig. 4 and Fig. 5).

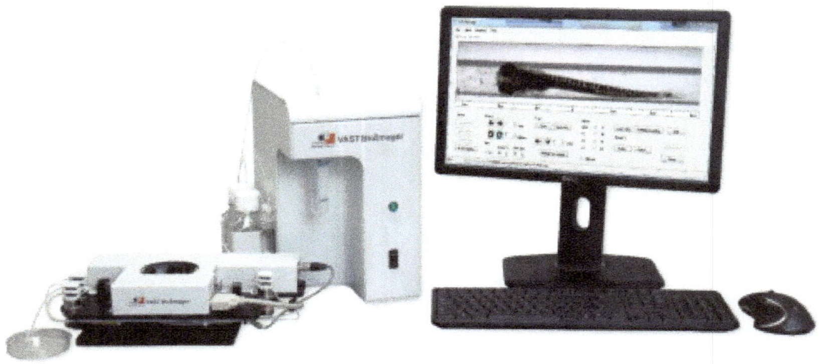

Figure 1: VAST BioImager is composed of two parts, the imaging unit (left) and the controller unit (center). A computer/monitor runs the software that controls the VAST BioImager and collects the larval images.

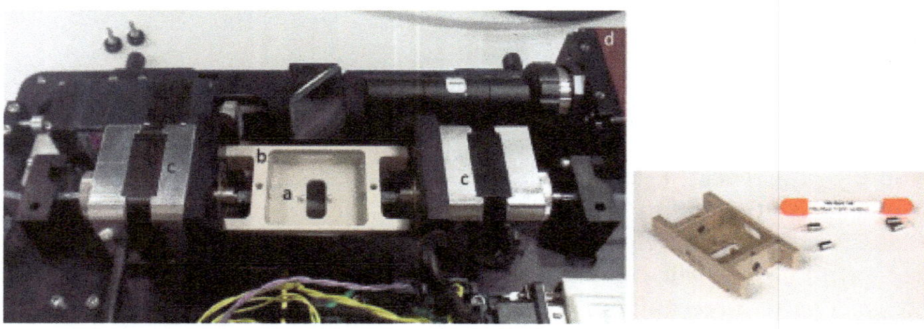

Figure 2: (left) VAST BioImager imaging unit showing the location of the capillary (a) in the water tray (b), flanked by the rotation motors (c). The on-board camera (d) captures images of the capillary off of the mirror. (right) Image of water tray and replacement capillary assemblies.

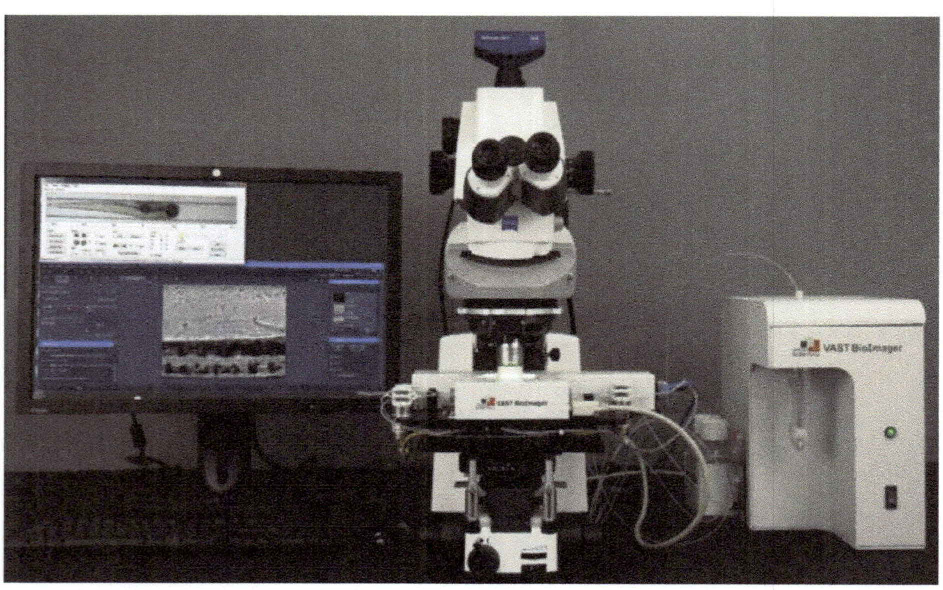

Figure 3: VAST BioImager attached to the stage of an upright microscope (Zeiss Axio) for imaging at a higher resolution and capturing fluorescent images.

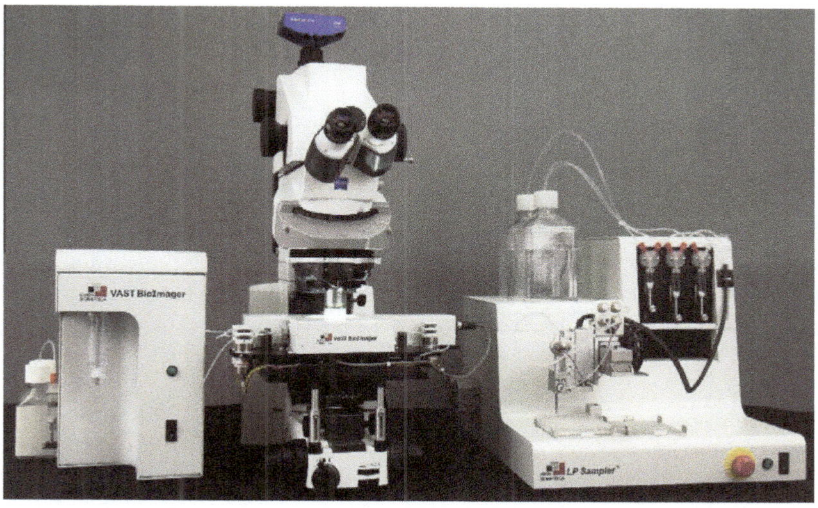

Figure: 4: VAST BioImager can accept larvae from wells of a 96-well plate using the LP Sampler.

Figure 5: VAST Pipettor is used to introduce user-selected larvae to the VAST BioImager for image collection.

DESCRIPTION OF VAST BIOIMAGER

The VAST BioImager is composed of two separate pieces of hardware, along with a computer and monitor. One of these, the imaging unit, contains the capillary in which the larva is held during imaging and the on-board color camera for initial larvae identification and orientation. The second is a controller unit that carries the electronics, the fluid syringe for pulling the larvae from the bulk reservoir and pushing them through to the capillary for imaging and fluid reservoirs used to accomplish this. Tubing and electronics connect the two units together, valves on the imaging unit open and close to control the flow of larvae and fluids through the system. A computer, monitor and software integrate the two units to achieve the automation of image collection. Several of the individual components are described in greater detail.

Fluid Syringe

A fluid syringe is responsible for moving fluids and larvae through the system. It is located on the controller unit and is connected to the buffer reservoir and the imaging unit of the VAST BioImager system by tubing. It functions in response to instructions from a script (controlled by software) to draw fluid from either the buffer reservoir or the sample line to load or unload the capillary with a fish.

Sample Cup and Probe

The VAST BioImager system can be configured to load sample fish from a 50 mL conical tube located on the controller unit. Larvae are drawn from the sample tube through a probe positioned within the 50 mL conical tube. The probe has a small propeller-type mixer attached to gently agitate and suspend the larvae in the sample. Users can control the speed of the mixer as well as the frequency to reverse mixing direction.

Sample Valves

A series of valves on the imaging unit allow for the controlled movement of fluids along specific paths to various parts of the VAST BioImager system. The sequence and timing of this is controlled by a script (software) but these settings can be modified if need be to better optimize a particular assay or procedure.

Water Tray

The imaging system is fitted with a plastic tray that holds clean water surrounding the rotating glass capillary. This configuration allows for the collection of images of the larvae contained in the capillary tube without the distortion which would otherwise occur due to differences in the refractive index between air and water. By submerging the capillary in water, this distortion is eliminated.

Glass Capillary

The glass capillary is a 3.5 cm long thin walled glass tube with a 600–700 μm inner diameter width range. It is housed in the water tray and is carefully positioned between two rotational motors. These motors allow the capillary to rotate 360° around its axis in order to position a certain area of the larva in the best orientation for imaging by the onboard camera and/or an external microscope mounted camera. The capillary is replaceable, if it should break or become dirty.

Rotational Motors

A pair of rotational motors suspends the glass capillary to allow full rotation around its axis. The system software automatically initiates rotational movement of the capillary in order to orient the fish. However, the user has manual control of these motors as well.

X-positioning Motor

The VAST BioImager system can be mounted on an upright microscope if the microscope has sufficient clearance for the imaging unit. The imaging unit can be fastened to the microscope's stage via an adapter plate. In order to image the desired part of the larva under higher magnification it may be necessary to move the sample filled capillary left or right under the objective. The user can set controls for this type of fine adjustment, either by automatic or manual *x*-axis motion control.

Fluid Buffer Container

The system draws fluid through various parts of the system from a 250 mL fluid buffer reservoir. The particular buffer used is determined by the researcher but for most applications Embryo media [9] is appropriate.

Waste Container

This container collects the fluid that has passed through the system. User may set conditions to automatically or manually output fish or other objects to a separate output container, such as a Petri dish or well of a multiwell plate.

Onboard Camera

The VAST BioImager system is equipped with an onboard color camera to capture images of the sample object. The onboard camera captures images at a high frame rate, greater than 100 fps, a requirement for rapidly identifying that a larva has passed through the capillary. The camera requires a specific PCI board on the accompanying computer.

Computer

IBM compatible PC utilizing 64 bit processer with PCI card for the camera, color monitor, and USB ports is required for use of the VAST BioImager system. This computer setup is an integral part of the VAST BioImager system.

External Microscope and Camera

VAST BioImager can be mounted to several different upright microscopes. With the external communications package, it will communicate with up to 2 external cameras and their accompanying software. The microscope, camera, and software of various manufacturers are compatible with the VAST BioImager. Not all microscopes are able to accommodate the imaging unit of the VAST BioImager system. Union Biometrica maintains a list of various compatible microscopes and their components.

LP Sampler

The LP Sampler is an autosampler that can work with the VAST BioImager to transfer larvae from wells of a 96-well plate directly to the imaging unit of the VAST BioImager.

VAST Pipettor

The VAST Pipettor is available for selectively loading fish from a Petri dish or other similar container. The researcher manually identifies the larvae to be imaged and uses the VAST Pipettor to introduce these, one at a time, to the VAST BioImager system.

Operating the VAST BioImager

The VAST BioImager is designed to automate the process of loading, identifying, orienting, and capturing images of a zebrafish larva, and then repeating the process with subsequent larvae. The larvae can be live, anesthetized, or fixed. The user is able to set certain system parameters that will affect the performance of the VAST BioImager. These parameters include fluid speed through the capillary, rotation speed of the capillary when a fish is being imaged, establishing image templates of the zebrafish, and image collection settings. Once these user defined features are set, the VAST BioImager is an automatic system. Several predefined experimental settings have been developed/tested and come with the VAST BioImager system. Researchers can use these predefined settings as starting points for their own experiments. Often no changes, or only a few, are necessary to start the automated collection of 2–7 dpf zebrafish larval images. What follows is a brief description of how a "typical" experiment would proceed. In this description larvae will be introduced from the 50 mL conical on the controller unit.

Start up

When the VAST BioImager system has been installed, with all electrical connections secured and all fluidic connections made, the first step would be to switch the VAST BioImager power button of the controller unit to the power on position.

Ensure fluid reservoir is filled with clean sample solution compatible for handling zebrafish larvae, such as Embryo medium [9].

Start VAST BioImager software. The VAST BioImager software screen opens onto the desktop of the computer. The VAST BioImager syringe will move to its home position (empty) and the onboard LED light turns on. This light illuminates the capillary for image capture by the on board high speed camera of the imaging unit.

Fill the capillary reservoir (water tray) with clean water. Check the image of the capillary on the software screen; if there are many particles on the outside of the capillary or in the fluid tray it may be necessary to clean the tray and/or use a small paint brush to gently wipe the outer surface of the glass capillary.

If using the 50 mL conical sample container, ensure the sample line is connected appropriately and place clean 50 mL conical on the sample holder on the controller unit.

Click the PRIME button (on the middle of the software screen, Fig. 6) to launch an automatic priming sequence. The syringe fills and empties in a series of steps to move clean fluid from the reservoir through the various parts of the system. If the fluid lines were empty, it might require up to four priming cycles to fill the lines with fluid and remove the air bubbles.

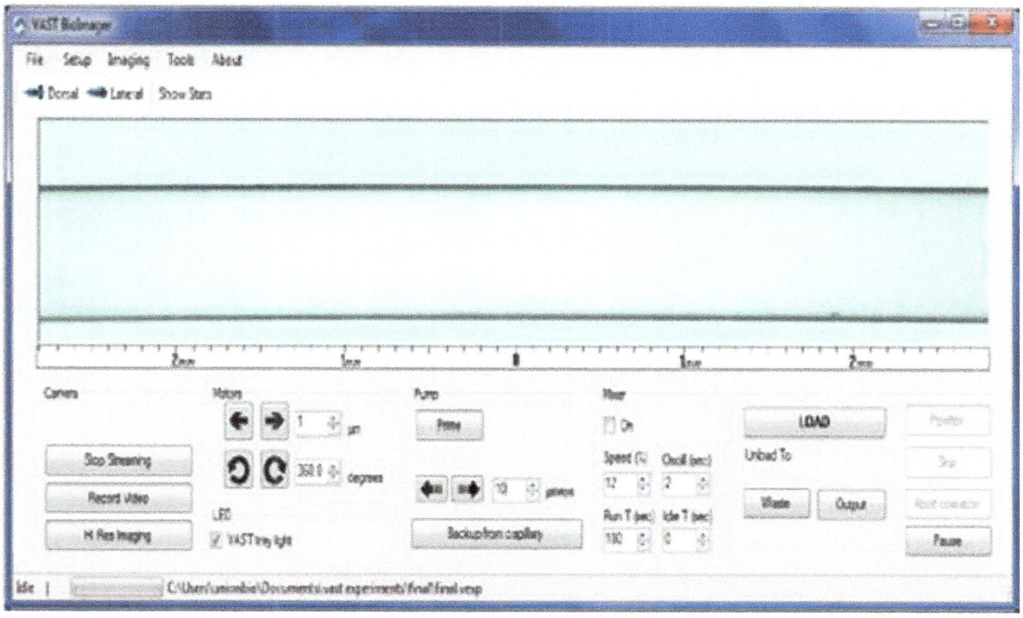

Figure 6: Example screenshot of VAST BioImager software showing various control features and an image of the empty capillary.

Once priming is finished, replace the fluid in the conical with fish larvae to be imaged. Adjust the speed and frequency of oscillation of the mixer. Control buttons for these manipulations are controlled by the software.

Go to the File menu and select an existing experiment including image templates. Alternatively, a new experiment could be set up but modifying a previous "standard" experiment is often the easiest approach.

Collect an image of the empty capillary. This is achieved by moving the mouse onto the image of the capillary and right click to open a menu, then select record empty capillary. This provides the background image for the camera and software to use with image recognition software to sense when a larvae passes through the capillary.

To utilize the automated rotational orientation feature of the VAST BioImager reference lateral and dorsal images are needed. These can be from a previous collection of larval images in the saved experiment, from alternative existing stored reference images, or the researcher could create new reference images. Specific larval ages or strains may require their own specific reference/template images for proper comparison and utilization of the automated rotational orientation feature.

In the object detection setup, click Show intensity on current image. Set the threshold value to be 20 or 30 below the average intensity. At this point the VAST BioImager is set to start loading larvae and collecting images.

Load a Larva

At the beginning of a planned collection of images of many larvae it is advisable to initially run the VAST BioImager system using manual controls to load and orient the fish to determine if the reference/template images and any high-resolution settings are set correctly prior to running in automatic mode. To set conditions for manual or automatic control of the sample and image capture, open the Operation mode feature found in the Setup menu and select the manual mode option.

Click the LOAD button on the software screen. This launches the load sequence where the system opens the sample valve while the syringe pulls fluid from the sample container through the system until an object passes in front of the camera and triggers the fluidics to stop. Features of the triggering object are analyzed using an image recognition algorithm to determine if the object resembles a zebrafish larvae or not. The fluids system then reverses direction to position the larva in front of the onboard camera.

Once the larva is loaded and in view on the camera, the motors will rotate the capillary while images are collected. These images are used to determine the orientation of the fish by comparison with the lateral and dorsal reference images the user has created or selected from previous experiments.

Capture an Image or Video from the On Board Camera

To capture an image of the larva once it is positioned in the view of the camera, the user will click the STOP STREAMING button to freeze the current image on the desktop. Minor adjustments can be made manually to better orient the larva before capturing an image. After clicking the Capture image button to capture the image, the software will prompt the user to designate a location to store images.

Additional images of the same larva are captured by activating the streaming feature of the camera, rotating to another position using the rotation buttons on the software screen (input the number of degrees to rotate each time the direction button is pressed), freezing the image by stopping the streaming to the camera and capturing the image.

If a video is required, the Record video button can be selected. This will initiate a video capture following the conditions set in the video setup menu. The time duration of the video as well as the frame rate can be adjusted and used for subsequent larvae.

Click the STOP STREAMING button when done with the image capture.

Unload the Fish

When operating the VAST BioImager system in manual mode and imaging of the current fish is completed, the user selects one of two unload destinations. The larva can be sent either to the waste container or by using the output feature to eject the current larva from the system to a separate collection vessel. This allows for the recovery of a larva that has just been imaged.

AUTOMATING IMAGE COLLECTION ON THE VAST BIOIMAGER

If the user is confident the conditions for object identification and image capture are set properly, they can set the system to automatically process multiple larvae from their samples. The user can define the rotational position and number of images to capture for each larvae, as well as the number of sequential larvae to image. These settings are input into an automated procedure using the AutoStore menu. Once these are set, the user can change the Operation mode found in the setup menu from manual to automatic and save these setting for capturing images of a collection of larvae in an automated process.

Clicking the LOAD button now will launch the automated VAST BioImager sequence to sequentially load a larva, position and capture images, store the images automatically, and move on to the next larva until the specified number of larvae are imaged.

Load a Fish from the LP Sampler

Fish can also be loaded into the VAST BioImager for image collection from wells of a multi-well plate. The LP Sampler is used for this process. When using the LP Sampler to load fish, a plate template is setup in the LP Sampler software. The user identifies the type of multiwell plate being used as well as which wells should be sampled. The process is controlled by a script of settings and instructions which takes into account the starting volume of fluid in the well. A collection of scripts have been developed and tested for a limited number of conditions, such as different volumes or plate types. These can be used as starting points and modified to accommodate different test conditions. To start moving larvae from wells for imaging, the users launches the process by selecting the Run Plate button. This launches the load sequence for the system to aspirate fish from the multiwell plate and send them to the VAST BioImager system. Then, the same sequence of events occur on the imaging unit. The sample valve opens while the syringe pulls fluid from the sample container through the system until a larva passes in front of the camera, the fluidics system stops and reverses direction of the flow to position the fish in front of the onboard camera, and image collection can occur. The stored images of larva are tagged with location information so that the user can track which images came from larvae from which well. This data management is essential for assays involving multiple conditions, where each well is a separate condition, such as might occur in chemical screening and toxicology studies.

The system will continue to load fish from the multiwell plate until the wells selected have been processed or the user chooses to abort the operation.

Load a Fish from the VAST Pipettor

Fish can be loaded manually from a Petri dish or other similar container using the VAST Pipettor. This approach to loading larva is ideal in those instances when the user wants to select from a group of larvae only those larvae they are interested in and not all of the fish in the collection. A change needs to be made by the user to the fluidics connections to convert sample introduction from the 50 mL conical on the controller unit to the VAST Pipettor. Then the user activates the Pipettor on the main screen of the VAST BioImager software to alert the system that the Pipettor will be used.

The user places the tip of the pipette probe into fluid near the zebrafish larva they desire to image. Loading is triggered when the top button on the VAST Pipettor is pressed and held down until the fish is pulled into the probe and travels into the tubing of the VAST Pipettor.

Once the fish has been aspirated into the VAST Pipettor, the user presses both buttons at the same time. This launches the load sequence causing the system to open the sample valve while the syringe pulls fluid from the VAST Pipettor through the system until the larva passes in front of the camera. The system then reverses direction to position the fish in front of the onboard camera.

After a fish has been imaged, the user can start the loading process for the next fish.

SETUP HI-RESOLUTION IMAGING (EXTERNAL MICROSCOPE AND CAMERA)

High resolution images and fluorescent images require an external microscope. The imaging unit of the VAST BioImager can be attached to the stage mount of numerous upright microscopes to allow for the collection of these types of images (high resolution and fluorescence images). Using the Hi-Resolution Imaging setup menu, the user can set the rotational and axial position conditions required to place the areas of interest under the microscope objective. If the external microscope supports automatic triggering for image collection, then the VAST BioImager system can be instructed to follow a script that automatically positions a specific region of the larvae to the focus of the objective and rotates the capillary to the desired orientation for image collection. For example, the user might want to collect two images per larva, such as one of the heart and a second of the vasculature in the tail. The VAST BioImager software is used to establish a reference location of the most anterior point of the image of the larva. The *x*-axis motor then moves the imaging unit of the VAST BioImager to a position along the larva appropriate for collecting images of the heart, captures these images, and then moves the imaging unit further and rotates the capillary to position the vasculature of the tail for imaging. These positioning steps can be saved in a script to be repeated for subsequent larvae. Multiple images from different parts of the larva can be collected, as well as the automated tomography image data collection. The VAST BioImager system connects to the external camera and microscope through standard TTL serial communication protocol. This allows for back and forth communication between the VAST BioImager and an external device such as a confocal or fluorescence microscope.

WORK AREA REQUIREMENTS

The VAST BioImager instrument can be set-up on most laboratory benchtops that measure at least 65–70 cm deep and a width of around 90–100 cm, with more space required if the LP Sampler will be used for introducing larvae for imaging. Additional space and access to the front and back of the various components makes work easier.

RESULTS AND DISCUSSION

The VAST BioImager system has been designed to allow end-user flexibility in controlling the conditions for loading and imaging zebrafish larvae by modifying the system software. We tested a number of different settings and settled on a specific collection of settings that achieved high efficiency in properly aligning a larva in the capillary for imaging and minimized events of missing a larva. A second goal was to reduce the speed for processing many larvae. These settings were tested and evaluated to determine the length of time to collect images of 100 larvae. The times using different sample introduction methods are compared. The results are presented in Table 1.

Table 1: Comparison of times to collect images from 100 larvae depending on the method used to introduce the larvae to the VAST BioImager imaging unit.

Sample introduction method		Time required to capture and store image(s) of 100 fish				
1 image with no rotation		# larvae required to capture 100 images	2 rotational positions imaged	5 s video (30 frames/s; stored to local drive)	2 linear positions with external device (1000 ms timeout delay; 2000 ms trigger in timeout)	
Bulk introduction (50 ml conical)	100fish in 25 ml fish water with tricaine	57 min	114	(+) 4 s/fish	(+) 5 s/fish	(+) 15.3 s/fish
VAST Pipettor		38 min	104			
LP Sampler						
Round well 96-well plate				or	or	or
1 per well (n1052)	150 µl volume with tricaine	83 min	103	(+) 6.5 min total for 100 fish	(+) 8.3 min total for 100 fish	(+) 25 min total for 100 fish
3 per well (n736)	150 µl volume with tricaine	58 min	121			
Square well 96-well plate						
3 per well (n200)	300 µl volume with tricaine	57 min	109			

The results in Table 1 are of larvae in Embryo media and anesthetized with tricaine. These data allow certain trends to be identified. It takes about an hour to collect images of 100 larvae from bulk introduction via the 50 mL conical sample cup. About 14% of the larvae could not be imaged. Most often this is because they overlapped with other larvae in a "traffic jam" and could not be separated for imaging. Table 1 reports that it takes about 114 larvae to get 100 images when sample is introduced in bulk from the 50 mL conical sample cup. There were no noticeable differences that could be attributed to age of the larvae although more testing needs to be done to assess if there really are no differences related to larval age. If we look at a different sample introduction mode, we see a difference in the time required to image 100 larvae. We also see small differences in the numbers of larvae that need to be handled to get 100 images, a comparative measure of how many larvae are missed. Data were collected by testing multiple 96-well plates determining the time to analyze the wells and the number of wells required to generate a dataset of 100 images of larvae. Using the LP Sampler and 96-well plates with one larva per well took longer, over 80 min, in comparison with bulk sample introduction (approximately 60 min). This increase in time is a result of the added time needed to move from one well to the next and the time required to wash the line between wells. However, the VAST BioImager in combination with the LP Sampler does allow for the automated analysis of all the wells. Furthermore, fewer larvae are missed because there is no opportunity for traffic jams to occur. Still, about 3% of the larvae were not imaged (about 103 wells needed to collect 100 images). The reasons for this loss is not completely understood although there are a couple possibilities. One is that a larva may fail to be recognized as a larva, by the image recognition software, when it differs sufficiently from the user-generated templates. A second reason might be that the larva gets stuck temporarily to the tubing or in the dead volume of the system and fails to be detected by the camera and is purged during the cleaning step. We know the larva are not being left behind in the wells, that they are being picked up by the probe of the LP Sampler and can be recovered from the output of the VAST BioImager. We continue to explore ways to eliminate this very low level of loss. If more than one larva is present in a well, the time to analyze a well increases because of the time needed to position each fish. Also, with multiple larvae per well the issue of traffic jams again occurs at about the same frequently that occurs when larvae are introduced in bulk from the 50 mL sample cup (over several trials, on average 109 larvae are needed to collect 100 images from the 96-well plate in comparison with 114 larvae to collect 100 images from the sample cup, although this difference may or may not be statistically significant). With three larvae per well we are able to collect images from at least one larva from the well greater than 99% of the time. The time to analyze a 96-well plate with three larvae per well takes about 120 min whereas a plate with one larva per well takes about 80 min. Having three larvae per well reduces the chances that a well fails to provide a larva for imaging (less than 0.1%, almost no incidence of missing all three larvae from a well in over several thousand wells tested). However, it takes about 50% more time to complete the analysis of an entire plate. The fastest method of introducing larvae to the VAST, one at a time, and imaging them is with the use of the VAST Pipettor. This requires the end user to manually select the larva for imaging. One hundred larvae can be imaged in less than 40 min. The end user can regulate the selection of larvae, insuring no traffic jams, and collecting images from each desired larva.

Once a larva is positioned in the capillary, the time to collect additional images or videos does not differs as a result of the method used to get it to the capillary. This is not unexpected as these two steps, that of loading a larva and then imagining a larva, are separate procedures. It is also clear that the rate limiting step in imaging larvae is to position them in the capillary tube.

Two different 96-well plate types were tested, round-well plates and square-well plates. The benefit of the square well plate over the round well plate is that the volume each square well holds is greater than the round wells. There was no discernible difference in the time needed to analyze from one plate type in comparison with the other. For both plate types it takes about 80 min to image 100 larvae.

The images are stored as bit-map (.bmp) images. This results in the collection of data sets that require enormous amount of storage space on the computer. For example, two images for each of 100 larvae can generate 100 Mbytes of data, or about 500 Kbytes per image (see Fig. 7). Videos require even more storage space.

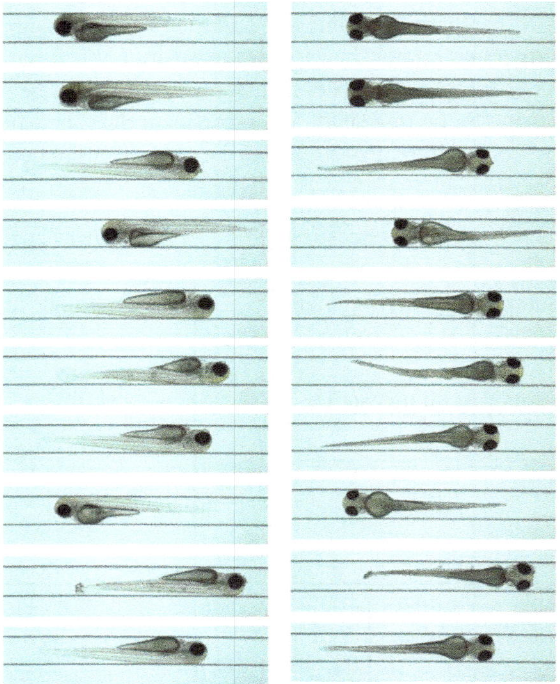

Figure 7: Example images from VAST. These are ten paired images of larvae loaded sequentially from a 50 mL conical sample cup, one lateral view (left) and same larva rotated 90° to collect image of dorsal view (right).

The VAST BioImager can be attached to an upright compound microscope to collect high resolution and fluorescent images. Transgenic fish expressing fluorescent proteins in specific cells and tissues are used to observe and study their development and physiology. Multiple high magnification images can be collected and stitched together to generate a composite high resolution image. Also, videos can be collected, allowing for analysis of certain physiological processes, such as heart beat rate and blood flow. Examples of high resolution images are shown in Fig. 8 and Fig. 9.

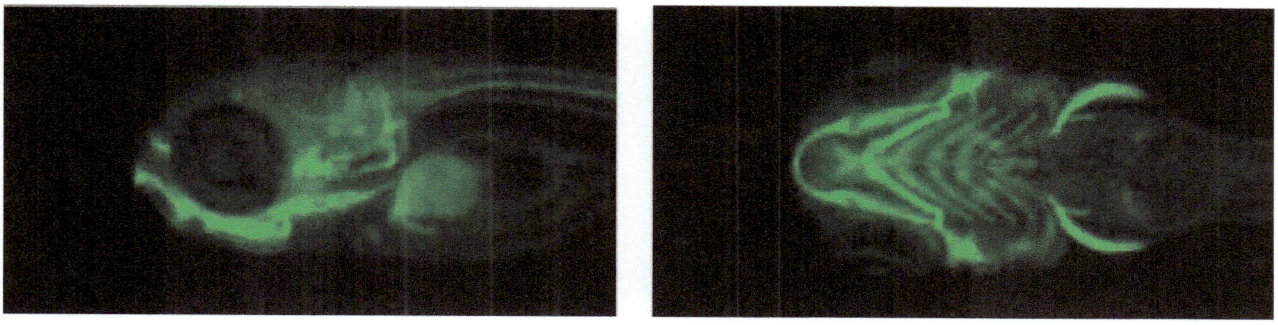

Figure 8: High resolution images (5X) of the head region of larva expressing GFP from sox-10 promoter, lateral view on left, ventral view on right.

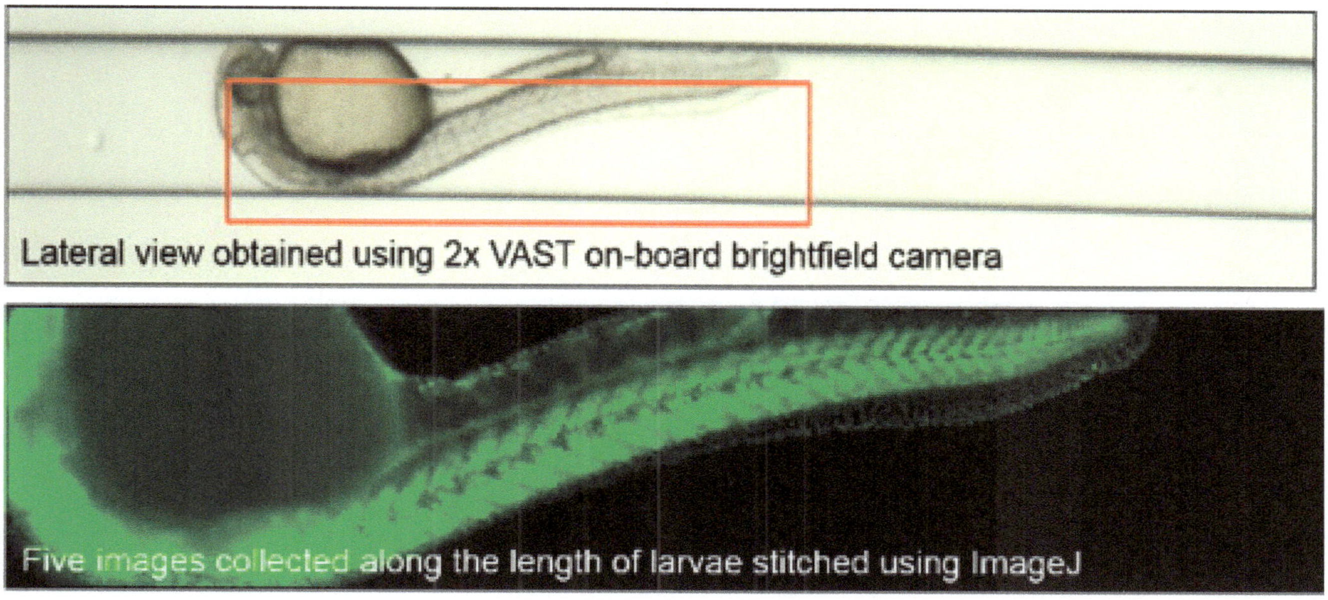

Figure 9: VAST BioImager-captured image (top) and corresponding high resolution images stitched together (below). GFP is expressed from a promoter active in all cells.

CONCLUSION

The VAST BioImager system provides a way of automating the collection of images from a large number of 2–7 dpf zebrafish larvae. Larvae can be transferred to the imaging unit sequentially from a bulk sample in a 50 mL tube, from wells of a multiwell plate using the LP Sampler, or selected individually by the researchers using the VAST Pipettor. The VAST BioImager automates larval handling, positioning and orientation tasks. High resolution and fluorescent images can be collected by mounting the VAST BioImager on an upright compound microscope such as a confocal or fluorescence microscope, to utilize their capabilities. The VAST BioImager automates the handling of larvae for images and thus allows for greater throughput of assays and screens requiring high-content image collection of zebrafish larvae such as might be used in drug discovery and toxicology studies.

ACKNOWLEDGMENTS

Helpful discussions, language, text and images were contributed by David Strack, Chris Bogan, Julia Thompson, Mikalai Malinouski, Zoey Shen, and Mariya Lomakina. Testing of software and establishing settings for experiments were performed by Julia Thompson, Mariya Lomakina, Mikalai Malinouski, Francis Smet, Peter McGuirk and Zoey Shen. Zoey Shen and Mikalai Malinouski collected data for Table 1. Thanks to David Strack, Chris Bogan, Rock N. Pulak, Alice Rushforth for proof reading.

REFERENCES

1. K. Dooley, L.I. Zon, Zebrafish: a model system for the study of human diseases, Curr. Opin. Genet. Dev. 10 (2000) 252–256.
2. G.J. Lieschke, P.D. Currie, Animal models of human disease: zebrafish swim into view, Nat. Rev. Genet. 8 (2007) 353–367.
3. R.T. Peterson, M.C. Fishman, Designing zebrafish chemical screens, Methods Cell Biol. 105 (2011) 525–541.
4. S. Lin, Y. Zhao, A.E. Nel, S. Lin, Zebrafish: an in vivo model for nano EHS studies, Small 9 (9–10) (2013) 1608–1618.
5. A.J. Rennekamp, R.T. Peterson, From phenotype to mechanism after zebrafish small molecule screens, Drug Discovery Today: Dis. Models 10 (2013) e51– e55.
6. A.J. Rennekamp, R.T. Peterson, 15 years of zebrafish chemical screening, Curr. Opin. Chem. Biol. 24 (2015) 58–70.

7. T.Y. Chang, C. Pardo-Martin, A. Allalou, C. Wählby, M.F. Yanik, Fully automated cellular-resolution vertebrate screening platform with parallel animal processing, Lab Chip 12 (4) (2012) 711–716.
8. C. Pardo-Martin, T.Y. Chang, B.K. Koo, C.L. Gilleland, S.C. Wasserman, M.F. Yanik, High-throughput in vivo vertebrate screening, Nat. Methods 7 (8) (2010) 634–636.
9. M. Westerfield, The Zebrafish Book, 4th ed., A guide for the laboratory use of zebrafish (Danio rerio), Eugene, University of Oregon Press, 2000, pp. 1.6–1.8.
10. M. Westerfield, The Zebrafish Book, 4th ed., A guide for the laboratory use of zebrafish (Danio rerio), Eugene, University of Oregon Press, 2000, pp. 4.4–4.5.

INDEX

G

H

I

L

M

N

O

P

R

S

T

U

V

X

Z